MW00844856

Introduction to Sustainability Analytics

Introduction to Sustainability Analytics

Ram Ramanan, Ph.D.

CRC Press
Taylor & Francis Group
Boca Raton London New York

CRC Press is an imprint of the
Taylor & Francis Group, an **informa** business

CRC Press
Taylor & Francis Group
6000 Broken Sound Parkway NW, Suite 300
Boca Raton, FL 33487-2742

© 2018 by Taylor & Francis Group, LLC
CRC Press is an imprint of Taylor & Francis Group, an Informa business

No claim to original U.S. Government works

Printed on acid-free paper

International Standard Book Number-13: 978-1-4987-7705-6 (Hardback)

This book contains information obtained from authentic and highly regarded sources. Reasonable efforts have been made to publish reliable data and information, but the author and publisher cannot assume responsibility for the validity of all materials or the consequences of their use. The authors and publishers have attempted to trace the copyright holders of all material reproduced in this publication and apologize to copyright holders if permission to publish in this form has not been obtained. If any copyright material has not been acknowledged please write and let us know so we may rectify in any future reprint.

Except as permitted under U.S. Copyright Law, no part of this book may be reprinted, reproduced, transmitted, or utilized in any form by any electronic, mechanical, or other means, now known or hereafter invented, including photocopying, microfilming, and recording, or in any information storage or retrieval system, without written permission from the publishers.

For permission to photocopy or use material electronically from this work, please access www.copyright. com (http://www.copyright.com/) or contact the Copyright Clearance Center, Inc. (CCC), 222 Rosewood Drive, Danvers, MA 01923, 978-750-8400. CCC is a not-for-profit organization that provides licenses and registration for a variety of users. For organizations that have been granted a photocopy license by the CCC, a separate system of payment has been arranged.

Trademark Notice: Product or corporate names may be trademarks or registered trademarks, and are used only for identification and explanation without intent to infringe.

Visit the Taylor & Francis Web site at
http://www.taylorandfrancis.com

and the CRC Press Web site at
http://www.crcpress.com

Dedication

This book is dedicated to my family Jana, Shivraj, Bala, Leena, Charu, and Robert for their love, encouragement, support, and inspiration.

Contents

Foreword

More than a century ago, Mohandas Karamchand Gandhi observed, "The earth provides enough to satisfy every man's needs but not every man's greed"[1]; he was one of the first to envision sustainable development and its quadruple dimensions. Extreme greed, whether for money or nature's resources, indeed has disastrous consequences. In the year 1930, Gandhi's friend and contemporary, Albert Einstein, provided the following guidance to scientists and engineers was as follows: "Concern for the man himself and his fate must always form the chief interest of all technical endeavors ... Never forget this in the midst of your diagrams and equations."[2] It was one of the first calls to scientists to be socially responsible. In 2015, Pope Francis, leader of the Catholic faith—with a following of more than one billion people, drew the world's attention to one of the mega issues of sustainability when he said, "Climate change is a dire threat that humans have a moral responsibility to address."[3]

The business leaders of today are challenged now more than ever by global sustainability issues. Author[4] emphasizes the need for sustainability education in business schools to train future leaders and decision makers in what he calls the *Clarion call for green MBA*, and says, "With increasing focus on sustainability factors from the marketplace (regulators, investors, financiers, and consumers), corporate sustainability disclosure is shifting from voluntary to vital. Recent advances in enterprise systems are making it feasible for corporations to track and transform sustainability performance. The materiality of these seemingly non-economic impacts is the critical link between sustainability and business strategy. Leaders need insight into how to determine which sustainability metrics are material to them and relevant to their business." This is the genesis of this book introducing sustainability analytics.

Preface

About a month before his death in 1955, Albert Einstein wrote: "Death signifies nothing ... The distinction between past, present and future is only a stubbornly persistent illusion."[5] Einstein's theory of relativity showed that even though space and time can independently differ for different observers, the four-dimensional space-time reality is the same for everyone; this implies events in space-time have a permanence to them that cannot be taken away. Once an event occurs, in essence it becomes part of the fabric of our universe. Every human life is a series of events, and this means that when we put them all together, each of us is creating our own indelible mark on the universe. Perhaps if everyone understood that, we might all be a little more careful to make sure that the mark we leave is one we are proud of.

Nearly half a century ago, my father gifted me a plaque on my graduation, etched with Einstein's quote: "The concern for man and his destiny must always be the chief interest of all technical effort." That drove me to pursue the environmental field. Several decades later, my children gave me a retirement gift, another plaque, this time with Gandhi's quote: "Earth provides enough to satisfy every man's need, but not every man's greed." This opened my eyes to the fourth dimension, ethics.

Over the past two decades, Dr. Ramanan has had the privilege of meeting with several Nobel laureates in the sustainability area.

In 2004, at a lunch with Nobel laureate Mario Molina, one who discovered the root cause of stratospheric ozone depletion, discussions turned to who parallels his discovery in the climate change arena. What surfaced quickly was the name of Nobel laureate Svante Arrhenius; indeed, his 1896 paper described how carbon dioxide could affect the temperature of the earth. Recent NOAA data shows a strong linkage between the surface temperature of the earth and the carbon dioxide level of the atmosphere. In 2011, during a conversation at an event, Nobel laureate and former vice president of the United States, Al Gore, opined that inaction on climate change can be reversed, if only we put a price on carbon in the markets and a price on denials in politics. Furthermore, he said, "I do not want our children asking, 'Why didn't you act? What were you thinking?' I hope they get to ask, 'How did you pick up the moral courage to rise and change, to do the right thing?'"

Two years later, at a dinner in 2013, Nobel laureate Rajendra Pachauri, chair of the Intergovernmental Panel on Climate Change (IPCC), said to Dr. Ramanan that scientific consensus among the thousand-plus scientists was a tough task—not as much because of differences in scientific views, but more because of the political pressure of the interest groups they served.

Because of the omnipresence of carbon in human life and its global impact on the most vulnerable people, climate change, *the two-degree classic,* truly tests how intergenerational equity and distributive justice are incorporated into making ethical choices. Distributive justice, an ethical mandate, requires that all human beings get an equal share of public goods, such as earth's atmosphere. Absent purpose as a moderator, powerful stakeholders could skew the objective through the inherent bias of self-interest. I hope to continue with my life's mission, which began half a century ago, with the arsenal I am best prepared with: dissemination of sustainability awareness and its quadruple bottom lines; promotion of startups with disruptive innovations; development of robust decision-making processes for corporate and policy stewards; and achieving intergenerational equity and distributive justice for all. This book, which introduces sustainability analytics, is another small step for me, and a path forward for our progeny.

Acknowledgments

I sincerely acknowledge the contributions made by the following people, to the enlightening examples included in the book. Thank you indeed.

ACUMEN[6]
1. Chris Bullard, Senior Post Investment Associate, ACUMEN, USA
2. Kat Harrison, Associate Director, Impact & Lean Data, ACUMEN, UK

BASF[7]
1. Bruce Uhlman LCACP, CLE, ISSP-SA, Manager, Applied Sustainability, BASF, USA
2. Peter Saling, Dr., Director, Sustainability Methods, BASF, Germany

CMU[8]
1. Allen Robinson, Department Head and Raymond J. Lane Distinguished Professor, Center for Air, Climate, and Energy Solutions and Department of Mechanical Engineering
2. Erin N. Mayfield, PhD Candidate, Department of Engineering and Public Policy
3. Jared L. Cohon, President Emeritus and Professor, Scott Institute for Energy Innovation, Department of Civil and Environmental Engineering

Chakr[9]
1. Kushagra Srivastava, CEO and Cofounder, Chakr Innovation, India
2. Arpit Dhupar, CTO and Cofounder, Chakr Innovation, India
3. Bharti Singhla, Head of Strategy, Chakr Innovation, India

IBM[10]
1. Andres Rodriguez, Program Manager IBM Corporate Environmental Affairs
2. Edan Dionne, Manager, IBM Corporate Environmental Affairs
3. Jay Dietrich, P.E., IBM Distinguished Engineer, Energy and Climate Stewardship
4. Wayne S. Balta, Vice President, Corporate Environmental Affairs and Product Safety

Author

Dr. Ram Ramanan has extensive global experience in sustainability; energy; environmental compliance; risk mitigation; ethics; and environmental, social, and governance (ESG) performance metrics. His diverse background includes management, evangelism, and advocacy leadership roles in corporate (ExxonMobil), consulting (AECOM, ICF), and academic (Illinois Institute of Technology Business School, Desert Research Institute (DRI), Nebraska, USA, Indian Institute of Technology (IIT), Bombay) environments. Also, as a US Department of State Environmental Leadership Fulbright Fellow, Dr. Ramanan has provided environmental thought leadership in India.

Dr. Ramanan currently serves as affiliate research professor at Desert Research Institute, a member of the editorial review board of the International Journal of Risk and Contingency Management, and Chair of the AWMA Sustainability Committee. Dr. Ramanan has previously served as industry associate professor and interim director of the Center for Sustainable Enterprise at the Illinois Institute of Technology Stuart Business School, where he was deeply involved at all levels of scholarship and leadership. He promoted R&D projects with industry, pioneered the annual green business roundtable, guided industry sponsored Capstone projects and developed curricula and taught graduate courses to 100+ MBA, PhD and JD students. Ram successfully led the effort to get the school program to be recognized as a top 25 Net Impact Green MBA program.

During his sixteen years at ExxonMobil, Dr. Ramanan focused on air compliance, developed and implemented risk-management strategies and programs, managed corporate liability at multiple superfund sites, and advised several multibillion-dollar ventures on environmental and social impact assessments in Asia, Africa, and the Americas. He was a member of the corporate EHS leadership team for the global implementation of best practice networks and CSR/Citizenship reporting. He also led industry advocacy efforts on air quality and risk-management issues, and actively participated in the development of environmental standards and regulations with the World Bank and federal and state agencies in the United States. Dr. Ramanan received Mobil's Pegasus Elite Award from the president of Mobil Oil for eliminating, with American Petroleum Institute, wasteful potential allocation of industry/society resources (~US$10 billion) from the risk-management program rules through an innovative, tiered approach for onshore oil and gas facilities.

Prior to joining ExxonMobil, Dr. Ramanan managed air quality, community right-to-know, and litigation support services to corporate clients at AECOM-ENSR. As regional leader, Dr. Ramanan also led the development of strategic direction and business plans. Post ExxonMobil, Dr. Ramanan served as the President of a Trinity Consultants' subsidiary and as Fellow at ICF International and provided enterprise environmental, health, and safety management information system (EMIS), corporate social responsibility, and climate change advisory services.

Prior to his doctoral work, for a decade, Dr. Ramanan has, as general manager/sales manager, managed the design, development, and installation of turnkey particle control systems, such as scrubbers, bag houses, and pneumatic conveyors for the steel, paper, and other process plants in India and Thailand, as well as the design and installation of turnkey edible oil complexes in India, and export to Tanzania and Zanzibar in Africa in collaboration with companies based in Belgium, Germany, and the United Kingdom. As a member of the Indian Standards Institution Committee, he actively participated in the development of emission standards for petroleum refineries and power plants.

Dr. Ramanan has served as chair of the Education Council and member of the International Board of Air and Waste Management Association (AWMA), where he led a group of more than fifty academics, including several tenured professors. Dr. Ramanan has also served as industry board advisor to the UT Austin Environmental Solutions Program at the University of Texas at Austin, where he promoted the concept of risk-based decision making that led to the establishment of the Risk-Management Center.

Dr. Ramanan holds a PhD (sponsored by AECOM/ENSR) from the University of Texas at Dallas, an executive MBA (sponsored by ExxonMobil) from the University of Texas in Austin, and a BS in chemical engineering from the Indian Institute of Technology. He is a professional engineer (Texas and India), a chartered engineer (UK), a qualified environmental professional, and a board certified environmental engineer of the American Academy of Environmental Engineers and Scientists. Dr. Ramanan is also an elected fellow of AWMA, Indian Institute of Chemical Engineers and the Institution of Engineers, India.

Dr. Ramanan has presented and published scientific, technical, and management papers, as well as executive briefings globally. He has provided executive briefings to environmental ministry experts from former Soviet Union nations (US Department of Commerce invitees), industry and government leaders of the Nigerian oil and gas sector, Reserve Bank of India research wing (as US Department of State Fulbright Fellow), the Conference Board USA, Northern Trust, and so on. Dr. Ramanan is a coauthor of the book *Environmental Ethics and Sustainability*, CRC Press, 2013, also published in Chinese by China Machine Press, 2017. More recently, Dr. Ramanan has authored two book chapters *Environmental Ethics and Corporate Social Responsibility* for the book *Spirituality and Sustainability: New Horizons and Exemplary Approaches*, Springer, 2016 and *Responsible Investing and Corporate Social Responsibility for Engaged Sustainability: Managing Pitfalls of Economics without Equity* for the book titled *Handbook of Engaged Sustainability*, Springer, 2018.

ENDNOTES

1. Singh, G., Mahatma Gandhi-A sustainable development pioneer, *Eco Localizer,* Accessed October 14, 2008 and available at http://ecoworldly.com/2008/10/14/mahatma-gandhi-who-first-envisioned-the-concept-of-sustainable-development/ in http://www.mkgandhi.org/articles/environment1.htm.

2. Einstein, A., Speech at the California Institute of Technology, Pasadena, CA, February 16, 1931, as reported in The New York Times, February 17, 1931, p. 6.

3. Pope Francis encyclical on climate change, On care for our common home, Accessed June 24, 2015 and available at http://w2.vatican.va/content/francesco/en/encyclicals/documents/papa-francesco_20150524_enciclica-laudato-si.html.

4. Ramanan, R., Need for green MBA and environmental economics education—Globally, *AWMA Annual Conference,* June 2008, Portland, OR; Ramanan, R. and W. Ashton, Green MBA and integrating sustainability in business education, *Air and Waste Management Association's Environmental Manager,* September 2012, pp. 13–15.

5. Bennett, J., Albert Einstein's special mark on the universe, Accessed February 11, 2016 and available at http://www.cnn.com/2015/11/25/opinions/bennett-einstein-theory-of-relativity/index.html.

6. ACUMEN, USA & UK, Accessed October 2017 and available at https://acumen.org/.

7. BASF, USA & Germany, Accessed October 2017 and available at https://www.basf.com/us/en.html.

8. CMU, USA, Accessed October 2017 and available at https://www.cmu.edu/.

9. Chakr Innovation, India, Accessed October 2017 and available at http://www.chakr.in/.

10. IBM Corporation, USA, Accessed October 2017 and available at https://www.ibm.com.

1 Sustainability Analytics and Social Responsibility

1.1 SUSTAINABILITY STEWARDSHIP AND CORPORATE SOCIAL RESPONSIBILITY

1.1.1 CHAPTER OVERVIEW

This introductory chapter defines sustainability and its quadruple bottom lines—profit, planet, people, and purpose. It takes the stewards on a journey through the sustainability mega-forces, risks, and opportunities; genesis of corporate social responsibility; and the significance of responsible investing for sustainable development. Then it identifies relevant stakeholders and underscores the untapped potential of applying analytics. Finally, the chapter covers pertinent key topics—metrics, materiality, indexes, regulatory frameworks, and integrated reporting.

1.1.2 SUSTAINABILITY AND THE QUADRUPLE BOTTOM LINE

Sustainability strives to protect the planet and ensure prosperity for all. A widely quoted definition of sustainable development comes from the Brundtland commission: "Development that meets the needs of the present without compromising the ability of future generations to meet their own needs."[11] From a corporate perspective, this could be seen as an effort to sustain production today without impairing the ability to produce in the future, even within the same generation. The objective has changed from conserving resources to continued productive use of resources that can be grown or renewed. Nobel laureate Robert Solow suggests that our actions now must ensure that the next generation lives as well as the current generation. However, he goes on to say that man-made capital (e.g., machines) are substitutes for natural capital.[12]

A frequently cited interpretation of what sustainability means is "Enlightened self-interest that achieves the triple bottom line of often competing economic, environmental, and social goals."[13] Elkington,[14] in 1997, argued that companies should produce and disclose *the triple bottom line* (*TBL*)—profit, people, and planet—to measure the financial, social, and environmental performance of the corporation over a period of time. TBL is one manifestation of the balanced scorecard.[15] In 1998, Ramanan,[16] taking a similar approach, presented a multidimensional value maximization framework with greater granularity. In addition to maximizing financial performance, this approach calls for, simultaneously, maximizing returns in environmental, (including health and safety) as well as social components (comprising employee, customer, and corporate reputation) (Figure 1.1).

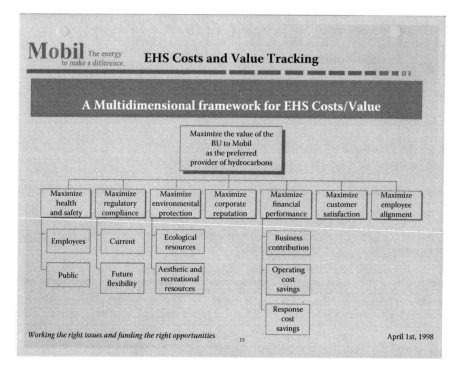

FIGURE 1.1   Multidimensional value generation and maximization framework.[17]

According to Mercer (2007),[18] "contemporary CSR includes corporate governance, employee relationships, customer relationships, environmental management, philanthropy, and community involvement." Based on their mission, concerned activist groups have promoted different aspects as the fourth bottom line: Future-orientation (by highlighting the significance of legacy and importance of the needs of future generations),[19] indigenous people (by differentiating their way of life and its harmony with the environment),[20] employee treatment (by justifying special consideration of people working within the organization),[21] and spirituality (by claiming it to be a monozygotic twin of sustainability—both having the human being at the center).[22]

However, organizations are run by people, and it is the individuals who are responsible for the actions and the consequent outcomes and impact on the corporations and society. Without purpose as a moderator, one could easily skew the objective through the inherent bias of self-or inner-group interest. For instance, US corporate law does not mandate maximizing share price or shareholder wealth, yet the myopic focus on short-term returns for the financial investor is driven by the skewed executive compensation system. In the sustainability arena, unethical behavior often manifests itself in the form of caring only for people who mimic us, protecting only parts of the ecosystem that overtly serve us, and, of course, and generating profits only for a subsection of the stakeholders. Examples of this kind of behavior include misleading investors through under-reporting material environmental risks from climate change, and misleading consumers through *greenwashing*, or over-claiming the environmental friendliness of products and services.

Ramanan[23] presents the inadequacy of triple bottom line and highlights the significance of ethics as the fourth bottom line of sustainability. He expands the triple bottom line context of people, planet, and profit with a fourth component—purpose—to emphasize the power of ethics as a balancing force to preempt the disastrous pitfalls of economics without ethics. Gandhi said, "The path taken matters." In a recent book, Taback and Ramanan[24] present a persuasive reasoning for incorporating ethics as a fourth bottom line to address sustainability bias issues and their impacts on all relevant stakeholders, often within governance, which leads to sustainability and the quadruple bottom line.

1.1.2.1 Profit (Economic or Prosperity)

Nobel laureate Milton Friedman's[25] stance in the context of social responsibilities of firms, that "there is one and only (one) social responsibility of business—to use its resource and engage in activities designed to increase its profits (maximizing shareholder value) so long as it stays within the rules of the game,"[26] dominates corporate ideology today. Intriguingly enough, the US corporate law does not mandate maximizing share price or shareholder wealth.

Profit, or the excess of income over expenses, is essential for longevity and scalability. Profit[27] is a financial benefit, a monetary metric, that is realized when the amount of revenue gained from a business activity exceeds the expenses, costs, and taxes needed to sustain the activity. Profit or loss go to the business's owners, or shareholders. Increasing profit signals increased corporate reinvestment, growth in retained earnings, or increased dividend payments to shareholders. Shareholders gain through profit sharing, dividend payments, or higher stock value.

Prosperity is a more moderate and inclusive term. It makes the economic bottom line also relevant to governments, nonprofits, and charities. United Nations Sustainable Development Goals[28] include good jobs and economic growth, innovation and infrastructure for the profit or economic or prosperity component. In addition to the traditional connotations of corporate profits and economic prosperity, this book will take a corporate perspective and focus on Global Reporting Initiative's (GRI) G4 Sustainability Reporting Guidelines,[29] which is one of the dominant sustainability reporting frameworks considering economic performance, market presence, indirect economic impacts, and procurement practices.

1.1.2.2 Planet (Environment)

Father of a nation of more than one billion people, Mohandas Karamchand Gandhi cautioned, "The earth provides enough to satisfy every man's needs but not every man's greed."[30] He was one of the first to envision sustainable development and its quadruple dimensions. In the words of the twentieth century poet and nature advocate Rachel Carson, "Only within the moment of time represented by the present century has one species—man—acquired significant power to alter the nature of his world."[31] Although, in this millennium, every day new stars and universes are being discovered, and colonization of Mars and the moon are being contemplated; we may have to continue to protect our one and only habitable planet for now. More recently, Pope Francis,[32] leader of the Catholic faith with a following of more than one billion people, has urged that, "Climate change is a dire threat that humans have a moral responsibility to address."

Anthropocentric activities, coupled with the vagaries of nature, such as volcanoes, forest fires, earthquakes, and tsunamis, are pummeling our earth's air, water, and biodiversity. Because of the planet's physical nature, there is an illusion of tangibility! The sheer vastness, variety, and variability make this dimension of sustainability very hard to fully comprehend.

Just the human activities that interact with the earth lead to hundreds of sub-components. For instance, air emissions alone are comprised of sulfur oxides, nitrogen oxides, carbon monoxide, particulates, ozone, volatile organic compounds, and several hundred toxic chemicals. Likewise, global warming encompasses greenhouse gases—water vapor, carbon dioxide, methane, nitrous oxide, ozone, and tens of other chemicals such as chlorofluorocarbons. Additionally, many activities are a result of these air emissions. Hundreds of new classes of materials—such as nano, bio, and genetically modified—are being invented today. Finally, the nexus of air, water, energy, food, and geopolitics very significantly compound the issue.

The United Nations Sustainable Development Goals[33] for the planet or the environmental bottom line include clean water and sanitation, climate action, life below water, life on land, renewable energy, responsible consumption, and sustainable cities. This book will take a corporate perspective and focus on issues covered by global treaties and pollutants covered by major regulations, which apply to select large sectors and are included within the GRI's G4 Guidelines,[34] one of the dominant sustainability reporting frameworks.

1.1.2.3 People (Social)

In the 1900s, Albert Einstein gave this guidance to scientists and engineers: "Concern for the man himself and his fate must always form the chief interest of all technical endeavors... Never forget this in the midst of your diagrams and equations."[35] This was one of the first calls to be socially responsible. The business ambiance has transformed significantly. A principal driver of this societal transformation is the recognition that business is no longer the sole property or interest of a very few. Notably, empowerment through synchronous interactive connectivity among stakeholders has had a significant role in this change. Today, the rise of new business models, such as philanthropic capitalism, venture philanthropy, mission-driven responsible impact investment, and private public partnerships (PPPs) to address global social issues, are all clear signals that the role of the corporation has transformed, and they must address social bottom lines.

The United Nations Sustainable Development Goals[36] has social bottom line components that include no hunger, no poverty, good health, quality education, gender equality, reduced inequalities, peace and justice, and calls for partnerships to achieve these goals. This book will take a corporate perspective and follow the GRI's G4[37] sustainability reporting framework definition of social bottom line and the relevant aspects under its subcategories—labor practices and decent work, human rights, society and product responsibility.

1.1.2.4 Purpose (Ethics or Governance)

People are inherently self- or inner-group serving. Without purpose as a moderator, one could easily skew the objective through the inherent bias of self- or inner-group interest. In the sustainability and environmental arena, unethical behavior most often manifests itself in the form of caring only for people who mimic us, protecting only parts of the ecosystem that overtly serve us, and, of course, generating profits only for a subsection of the stakeholders (such as misleading investors through under-reporting material environmental risks from climate change and misleading consumers through *green-washing*, or over-claiming the environmental friendliness of products and services)."Ethics is the difference between what a person has the right to do and the right thing to do."[38]

Ramanan[39] expands the triple bottom line context of people, planet, and prosperity (or profit) with a fourth component—purpose—to emphasize the power of ethics as a balancing force to preempt the disastrous pitfalls of economics without ethics. Long overdue, only now is governance, incorporating ethics and integrity metrics, becoming a regular part of sustainability reporting under several dominant frameworks like GRI[40,41] and Sustainability Accounting Standards Board (SASB).[42]

1.1.3 SUSTAINABILITY MEGA-FORCES, GLOBAL RISKS, AND TRANSFORMING BUSINESS AMBIANCE

1.1.3.1 Sustainability Mega-Forces

KPMG[43] highlights ten sustainability mega-forces that will impact each and every business over the next twenty years. Governments are intervening to manage externality associated with each mega-force and their intertwined cascading effects aggressively.

1. Climate Change: For example, mean temperature rise is expected to be 0.5°C–1.0°C
2. Energy and Fuel: For example, energy consumption is up 47% while population is up only 26%
3. Material Resource Scarcity: For example, raw materials extraction is up by 55%
4. Water Scarcity: For example, four billion people live now in water-stressed or water-scarce regions
5. Population Growth: For example, total population is now up to seven billion, and percentage of people above the age of 65 is expected to go up by 65%
6. Urbanization: For example, urban dwelling is going up by 44%
7. Wealth Concentration: For example, the richest 1% now has as much wealth as the rest of the world combined,[44] and in 2015 just 62 individuals had the same wealth as 3.6 billion people—the bottom half of humanity[45]
8. Food Security: For example, aggregate demand is going up by 50%, and "the total incidence of undernourishment in the developing countries was 827 million persons in 2005/2007"[46]
9. Ecosystem Decline: For example, ecological damage was estimated to be US$2.15 trillion in 2008
10. Deforestation: For example, net forest cover is going down by 13%

1.1.3.2   Sustainability Global Risks

The World Economic Forum (WEF)[47] defines global risk in this report as an "uncertain event that could cause significant negative impact across multiple nations and possibly industries over the next decade." Risk is a function of likelihood and consequence. It can be mitigated either by reducing the probability of risk or by reducing the potential impact. Climate change, water crises, and large-scale involuntary migration appear on both the impact and likelihood top-ten lists of WEFs 2016 Global Risk Report, making them the highest priorities that deserve attention.

Involuntary migration tops the list of most likely risk, and is complemented by three climate change-related environmental risks: extreme weather events, failure of climate change mitigation, and adaptation and major natural catastrophes. Climate change tops the list of most negative impact risks, with a potential for cascading effect leading to exacerbation of water crises, food security, and agricultural production across geographies, involuntary migration, and social instability. Impact of geopolitical volatility on global fund flow and cooperation makes climate change resolution significantly more challenging.

Finally, it is worth noting that data fraud and theft and illicit trade make it to the top-ten likely risks while fiscal crises and asset bubbles are on the top-ten highest negative impact risks. Failure of national governance is the driver for all four risks, thus making the case for ethics to be the fourth bottom line of sustainability.

1.1.3.3   Transformed Business Ambiance

The business atmosphere and social contracts have changed significantly in recent decades.[48] In the sustainability context, governments and corporations are facing a plethora of ever-expanding management challenges related to issues spanning physical, financial, geopolitical, social, and environmental concerns. Corporations must recognize the strong connection between competitive advantage and sustainability issues. Additional complications arise from the cascading effect of some of these global risks; for instance, climate change has the potential to exacerbate water crises, food security, and agricultural production across geographies, as well as large-scale involuntary migration and social instability. Company productivity and profitability are closely related to environmental impacts (e.g., where the company's operations are dependent on natural resources, supplies may not last if they indulge in unsustainable fishing or logging for timber where nature may not be able to replenish these resources at these excessive rates of depletion), supplier access and viability, employee skills, health and safety, and resource use (e.g., water, energy, rare earths).

Moreover, as the world becomes interconnected, it draws a large population of poor and untapped human talent into the marketplace and presents an opportunity to address the bottom of the pyramid. Not only is addressing the bottom of the pyramid the right thing to do, it is also an opportunity to extract the treasure. Growing disparities in wealth increase risk, while being inclusive may significantly enhance global harmony and growth. Being inclusive is not merely driven by the goodness of heart or a pure desire to be good citizens, but has become imperative for long term survival and it is a sound business practice. This is also an opportunity for governments and private corporations to partner to serve the people prosperously or profitably.

1.1.3.4 Bottom of the Pyramid[49]

Since the 1990s, the late professor C. K. Prahlad and his fellow researchers at the Ross School of Business at the University of Michigan argued for corporations to find their fortune at the bottom of the pyramid (BOP) and eradicate poverty and hunger. There are more than four billion people who live under the poverty line but do have need for products and services. Prahlad said that "low-income markets present a prodigious opportunity for the world's wealthiest companies—to seek their fortunes and bring prosperity to the aspiring poor."[50] Global overcapacity and intense competition in other higher tiers of prosperity make this a more attractive opportunity. The fact that people in the poorest rung of income and wealth are becoming aware of and aspiring to obtain products and services has huge implications for multinational corporations (MNCs). This is a huge untapped market, and MNCs need to create affordable products and services for these populations.

Strategies for MNCs to make a fortune serving the BOP include a radically different price performance and a call for innovation across the board, new product development, manufacturing, and distribution given the need for a new price structure based on lower purchasing power. Quality has to be viewed in the profoundly different context of robust products for harsh conditions and drastically lower prices. Lower price and lower quality with a price-quality trade-off is an option, but ethical behavior calls for transparency. Some products may not need to last as long. Profitability has to be revisited from the perspective of low margins, very high volume and high investment intensity. Finally, consistent with the other changes occurring globally, sustainability must be considered—a reduction in resource intensity and use of recyclability and renewable energy.

Countering Prahlad's radical approach, Karnani states that MNCs are to seek fortune at the BOP and eradicate poverty. Karnani says[51] that the poor are vulnerable, lack education, are ill informed, and often make choices that are not in their interest. The poor are exploited, and he recommends alternate strategies for the BOP. He sides with Nobel prize-winning economist Amartya Sen's view, "A person's utility preferences are malleable and shaped by one's background and experience, especially so if one has been disadvantaged. We need to look beyond the expressed preferences and focus on people's capabilities to choose the lives they have reason to value."[52]

Karnani further recommends that the role of the private sector be to view poor people as producers and create labor intensive jobs; and not to exploit them as consumers. This is a case of market failure, and this externality needs intervention through legal, regulatory, and social mechanisms to protect consumers. "The BOP approach relies on the invisible hand of free markets to eradicate poverty ... instead, the State should extend a very visible hand," says Karnani.[53] The growth of private-public partnerships to build infrastructure, especially in underserved markets, is a very welcome global trend.

1.1.4 GENESIS OF CORPORATE SOCIAL RESPONSIBILITY

Einstein's contemporary, and the father of a nation of more than one billion people, Mohandas Karamchand Gandhi cautioned, "The earth provides enough to satisfy every man's needs but not every man's greed."[54] He was one of the first to envision sustainable development and its quadruple dimensions. In 1953, Howard Bowen[55] wrote, "Corporate social responsibility (CSR) expresses a fundamental morality in the way a company behaves toward society," and birthed the term Corporate Social Responsibility (CSR).

In 1994, John Elkington[56] coined the term *triple bottom line*. In 1998, Ramanan[57] presented a multidimensional framework for tracking environmental cost and value at the US conference board, calling for value generation in addition to financial performance, through ecological resource protection, employee alignment, customer satisfaction, and corporate reputation. In 2000, Jensen[58] advocated a single integrated objective function for the corporation and said, "Enlightened stakeholder theory adds the simple specification that the objective function of the firm is to maximize total long-term firm market value. The corporate social responsibility debate has intensified greatly in the last decade. A recent edict by Pope Francis[59] on mitigating climate change risks as a moral duty was followed by his urging of the president of the United States, Donald Trump, to not withdraw from the Paris Climate Change Agreement."

Success of a firm should be measured by changes in total long-term market value of the firm. Maximizing a firm's long-term market value precludes mistreating any important stakeholder. However, this calls for aggregating the multi-capital performance measures of a firm, including several intangibles, into one common denominator, defining a common long-term duration and discount rate.

1.1.4.1   Why Should Corporations Be Socially Responsible?

Porter[60] lays out a simple description of the evolution of responses to the question, "Why should a corporation be socially responsible?" The early arguments calling it a *moral obligation to be a good citizen* and do the right thing is not tenable, as they call for individual corporations to make decisions by balancing competing values (e.g., whether to subsidize today's medicine or to invest in tomorrow's cure). He goes on to point out the difficulty with the *license-to-operate* argument as well. While it is obvious that communities do provide permission to the corporation to operate within their confines and utilize their resources in exchange for employment, tax revenues, products, and services, this approach has serious limitations. Government intervention addressing external elements through a permit process, such as requiring environmental and social impact assessments, has been moderately effective. An inherent part of the process is consulting and engaging stakeholders. However, the process of obtaining stakeholder permission by meeting their needs has a great inherent danger: it could easily lead the corporation and often the relevant community to cede their CSR agenda to the most vehement stakeholder or group. A well-organized and articulate interest group could easily get its way at the expense of what may be more socially equitable.

Building a corporate image and brand loyalty is invaluable in the marketing space. It is recognized as at least a moderately successful technique for promoting the value of corporations' socially responsible behavior. In the era of marketing, in which perception is promoted as reality and reputation and brand are pushed as factors that mattered most, CSR arrived as a gift from heaven to public relations. At this promotional stage of CSR, public relations are used to reach the public and, in particular, consumers and investors. Franklin[61] says, "Many companies pretend their sustainability strategy runs deeper... Need less misdirection and more redirection." Karnani[62] notes that corporations resist fundamental transformation and often indulge in cosmetic campaigns and green-washing.

Green-washing, which in one form is lobbying against public interest, took on a life of its own. A classic case of ethical failure was the use of smoke (pun intended) and mirrors by the tobacco industry to hide the negative health impacts of cigarettes by showing macho athletes smoking. Another example is the campaign by contrarian scientists and industry-sponsored think tanks to sow doubt about climate change. Not surprisingly, many of the same *expert advisers* were involved in both tobacco and climate change issues in the US.

Getting consumer purchasing preference as insurance to temper public wrath in case of a company-caused crisis has been frequently observed and even documented in both academic and industry trade literature. There have been cases of tainted products reaching the consumer market; this was the case with Tylenol, but the well-developed public perception of Johnson & Johnson being a responsible company tempered negative reaction from the consumer. While there are numerous anecdotal events to support this kind of tempered public reaction, it still suffers from lack of conclusive correlations to good deeds. These correlations to show that good deeds help temper negative reactions in case of a company-caused crisis are nebulous at best, and there is no conclusive evidence that shows causality. For instance, it could not fend off the recent jury award of US$72 million against Johnson & Johnson on talc causing ovarian cancer; an example of unethical behavior by the company in the face of their 1997 internal memo from a company medical consultant, who said "anybody who denies" the link between use of hygienic talc and ovarian cancer is "denying the obvious in the face of all evidence to the contrary."[63]

Finally, corporate sustainability has been promoted as an act of enlightened self-interest to secure long-term economic performance by avoiding short-term social or environmental cost cutting. A frequently cited interpretation of what sustainability means is *enlightened self-interest that achieves the triple bottom line of often competing, economic, environmental, and social goals.* Porter criticizes the term as vague and lacking clarity because the required trade-offs remain undefined. Sustainable actions by corporations occur successfully when the triple bottom line goals are congruent—for example, energy conservation saves money, reduces emissions, and serves more members of the society—affordable, and available. But there is a huge gap in this model. Porter says there is no defined framework to validate trade-offs and optimize value. Furthermore, Ramanan[64] suggests absent ethics as the fourth bottom line; in other words, sans purpose as a moderator, even congruence of the other three may be skewed, and not incorporate issues such as distributive justice or intergenerational equity in the decision-making process.

1.1.4.2   Advent of Strategic Corporate Social Responsibility

The management age of business in relation to CSR called for aligning priorities with companies' core operations and competitive strategies. The concept was to embed CSR in management systems, let the voluntary CSR metrics set the bar, and let divisions and the managers make a business case for CSR. This approach made the shareholders and the aggressive NGOs the target stakeholders.

Porter[65] suggests that corporations should prioritize the social issues they choose to address based on their relevance to corporations' core business and the value they add and should only indulge in social issues that are congruent with a competitive advantage for them. Generic social issues—those that are not significantly affected by a company's operations and the resolution of which do not materially affect its long-term competitiveness (such as carbon emission for a bank)—should not be addressed. On the other hand, value chain social impacts are more relevant. They are significantly affected by a company's operations in the course of ordinary business (e.g., carbon emissions of the UPS delivery fleet). Further up on the totem pole are the social dimensions of a competitive context—where issues in the external local environment such as consumer preference or local regulations may greatly affect a company's current or midterm competitiveness (e.g., carbon emissions from Toyota's automobiles).

Having identified the issues of importance, the implementation of strategic CSR—in which good citizenship mandates efforts to do no harm and mitigate harm that occurs—is a given. This could go beyond legal compliance. For instance, a corporation may apply their more stringent environmental standards to operations in regions with lower requirements. This is common among major corporations in the oil and gas sector, as well as, many large multi-national corporations. This is congruent with their need to protect their image and minimize liability. The transformation from a defensive stance to a strategic CSR is the effort focused on a win-win, which benefits society while reinforcing strategy—investing in social aspects that enhance competitiveness. Karnani,[66] however, goes on to say that this is clearly a business imperative, and calling it strategic CSR is a misnomer.

1.1.4.3   Transforming to Systemic Corporate Social Responsibility

Visser[67] notes that all that happened with the advent of strategic CSR was that it moved from peripheral to incremental to uneconomic. However, more of the same CSR will not meet the challenge, and he calls for systemic CSR in this business age of responsibility. At a macro level, CSR shifts from philanthropic, image-driven, and standardized to collaborative, reward-based, and scalable. It includes innovative partnerships, stakeholder involvement, social entrepreneurship, and real-time integrated reporting. At the micro level, the shift will be from charity projects, product liability, and ethical consumerism to serving the bottom of the pyramid, social enterprise, and ratings and data streams for performance tracking and transparency.[68]

1.1.4.4   Morphing from Strategic/Systemic Corporate Social
Responsibility to Creating Shared Value

The previous sections present how CSR morphed from a basic defensive necessity to a strategic need on to a systemic imperative. It highlights the link between competitive advantage and corporate social responsibility and shows how the emergence of an integrated systemic responsibility approach drives corporate value. Porter,[69] guru of strategy from Harvard, has introduced the big ideas of "creating shared value" and a "higher form of capitalism." Increasingly, leaders recognize that purpose matters and that there is a higher form of capitalism. At the same time, the role of business is transforming from meeting a social contract to realizing tangible economic gains by creating shared value. Porter calls for transformation from strategic CSR and redefines the purpose of the corporation as creating shared value, not just profit, per se, and presents a new bottom line for corporations to pursue—profits involving a social purpose, one that represents a higher form of capitalism.[70] He defines the creation of shared value concept as "policies and operating practices that enhance the competitiveness of a company while simultaneously advancing the economic and social conditions in the communities in which it operates."

Creating shared value (CSV) is not simply a matter of philanthropy, sustainability, or executing social responsibility but is a new way to achieve economic success in which societal needs, not just conventional economics, define markets, and social harm is recognized in order to create internal costs for firms. A basic tenet of creating shared value is that both social and economic progress must use value principles, where value is defined as benefits relative to costs. There are numerous ways in which corporations could create value by reconceiving products, markets, and business models. Examples include offering more nutritious food; harnessing digital intelligence (e.g., IBM) to conserve consumer power usage and cut costs (e.g., using differential peak rates); using less energy, natural resources, and water; producing healthier and eco-friendly products; delivery of books, magazines, software, music, and movies (e.g., Netflix) via the Web, thus eliminating materials and reducing costs. In addition, business practices that encourage the ethical creation of value across the value chain could eradicate social harm and simultaneously eliminate costs for the corporation. Examples include not marginalizing small suppliers (eliminating them would hurt company procurement), providing living wages, building supportive industry clusters at the company's locations, and creating open/transparent markets. Employing these strategies provides better incentives for quality and efficiency and ensures the long-term sustainability and reliability of supplies. Creating these positive cycles of economic and social development also leads to seeding new companies and an increased supply of skilled workers.

Porter nicely contrasts CSR and CSV: Value gets redefined as creating economic and societal benefits relative to cost rather than doing good. The goal moves from citizenship to philanthropy, sustainability to joint company and community value creation. The driver for action changes from response to external pressure to one that is an integral part of competitiveness. The activity becomes integral in profit maximization. Also, the impact is no longer limited by the corporate footprint and CSR budget, and it realigns the entire company budget. Finally, external reporting and personal preferences do not determine the agenda; the agenda is company-specific and internally generated. Governments, with the right kind of regulation, could encourage companies to create shared value and avoid unhealthy trade-offs between economic and social goals.

1.1.5 SUSTAINABILITY TSUNAMIS AND THE NEW SOCIAL CONTRACT

1.1.5.1 Sustainability Tsunamis[71,72]

Financial tsunamis and environmental disasters that surfaced in the last two decades were unprecedented. Multiple environmental disasters occurred such as the US Environmental Protection Agency (EPA) Anima River toxic spill, BP Gulf of Mexico oil spill, and the devastating storms Sandy, Katrina, and Allison, all commonly attributed to climate change. In Rachel Carson's now famous words, "Only within the moment of time represented by the present century has one species—[hu]man—acquired significant power to alter the nature of his world,"[73] especially when seen in the context of climate change. Concomitantly, several "too big to fail" financial institutions, such as AIG, Fannie Mae, Freddie Mac, Lehman Brothers, Merrill Lynch, and Washington Mutual, fell like dominoes—some of the largest failures in US history. These were many times faster, much more obscure and global in reach compared to any, including the Great Depression of 1929.[74]

The speed and scale of these greed-driven and callous actions make them potentially catastrophic; the penalties for ignoring apparently nonfinancial environmental, social, and governance (ESG) issues are severe. The following examples cover a spectrum of colossal unethical behavior: BP cutting corners on required safety and maintenance expenses to achieve next quarter returns; Enron executives creating complicated illusory and fraudulent accounting to mislead shareholders and lenders and taking an average executive bonus of $50 million, just prior to the failure of the company, while firing employees with a severance of $50,000; Bernard Madoff's downright predatory cheating of uninformed speculators in a Ponzi scheme of investments; deplorable working conditions and human rights violations at Apple's manufacturer Foxconn in China, which halved its market capitalization; and UBS's manipulation of the London Interbank Offered Rate (LIBOR), a key global interest rate which led to $1.5 billion in fines.[75] In each case, the company prioritized financial over ESG performance, and to varying degrees, one or more of the three bottom lines—economic, environmental, or social—were overemphasized to the detriment of ethics that led to devastating results. The Compliance and Ethics Leadership Council[76] observed that companies with weak ethics cultures had five times as many incidents of misconduct as those with strong ethics cultures, and this once again highlights the need for ethics as the fourth bottom line of sustainability.

Opacity is the arsenal of deception in the financial sector. The tsunami of toxic or troubled assets highlights predatory lending; greed at its peak. Toxic assets, because of significant drop in value, lack an active market to trade—for example, subprime mortgages. Speculators created financial derivatives, a security whose price depends on or derives from one or more underlying assets. The author terms them *Wall Street neutrons*[77] to indulge in speculative bets in which most trades are speculations outside real economy. As Visser[78] puts it, "Speculators may do no harm as bubbles on a steady stream of enterprise. But the position is serious when enterprise becomes the bubble on a whirlpool of speculation." Once again, a parallel could be drawn between unethical companies that would rather not disclose their material environmental risks and liabilities to those that would hide their large financial misappropriations and losses, both for fear of its impact on the next quarter stock price. Transparency is the only known antidote to fraud veiled in complexity.

1.1.5.2 The New Social Contract

Corporations and their contracts with society and the relevance of stakeholders have been debated, from Nobel laureate Milton Friedman's[79] stance on the social responsibilities of firms: "There is one and only social responsibility of business—to use its resource and engage in activities designed to increase its profits (maximizing shareholder value) so long as it stays within the rules of the game,"[80] to Edward Freeman's[81] view on how firms can take into account the interests of several stakeholders on to the *new social contract*, defined as "where business is one thread in the complex web of interwoven society … [and] often becomes responsible, cradle to grave, for not just its inanimate inputs and outputs, but for all related human and environmental interactions."[82]

Capitalism in general and the American dream in particular interprets greed to be a healthy trait. "Greed is good—[it] has marked the upward surge of mankind,"[83] is the mantra of the early business age. The only target stakeholder for a business in the pervasive greed stage is the shareholder. This early business age mantra, along with an obsession with the primacy of shareholder interests, has driven most early entrepreneurial efforts to privatize gains and socialize costs. Defining US corporations as legal persons mandated to relentlessly pursue shareholder benefits exemplifies corporate greed, especially given that there is no such mandate in US corporate law. Greed became pervasive in business among executives, corporations, banks, and financial markets. Costs of external consequences such as pollution or human rights violation were pushed to be borne by society while the shareholders benefited. Initiatives that help society such as emission reduction or above sustenance wages were seen as costs to the company.

Business slowly morphed in part from pure greed to some form of philanthropy. The realization was that with great wealth comes great responsibility. Business should give back to society to let economic benefits trickle down. But the motto was *first get rich and then get generous* (the power of largess). Several organizations followed Carroll's[84] weighted CSR pyramid: be profitable, be legal, be ethical, and be generous. Over the past century, there have been many iconic charitable leaders. John D. Rockefeller Sr., the founder of Standard Oil, believed in giving back to society and sharing the fruits of success. His approach was called post-wealth generosity—making lots of money first and then dedicating oneself to distribute those riches and leaving a legacy. Andrew Carnegie, the steel magnate, had a three-part dictum: first, to get all possible education one can; second, to make all the money one can; and third, to give it all away to worthwhile causes. Institutional philanthropy—charitable donations directly from business profits to causes such as United Way—has been around since the 1900s. This stage targets communities as stakeholders through charitable donations.

"The corporate scandals and implosions of the past decade, climaxing in the recent global financial crisis and environmental disasters, have highlighted how critically ethically, environmentally, and socially responsible decision making and leadership are to the long-term survival and success of both individual businesses and society."[85] Today, with greater recognition that shareholders are only one of many stakeholders, there is a consistent and continual move from shareholder primacy, towards stakeholder primacy, and it is impossible to ignore the changing business ambiance and social contract under which corporations have to operate. A principal driver of this societal transformation is the recognition that business is no longer the sole property or interest of a very few. Notably, empowerment through synchronous interactive connectivity among stakeholders has had a significant role in this change.

Taback and Ramanan[86] capture the impact of these developments as, "In today's global environment, societal needs are defining markets, and business leaders have to address a range of issues from poverty and hunger to sustainability and ethics. Ethical issues include bribery, fraud, green-washing, and a culture of corruption. Corporations and leaders have to manage corporate social responsibility and integrate it into their global strategy instead of treating it as just a moral obligation or a risk/reputation management exercise. They also have to endeavor to build new competencies in managing transparency, accountability, stakeholder engagement, ethics culture and social innovation, which are critical for business success in the Next Economy."

The role of the public corporation and the nature of the *social contract* have been changing over the past two centuries but have changed at a faster pace in recent decades. Business used to be charged with producing a functioning product and selling it at a profit, often at the head of a linear chain. Today, business often becomes responsible from the cradle to the grave. Is the offering safe throughout its life cycle? Does the process violate any ethical principles on its path? Does it meet the most pressing social needs, and if so, does it do so effectively and responsibly? The process brings in the issue of equity, which was for too long a missing component of market efficiency. Water is no longer just a beverage; the question today extends to whether the farmer was equitably compensated for the diverted water. Was it a fair trade? Did it avoid increasing stress on community access to water? An attitude of "that's not my job" is not an option anymore. Companies, in particular major corporations, can no longer outsource negative societal consequences. Leaders will increasingly be assessed on long-term impact on societal well-being.[87]

1.1.5.3 Expanded Fiduciary Duty

The financial investment activity, as the economic engine of all businesses, has a very unique role in sustainable development and risk management. Its stewards, institutions, private equities, banks, project lenders, governments, venture capitalists, and individual equity investors all have the equivalent of a fiduciary duty to channel investments into operations that are responsible from longevity, risk, and reward perspectives. The European Union has adopted Directive 2014/95/EU[88] on the disclosure of environmental, social, and governance metrics and diversity information from publicly listed companies with more than 500 employees. The directive is under transposition to national law by EU member states. Multiple stock exchanges and governments worldwide have mandated significantly greater transparency in reporting material risks to protect individual equity investors. For instance, fiduciary duty of investors calls for climate change to be treated like other highly modeled investment/longevity risks. In recent years, the insurance and reinsurance sector have taken substantial steps in dealing with environmental callousness. Swiss Re and Marsh offer carbon risk insurance products; however, they have cautioned clients that senior executives could lose insurance protection against climate change-related liability claims by shareholders if they fail to adopt adequate risk mitigation policies.

These developments clearly define the direction of the new metrics by which public policy stewards, corporate leaders, and institutional investors are increasingly likely to be measured, and which will potentially determine their success, longevity, and compensation in years to come. They include, without a doubt, a call for sustainability stewardship—with an expanded fiduciary duty, which could benefit greatly from the application of sustainability analytics.

1.2 SUSTAINABILITY ANALYTICS, STAKEHOLDERS, AND SUSTAINABLE DEVELOPMENT

1.2.1 CHAPTER OVERVIEW

This chapter begins with an introduction to the key capital drivers of organizational success and introduces the potential of sustainability analytics. Then, following a review of sustainable development goals and the relevant stakeholders, it presents a primer on sustainable development economics, public goods, externality, public policy, and the role of advocacy. It also provides a discussion on economic instruments and regulatory policy elements available to governments to intervene and ensure sustainable development. The final subsection highlights the emergence of public–private partnership (PPP) for sustainable growth.

1.2.2 KEY CAPITAL DRIVERS OF ORGANIZATIONAL SUCCESS

The success or failure of an organization is driven by six types of capital:[89] financial, physical, intellectual, human (employees), social, and natural (environmental). Usually, the financial and physical, and occasionally the intellectual and human capital performances, are captured in the traditional organizational reports. The governance or ethics capital component is conspicuous by its absence. Barring some exceptions, until recently, the social, natural (environmental), and governance or ethics capitals have for way too long been treated as issues external to the organization's performance. Organizations, to ensure long-term viability of their business model, must also recognize that the foundation of their value creation potential depends on their ability to access resources, attract and retain talent, generate ideas for innovation, and build strong brands, customer trust, and loyalty.

1.2.3 SUSTAINABILITY ANALYTICS—THE UNTAPPED POTENTIAL

Business analytics and big data tools enable organizations to convert the raw data into actionable insights to achieve their sustainability goals. SAS[90] reports that almost 46% of C-level executives believe that analytics can have an impact in the area of sustainability. In the private equity industry, per Insead Global Private Equity Initiative (GPEI),[91] while finding a comprehensive way to measure impact of operations and ESG actions is a challenge, "Solid data and targeted analytics can be a critical driver for motivating and directing a company's efforts."

Deloitte uses the term *sustainability analytics* as a new field of expertise that integrates big data analytics with sustainability. They define sustainability analytics as "an approach that aims to effectively use technology to collect, disseminate, analyze, and use sustainability-related information across the enterprise."[92] Ernst & Young (EY)[93] states that environmental, health and safety (EHS), and sustainability analytics programs help companies reduce risks and drive cost savings. The first three components listed by EY, summarization, visualization, and statistical econometric analysis, are elements of more traditional analytics. The next three analyses, comprising of spatial, human-driven algorithmic, and heuristic machine learning represent the more advanced analytics applications today.

The right relevant measures that relate active ESG management to monetary value help integration and enable stewards focus on the key performance indicators (KPIs) that drive value. A few sustainability analytics applications[94,95] are now on the market with niche capabilities to analyze the data and provide insights that help transform sustainability information into action across the value chain and life cycle. The untapped potential of sustainability analytics is only now beginning to be recognized and used.

1.2.4 SUSTAINABILITY AND RELEVANT STAKEHOLDERS

All living species and inanimate objects are impacted by sustainability. Stakeholders are individuals or groups that have an interest in the actions and impacts of an organization (e.g., customers, suppliers, shareholders, employees, communities, regulators, special interest groups, and nongovernmental organizations [NGOs]).[96]

The UN[97] identifies several major groups as stakeholders for sustainable development. Stakeholders relevant to sustainability are participants, influencers and the vulnerable. Participants are directly involved in the commercial exchange process of business and industry and include shareholders, corporations, financial institutions, state owned enterprises, supply chains and consumers. Influencers are instrumental in the development of public opinion and policy and include local authorities and NGOs such as regulatory agencies, scientific communities, public interest activists, industry trade groups, and the media. The third group comprises of the vulnerable sections of society that require special protection from exploitation, including employees, women, children, and select socioeconomic groups. The following representation is a regrouping in accordance with their role in sustainability and potential utility of analytics.

1. *Participants—business and industry*: Shareholders, corporations, financial institutions, government enterprises, supply chains, and consumers
2. *Influencers—local authorities and nongovernmental organizations and the scientific community*: Regulatory agencies, academics, consultants, researchers, public interest activists, industry trade groups, and the media
3. *Vulnerable and special social groups—workers, women, children, and special social groups*: Employees, contractors, trade unions, women, children, and select socioeconomic strata.

While every group is either impacted or impacting the others through their exchange interaction, groups A & B have a more significant role in managing the sustainability impact on the other groups and themselves. The stakeholders in group C are usually the impacted and often play a defensive activist role in protecting their interests. Equity and institutional investors as well as corporate and public policy stewards, clearly have a dominant role in contributing to sustainable development decisions and accomplishing the quadruple bottom line. This makes them particularly attractive as the key users and beneficiaries of sustainability analytics applications.

1.2.5   Sustainable Development Goals—UN Agenda 2030

In 2015, the UN has adopted the following as the sustainable development goals (SDGs),[98] which replaced the millennium development goals (MDG).[99] On January 1, 2016, the 17 sustainable development goals (SDGs) of the UN agenda 2030[100] for sustainable development, which was adopted by world leaders in September 2015, officially came into force. The seventeen SDGs seek to end poverty, protect the planet, and ensure prosperity to realize the human rights of all and are intended to be integrated, indivisible, and targeted to balance the triple bottom line of sustainable development: economic, environmental, and social. These 17 SDGs may be categorized under the triple bottom line of sustainable development as:

1. *Economic*: Good jobs and economic growth, innovations, and infrastructure
2. *Environment*: Clean water and sanitation, climate action, life below water, life on land, renewable energy, responsible consumption, and sustainable cities
3. *Social*: No hunger, no poverty, good health, quality education, gender equality, reduced inequalities, peace and justice, and partnerships for these goals

However, the governance or ethics component is conspicuous by its absence (Figures 1.2 and 1.3.).

1.2.6   Sustainable Development Economics and Public Policy

1.2.6.1   Economic Theories and Public Goods

Sustainability economics is an applied field of economics and covers the study of the impact of the economy on the environment and society, and the significance of the environment and society to the economy. It differs from ecological economics, which takes a biophysical view, and also from resource economics, which deals with the production and use of natural resources. Environmental and social issues are often tough sociopolitical choices, and economists continue to battle several market myths in these highly charged debates of public opinion. It is worth reviewing how economists are perceived to see sustainability, especially environmental and social equity issues. Let us explore some myths.

Goal 1	No Poverty	End poverty in all its forms everywhere
Goal 2	No Hunger	End hunger, achieve food security and improved nutrition, and promote sustainable agriculture
Goal 3	Good Health	Ensure healthy lives and promote well-being for all at all ages
Goal 4	Quality Education	Ensure inclusive and equitable quality education and promote lifelong learning opportunities for all
Goal 5	Gender Equality	Achieve gender equality and empower all women and girls
Goal 6	Clean Water and Sanitation	Ensure availability and sustainable management of water and sanitation for all
Goal 7	Renewable Energy	Ensure access to affordable, reliable, sustainable, and modern energy for all
Goal 8	Good Jobs and Economic Growth	Promote sustained, inclusive, and sustainable economic growth, full and productive employment and decent work for all
Goal 9	Innovation and Infrastructure	Build resilient infrastructure, promote inclusive and sustainable industrialization, and foster innovation
Goal 10	Reduced Inequalities	Reduce inequality within and among countries
Goal 11	Sustainable Cities and Communities	Make cities and human settlements inclusive, safe, resilient, and sustainable
Goal 12	Responsible Consumption	Ensure sustainable consumption and production patterns
Goal 13	Climate Action	Take urgent action to combat climate change and its impacts
Goal 14	Life Below Water	Conserve and sustainably use the oceans, seas, and marine resources for sustainable development
Goal 15	Life on Land	Protect, restore, and promote sustainable use of terrestrial ecosystems, sustainably manage forests, combat desertification, halt and reverse land degradation, and halt biodiversity loss
Goal 16	Peace and Justice	Promote peaceful and inclusive societies for sustainable development, provide access to justice for all, and build effective, accountable, and inclusive institutions at all levels
Goal 17	Partnerships for the Goals	Strengthen the means of implementation and revitalize the global partnership for sustainable development

FIGURE 1.2   United Nation sustainable development goals (SDGs).[101]

1.2.6.2   Myth of Markets

The myth of the universal market is that an invisible hand of market efficiency solves all problems. Private markets are perfectly efficient with no government interference, if they meet certain conditions. The fundamental welfare theorem of economics states that under certain conditions, market outcomes are efficient. There should be no public goods, no externality, no monopoly buyer or seller, and no unlimited scale of economy. Stock markets meet most conditions. However, some of those *required conditions* might not be met for environmental and social equity problems, and unless the market can self-regulate these problems, a government intervention

FIGURE 1.3   United Nations sustainable development goals icons.

may be warranted. For instance, environmental pollution has a negative externality, and natural resources are public goods. Economists do understand this. They know that universal market does not solve all problems, and they do recognize that laissez-faire policy for market externalities (e.g., environmental pollution) leads to social inequities and requires government intervention to ensure fairness and justice for all people.

The myth of market solutions assumes that economists will always recommend market solutions—this is not true. Economists recognize that tradable permits or emission to cover social costs may work for uniform mixing pollutants (e.g., acid rain causing sulfur dioxide from power plants) because they meet most conditions necessary for market efficiency. The same market-based solution will not work for local impact hazardous air pollutants (e.g., benzene) because that could lead to hot spots at high-impact locations. Rather than control or minimize benzene emissions from the manufacturing facility, it is easier to buy benzene emission credits from a large number of low benzene emission impact locations and satisfy government regulatory requirements to get operating permits. This often leads to the environmental regulatory authorities permitting unusually large numbers of manufacturing facilities that emit benzene in one concentrated area (creating a neighborhood that has very high levels of benzene in the air, referred to as hot spots for benzene). These in turn result in very high levels of benzene levels in the air in lower eco-strata locations, for people living closer to those high benzene emission manufacturing plants. This is referred to as environmental injustice.

The myth of market prices blames economists for using market prices to evaluate even non-market solutions. They realize that the economic value of damages to human health are much higher than the sum of health-care cost and lost productivity. They recognize that human suffering in terms of stress and pain of trauma to self or to a loved one far exceeds the obvious tangible economic impacts. Similarly, the economic value of the Amazon forest is much greater than just the aggregate market value of future medicines in yet to be discovered herbs, value of millions of trees that serve as carbon sinks, and the value of lost ecotourism. The forests have a non-use value based on just being there—wilderness as a world heritage for future generations of humanity. However, for the purpose of prioritization of resource allocation, aggregation of these factors becomes unavoidable.

Finally, the myth of market efficiency blames economists for being concerned only with market efficiency and ignoring equity or distribution effects. Distributional equity is difficult to value. Environmental and social regulations are not yet good at resolving this issue. Environmental justice in the US is a first step, and lawsuits and activism continue to push it. Economists are only now learning to incorporate these equity or distribution effects through multi-attribute trade-off analysis.

1.2.6.3   Market Externality

In 1920, Arthur Pigou, a Cambridge University economist, developed a concept of economic externalities. Market exchange generally involves a supply and a demand side, and together in a free market situation, market efficiency is achieved. However, if the seller and the buyer market exchange process impact an external entity that has no say in setting the exchange price, a market externality is created. For instance, a producer continues to produce a product and does not incur the cost of pollution. The producer, given the lower cost of production incurred, can sell the product cheaper and yet make a profit; and the buyer will increasingly consume more of that product at a lower price. However, the community impacted by the pollution from the producer has no say in the pricing process, and that creates a negative externality. Positive externality is where there are public benefits from a market activity, and government subsidies are needed to avoid undersupply of a product. The US Medicaid program to provide health care to low economic strata individuals is an example of this.

In the environmental context, tax incentives for renewable energy projects are a prime example. Negative externality is where the social cost of market activity is not covered by the economic exchange, and this leads to product overconsumption. A tax equal to the value of the negative externality results in market efficiency. The social cost of pollution is often greater than the private cost to the polluter, and the government needs to intervene with a tax on pollution to minimize the impact on external constituents. These constituents are often outside the market exchange, but get exposed to the pollution generated by the market activity. These are Pigovian taxes or fees, generally paid to the government. It is the marginal savings from pollution at the optimal pollution level or "a fee paid by the polluter per unit of pollution exactly equal to the aggregate marginal damage caused by the pollution at efficient level of production."[102] The equi-marginal principle, in controlling emissions from several polluters, requires that the marginal cost of control be equated across polluters to achieve emission reduction at the lowest possible cost. Pigovian fees satisfy the equi-marginal principle, since all polluters set their marginal savings at the same value, that of the Pigovian fees. It is worth noting the difference between effects of fees versus subsidies on market efficiency. In the environmental context, the polluter pays fees or gets tax subsidies for installing pollution control; are these similar? If not, which is better? Subsidies breed inefficiency. A subsidy may allow firms to continue, while if a company has to pay fees, it may cease to exist.

Subsidies make polluting processes seem more attractive. Fees have the opposite effect, promoting innovation or efficiency.

1.2.6.4 Classification of Goods

Classification of goods is a fundamental concept in economics and quite relevant to the environmental space. Excludability, a basic trait of a product that allows assignment of private property rights, is a necessary condition for the invisible hand of market efficiency to work. A valuable product is excludable if it is practical and feasible to selectively allow consumption of goods (e.g., cars, food, movies). A harmful product is excludable if it is practical and feasible to selectively allow consumers to avoid consumption (e.g., cigarettes). Ethics plays a significant role in making choices between valuable and harmful products; most products offer both benefits and risks. A common example is a medicine that cures and yet has undesirable side effects. Should it be permitted, what levels of constraints are needed to prevent misuse (e.g., a habit-forming prescription drug for instance, opioids that help manage pain but are habit forming or addictive). Similarly, rivalry is another characteristic of a product where consumption by one individual eliminates the possibility of another to consume or enjoy the same. For instance, beverages, food, clothes, and airline travel are classified as rival products. In these situations equitable production and distribution have a significant ethical component. However, many public goods such as defense, public parks, cyberspace, and urban air (both clean and polluted) are not excludable, and this leads to externality and market failure.

1.2.6.5 Tragedy of the Commons

In 1968 Garrett Hardin highlighted the problem with the application of invisible hand of market efficiency-based solutions to common goods.[103] In the well-articulated context of the food basket, he describes an open-to-all grazing pasture where anyone may keep as many cattle as desired. But when its grazing capacity is exceeded, a tragedy occurs—all the cattle die. When food baskets are created, that is, parcels of pasture are assigned private property rights, the grazing is limited to match nature's ability to replenish, and that averts the tragedy. In the context of environmental pollution, instead of taking food from the pasture, air or water pollutants are discharged into the common areas at a rate that is way beyond nature's ability to recycle. But unlike the food basket, air and water cannot be readily boxed: they are public goods. Assigning private rights to public goods like parcels of air and water, more correctly termed air and water sheds, is not practical, but more importantly, the polluter has no incentive to limit pollution. Thus, putting into practice the private property solution from the context of the food basket tends to favor more pollution. This is market externality, discussed in Section 1.2.6.3, and requires government intervention to ensure clean air and clean water sheds to protect public health and provide equitable access. This calls for a redefinition of property rights.

1.2.6.6 Property Rights to Overcome Externality

In 1960, Nobel laureate economist and centenarian Ronald Coase demonstrated theoretically that well-defined property rights could overcome the problem of externality.[104] He demonstrated how enforceable property rights make goods excludable and allow markets to work. He proposed that if stealing is not illegal, nearly all goods are non-excludable. Any common privately owned property such as a car is excludable and belongs to the owner, but only if the owner's right is enforceable. In the context of pollution, one could think of the "right to no pollution" or the "right to compensation." This establishes the "right to pollute as a property."[105] This assigns property rights to an air shed or water shed. If this right to public goods like clean air or clean water could be enforced, these goods become excludable. But is there a right to pollute? Who has the right—the polluter or the victim? Coase showed that if there are no trading barriers, trading of this right to pollute leads to market efficiency regardless of whether the polluter compensates the victim or the victim pays the polluter to not pollute (regardless of how abhorrent the concept is) makes no difference to market efficiency.

Per Coase, although counterintuitive, initial assignment of property rights does not matter for economic efficiency. However, this is true only if it meets several conditions, provided that: everyone has perfect information, consumers and producers are price takers, court systems in place for enforcement purposes are costless, producers maximize profits, consumers maximize utility. Also, there are no transaction costs and there are no income or wealth effects. But if any of these conditions fails, initial assignment of property rights does matter.[106] Transaction costs matter—costs of trading rights must be minimized. If impediments to trading exist, initial distribution matters, and makes a difference to the involved parties. Therefore, vesting pollution rights through allocation of emission credits, for example, is like handing out cash. Also, victims should not be loath to pay, which is a psychological barrier and not an economic issue. While it rarely happens, Santa Maria, CA citizens paid to relocate cattle feed lot to eliminate pollution. Market efficiency was attained, although the victims paid for it, in that case the right to no-pollution was worth more to the victims.

Finally, markets become efficient with no concern for equity, and this has political significance. Take this political economy model, for example: The government induces polluters to take socially desirable actions, which may not seem to be in the polluters' interest. Government intervention may be in the form of environmental regulations, economic incentives, or trade restrictions. But determining exactly the level of pollution best for society is a complex problem. These are discussed in greater detail in the following section on economic instruments and regulatory policy mechanisms that the government may use for intervention.

1.2.6.7 Economic Instruments and Regulatory Policy

How much environmental protection is appropriate? What is the right balance between sustainability and use of natural resources? This is a normative sociopolitical question. Normative economics, however, is value-laden. Individual/social preferences can be biocentric, anthropocentric, or centered on sustainability. Biocentrism has the biological world as the center, focused on the intrinsic value of life, and does not consider usefulness to human beings to be one of its instrumental values. Anthropocentrism assumes that the environment is there to provide material gratification to humans. Sustainability strives to preserve the integrity of ecosystems. A widely quoted definition of sustainable development comes from the Brundtland commission: "development that meets the needs of the present without compromising the ability of future generations to meet their own needs."[107]

There are, broadly, four major regulatory mechanisms that the governments can utilize to intervene with an optimal level of regulation and manage sustainability issues. These are command and control, liability through law suits (also known commonly as toxic torts), economic penalties or incentives such as, emissions fees or tax credits, and market mechanisms such as emissions credit trading. Command and control is the most dominant form of regulation, in which specific pollution control equipment, technology, or emission limit for type of plant or specific pollutant as prescribed by regulation. These regulations carry significant financial and personal criminal liability/penalties for noncompliance. Goals, such as equitable access to electricity, is accomplished through price or rate control for regulated monopolies like utilities. Likewise, social equity and justice are enforced through the legal framework and courts. The *liability* approach—polluters are responsible for the consequences, so polluters pay for all damages—creates incentives for the polluter to take precautions, as this approach can lead to toxic torts. Some jury awards may be significant. Because of their ease of implementation, these two—command and control and liability approach—are more commonly used by governments as regulatory mechanisms. However, they do not offer incentives to promote free market efficiency, entrepreneurial ventures, or innovation.

The other two mechanisms offer more direct economic incentives. For instance, *emissions fees* call for a charge per unit of pollution—it is in the polluter's interest to reduce pollution to lower the fees they must pay. This approach could achieve predefined environmental standards at the lowest possible cost. But the control authorities often do not know the correct fee to charge in order to reach the optimum pollution for market efficiency. Also, this method works better when tradable permit transaction costs are high. "Marketable permits/emissions credits trading," on the other hand, allows polluters and speculators to buy and sell rights to pollute. It separates who pays and who installs controls.[108] Making credits transferable makes them private property and tradable. Trading changes the permit to pollute (a command and control approach) into an economic incentive. A polluter may install excess controls that yield emission reduction credits (ERCs), and use the revenue from the sale of ERCs to pay for the controls. Efficiency is achieved through the use of purchased or internally generated ERCs to avoid controls for operating facilities and equipment units that are more expensive to control. This approach works better for uniform mixing pollutants where no hot spots are created. For example, it has reduced acid rain pollutant compliance costs by $10 billion in the US. However, spatial distribution and the consequent hot spots, or initial allocation of the right to pollute and the consequent impact on equity, are continuing issues. Per Coase's theorem, both marketable permits and emissions fees could allocate controls in a cost-effective manner to precisely meet predefined pollution levels.[109,110]

1.2.6.8   Government Intervention and Interest Group Advocacy

Political economy model is often the basis for government intervention or action to promote underserved areas and sustainable development, and prevent undesirable products or irresponsible operations for public health and safety. Financial incentives and regulatory mandates with penalties for noncompliance are typical features of these efforts. Many government programs worldwide encourage the development of renewable energy, while others mandate the enhancement of fuel efficiency standards for vehicles. Some government mandates require more stringent emissions control requirements while others incorporate emissions fees or taxes to encourage reduction. Government agencies that manage financial sector operations, such as the stock exchanges, require publicly listed companies to report their sustainability performance as well disclose their material sustainability risks in their operations, in both the near-term and long-term.

For the intervention process to be effective, the public policy stewards must identify risks to revenue and profit growth resulting from specific environmental, social, or governance factors, as well as plan and implement a data-collection strategy to gather the information necessary to evaluate an investment and make a persuasive, balanced, and evidence-based case for sustainable investing within their constituencies and implement.

First, the players in the ecosystem and the incentive structures that link finance, economy, and the energy and resource base are identified. Second, the techniques of investment analysis and due diligence methods and financial statement analysis, applicable across asset classes, geographies, and investment horizons, are understood. Finally, studies linking corporate financial performance and indicators of excellence in environmental, social, and governance factors are evaluated for related payoffs, such as the potential of environmental policy to create business opportunity, the legal and financial risks related to human rights violations that can result from unregulated corporate behavior, the elevation of sovereign risk that is associated with social injustice, and the business impact of often overlooked natural hazards.

Today the intervention process has taken gargantuan proportions and has even become the staple of presidential debates of the most powerful nations in the world on issues like climate change science! Changes in political leadership often reverse the course to protect constituent interests, such as the recent withdrawal of the US from the Paris Climate Change Agreement by President Donald Trump.

Taback and Ramanan[111] present a succinct summary of the government regulatory intervention and interest-group interaction process. "The genesis of every significant step forward towards government intervention can typically be traced to the occurrence of a visible event with an undesirable social consequence linked to unacceptable corporate behavior. Governments, activists from nongovernment organizations and the media hold companies accountable for these social consequences. Synchronous connectivity creates rampant awareness and empowerment among the public at large and leads to responses in the form of consumer boycotts, lawsuits, and shareholder resolutions. Communities call for action and governments mandate corporate disclosure of ethical, social and environmental risks and action to prevent recurrences and mitigate undesirable social consequences." Affected interest group advocates participate in the development of regulatory policy and compliance standards. Facebook storm - lack of data privacy protection, and the related congressional and media calls for more regulations and oversight is a classic case.

"Environmental advocacy is presenting information on nature and environmental issues that is decidedly opinionated and encourages its audience to adopt more environmentally sensitive attitudes, often more biocentric worldviews."[112]

One of the earliest examples of activism and advocacy was in the 1970s, a movement called *Chipko*, in the Himalayas of India by a group of nearly illiterate poor women who placed their bodies between trees and contractors' axes to protect them from being cut down, and launched a crusade against deforestation, in the process becoming the first *tree-huggers*. In the US, Love Canal led to superfund laws demanding the cleanup of contaminated sites. Activism, which carries disruptive connotation, is not limited to passionate individuals or nongovernmental agencies. Many are driven by political leaders to protect the people in their constituencies or organizational leaders to protect their long-term value. A classic example is Paul Polman of Unilever. "In January 2009, one of his first actions as CEO was to abolish quarterly reports and earnings guidance,"[113] deviating from the industry norm to stress longer-term vision.

Economic theory of externalities predicts that private actors will ignore the external costs borne by the environment, and national governments are the likely stewards of natural capital. However, today, many megacorporations, such as Walmart, ExxonMobil, and Apple (with 2016 revenues of US$480, $240, and $240 billion, respectively), and parts of nations (e.g., California in the US, with a GDP of US$2.5 trillion; Uttar Pradesh in India, with a population of more than 300 million) are larger than most countries in GDP or population. Impetus for action by organizational stewards come from the recognition of the potential impact on image and stock prices because of society's growing expectation of responsible action on the part of businesses, including society's demand for ever-increasing transparency on performance. Also, while cleanup costs after pollution may have seemed optional, there is growing cognition that post pollution cleanup is first not optional any longer, and is often an order of magnitude more expensive. Also, there is increasing realization that exhausting or wasting a key resource is a cost that is not prudent.

Effective activism and advocacy comprises collecting relevant information, getting the message out, building alliances, going public with the issue, organizing demonstrations, participating in rule-making, and, if necessary, boycotting brands or exploring other legal recourses. The stakeholders who impact or are impacted by the business and activities include employees, customers, shareholders, investors, communities, local governments, nongovernmental organizations, suppliers, their employees and communities, and the environment in which these products and services are sourced, made, sold, used, or disposed of. Sustainable development brings in future generations who do not have a voice in constructing or directing today's decisions. The actions of stakeholders have grown multifold, with significantly greater participation by a wider set of constituencies driven by the growth of the internet, providing real-time access to substantial data globally for scrutiny, many-to-many synchronous interactivity with very few barriers, and the ability to connect communities at unprecedented speeds.

1.2.6.9 Sustainable Development and Public–Private Partnership[114]

Public–Private Partnerships (PPP) are typically between a government agency and a private sector entity to finance, build, and operate projects, such as public transportation networks, sustainable development of an underserved region of the world, or accomplish elimination of avoidable infant mortality. The government agency could be a federal, state, or municipal authority of a country, or it could even be a funding agency such as the US Agency for International Aid (USAID) or International Finance Corporation (IFC), the financial arm of World Bank, or the United Nations Sustainable Development Program (UNSDP). The private sector entity could be an entrepreneurial venture, a for-profit company, or one of several emerging responsible investment business models for sustainable development, such as philanthropic capitalism, venture philanthropy, mission-driven charitable foundations, and impact-driven high-net-worth individuals with patient capital. Risks and returns are shared between the partners according to the ability and missions of each.

While financing could come from either or both partners, it requires repayments from the public sector, users over the project's lifetime, or both. For instance, for a wastewater treatment plant, payment comes from fees collected from users. Toll-based bridges, tunnels, and highways have been following the public–private partnership models for more than a century. Need for large investments, long-term patient capital, and often a passion for social responsibility makes the Public–Private Partnership (PPP) model rather attractive for sustainable development projects. One partner's authority to enforce long-term repayment by a large, captive set of users, coupled with the other partner's desire and drive to make an impact is a powerful recipe. Private-sector innovation could provide operational efficiency. However, they bear the burden of project delay, budget exceedance, and insufficient demand. In case of sustainable development projects for underserved regions, geopolitical risk may be overwhelming.

1.3 SUSTAINABILITY—EVOLVING REGULATIONS AND RESPONSIBLE INVESTMENTS

1.3.1 CHAPTER OVERVIEW

This chapter provides an overview of sustainability regulations that are evolving globally. Traditional financial reporting does not adequately account for how corporate sustainability performance can enhance or impede both shareholder and stakeholder value. The chapter also shows how *integrated corporate reports* that combine financial and sustainability reporting could close the gap by incorporating externalities and other intangible assets by capturing intrinsic values. This enables investors to make better informed decisions. Company reported nonfinancial information coming directly is more likely to be valued by investors.[115] Finally, this chapter traces the evolution of responsible investing to its current forms and demonstrates how CSR; environmental, social, and governance (ESG); and mission or principle issues have become financially material and have a direct impact on risk exposure and goal accomplishment of public, private, and government investments. It also highlights some of the broadly accepted voluntary principles.

1.3.2 SUSTAINABILITY REGULATIONS—EVOLVING GLOBALLY

As discussed in Section 1.2.4, stakeholders relevant to sustainability are participants, influencers, and the vulnerable groups.'

While every group is likely to benefit long term from sustainability regulations, the requirements and impact of regulations vary by group. Influencers formulate regulations and monitor compliance, participants comply, the vulnerable are protected and the society at large benefits. Equity and institutional investors, as well as corporate and public policy stewards, clearly have a dominant role in contributing to sustainable development decisions and accomplishing the quadruple bottom line. Pension funds and institutional investors often file corporate shareholder resolutions seeking data on companies' risks and initiatives related to climate change (e.g., policy, emission reports, and mitigation plans). Equity investors are equally concerned about environmental and other sustainability risks and the longevity of corporations.

Emerging regulations seeking disclosures emanate from government departments of environment, trade and commerce, and finance and treasury to ensure sustainable development, to protect investors, and to collect fair share of taxes. Lenders and institutional investors are increasingly required to disclose, through integrated reporting, how their investments are channeled into responsible operations from the perspectives of longevity, risk, and reward. Stock exchanges are recognizing the need for transparency on corporate sustainability strategy. The US Securities and Exchange Commission (SEC) and several stock exchanges across the developed world call for the reporting of material risks in their operations as part of their annual financial reports.

The government has a dual role of leadership in sustainability reporting through state-owned enterprises, as well as mandating and monitoring sustainability reporting through their public governance arm. State-owned enterprises have a natural stewardship role in progressing sustainability reporting. They serve as pilots and role models. Their development of metrics and measurement of sustainability help advance government mandates for sustainability reporting across all sectors. Some European nations such as France, Spain, and Sweden, as well as all the BRICS nations, Brazil, Russia, China, India, and South Africa specifically target and mandate sustainability reporting from state-owned enterprises.[116] Many country governments and stock exchanges seek third-party verifications for assurance. Likewise, because of the growing linkage of sustainability impacts to financial performance, multinational companies and global investors want qualified and vetted third party verification assurance. Global assurance standards available today include International Standard on Assurance Engagements (ISAE) 3000[117] of International Auditing and Assurance Standards Board of the International Federation of Accountants and ISO 14064-3 for greenhouse gas (GHG) assertions.[118] Some country-level assurance standards for verification of sustainability reports are included in the discussion of the relevant country or region regulations.

1. Since 2001, companies listed on the stock exchange in France are required to include social and environmental impacts in their annual reports. The 2010 Grenelle II Act of France expanded the mandate beyond environmental and social performance reporting, and requires third party verification.[119] The assurance of verification related to company's transparency obligations on social and environmental matters is mainly designed to comply with ISAE 3000[120] of International Auditing and Assurance Standards Board of the International Federation of Accountants and French professional standards.

2. The 2007 Environmental Information Disclosure Act of China mandates public disclosure of compliance and serious releases. Incentives such as grant priority are offered for voluntary disclosure of environmental information on resource use, emission level and reduction targets, and so on. The 2008 Green Securities Policy adopted in China requires several highly polluting industry sector companies listed on the Shenzhen and Shanghai stock exchanges to disclose environmental information to the public. China Ministry of Finance issued China Certified Public Accountant Practicing Standard, CAS3101, which follows ISAE 3000 but requires sign off by a certified practitioner.

3. China Stock Exchanges Shanghai[121] and Shenzan,[122] for over a decade, required all companies listed on the stock exchange and all companies listed in the SSE Corporate Governance Index 240 to provide an ESG report.[123] Hong Kong Stock Exchange's new HKEx ESG Reporting Guide[124] came into effect on January 1, 2016.

4. The 2012 requirement of India's Securities and Exchange Board calls for business responsibility reports from the top one hundred companies. A unique feature of the Companies Act 2013 of India is that it requires companies, beyond a certain size of operation, to set up a Corporate Social Responsibility board committee to develop socially responsible policies and to ensure allocation and application of "at least 2% of the average net profits of the company made during the three immediately preceding financial years" on CSR activities.[125] If the company fails to spend this amount on CSR, the board must explain why, in its annual report. The act defines CSR as activities that promote poverty reduction, education, health, environmental sustainability, gender equality, and vocational skills development.

5. The Australian Stock Exchange (ASX) requires listed companies to disclose material sustainability—economic, environmental or social risk, and mitigation plans. DR03422 General Guidelines on the Verification, Validation and Assurance of Environmental and Sustainability Reports 2003 was issued by Standards Australia.

6. Since 2012, the UK Department for Environment, Food & Rural Affairs (DEFRA) requires all companies listed on the London Stock Exchange to report their greenhouse emissions in their annual reports. One very interesting feature is the requirement to include at least one ratio that

relates reported GHG emissions to company activity, such as carbon intensity. AA1000 Assurance Standard issued in 2008 by UK-based AccountAbility is designed to help ensure that sustainability reporting and assurance meets stakeholder needs and expectations.

7. The European Union adopted Directive 2014/95/EU[126] on the disclosure of nonfinancial and diversity information by organizations with more than 500 employees. They must include in their management reports policies, main risks and outcomes on environmental, social, and employee aspects, human rights, anticorruption and bribery, and diversity. The directive is under transposition to national law by EU member states.

8. In the US, the need for mandated corporate transparency is becoming acknowledged at a steady pace, with regulations on the rise. US investment banks are required to conduct due diligence for material risks, including environmental liabilities prior to preparing prospectus for any new initial public offering (IPO). The US financial reform regulations hold banks responsible for their actions long past the date of transaction. The Dodd Frank Act[127] requires reporting of conflict minerals. The US SEC, in order to protect equity investors investing in publicly listed stocks, has guided listed companies to manage climate risk like any other business risk. In 2010, the SEC created guidelines for companies to report on climate risks in their proxy statements, which accelerated the integration of ESG factors in mainstream financial risk disclosures for US companies. They suggest that the climate risk mitigation may require internal capacity-building and stakeholder and community engagement and warn that uncertainty is not a reason for inaction."[128]

1.3.3 SUSTAINABILITY AND INTEGRATED REPORTING

"A sustainability report is a report published by a company or organization about the economic, environmental and social impacts caused by its everyday activities."[129] The quality of the report will depend on the organization's size, sector, global reach, geopolitical significance, consumer and investor demand, and management plans to use sustainability strategy as a differentiator to propel ahead of the competition. Jeanne Chi Yun Ng, director, Group Environmental Affairs, China Light and Power, says that integrated reporting of financial and nonfinancial issues (e.g., environmental, social, and governance risks) has provided multiple benefits for the company.[130]

Per EY,[131] sustainability reporting appears to be reaching a *tipping point*. Once reporting becomes standardized, sustainability performance indicators will become as important as financial performance metrics. Integrated financial and sustainability reporting will drive the need for high quality and comparable and measurable data on sustainability performance and impacts to ensure maximum transparency and ease of comparison. The mainstreaming of sustainability reporting is driven by business consumer demand. For instance, Oster observes that, "For economists, CSR reports are signals. This is one reason that Apple's opposition to a shareholder petition to require a CSR report in 2009 dismayed some of the firm's supporters."[132]

1.3.3.1   Status of Sustainability and Corporate Social Responsibility Reporting

There are number of recent studies on progress of corporate sustainability and reporting. They include Bloomberg,[133] KPMG[134] and Klynveld Peat Marwick Goerdeler (KPMG),[135] Ernst and Young (EY),[136] Massacussetts Institute of Technology (MIT) and Boston Consulting Group (BCG).[137] One recent study[138] shows that based on market value of stocks traded, about one tenth of the publicly listed companies in the United States generate CSR reports. The study also finds that firms with high costs of equity capital tend to release corporate social responsibility reports, and among those reporting, firms with superior social responsibility performance attract dedicated institutional investors and analyst coverage. "Although not mandatory in most countries, CSR reporting has continued to grow; just over the past three years, of the largest 250 global companies, CSR reporting has grown from 80% to 95%, with reputation and ethical considerations topping the list of drivers."[139] As one may expect, companies with higher performance on sustainability metrics are the first ones to report. These could also be the result of a firm's attempt at reducing cost of capital and attracting socially responsible, and in particular, institutional investors.

1.3.3.2   Integrated Financial and Sustainability Reporting

Integrating sustainability and financial reporting adds further credibility to sustainability disclosure and helps communicate business value of sustainability using analytics to external stakeholders. For instance, SAP integrated reporting is based on the interconnection of social, environmental, and economic performance, and each creates a tangible impact on the others. SAP puts a monetary value to employee engagement, healthy business culture, and carbon emission reduction benefits.[140]

Ernst & Young suggest that integrated reporting of financial and ESG factors provides a more comprehensive long-term reflection of the value of the firm.[141] It should connect sustainable business practices, tangible and intangible material assets, capital risks and opportunities, and short- and long-term value creation objectives. Despite the potential influence of the nonfinancial factors on future value creation, its integration into decision making remains a challenge. Investors are turning to integrated reporting for a better comprehension of the whole picture. The complete range of resources and relationships that affect future value creation are best captured and disclosed through the multi-capital approach in integrated reporting.[142]

The International Integrated Reporting Council (IIRC) is a global alliance of regulators, investors, corporations, accounting professionals, and NGOs. The IIRC's vision is to align capital allocation and corporate behavior to financial stability and sustainable development through the integrated reporting and thinking. It is a process based on integrated thinking that leads to periodic integrated reports and related concise communication about how an organization's strategy, governance, performance, and prospects, in the context of its external environment, lead to the creation of value in the short, medium, and long term."[143] IIRC's[144] purpose of creating an integrated reporting framework (IR) is to enhance accountability, stewardship, and trust; to harness flow and transparency of information; and to provide investors with the information they need to make more effective capital allocation decisions to facilitate better long-term investment returns. The integrated report will also bring consistency with a number of corporate reporting trends across the globe. Finally, the integrated report will meet a range of market drivers, which are not being satisfied currently by complex and dated reporting methods. These include opportunities afforded by new technology, and the need for transparency, inclusiveness, and more information that is material to modern business.

1.3.3.3 Emerging Global Trends in Sustainability and Integrated Reporting

An Ernst & Young survey[145] of 272 sustainability executives in 24 industry sectors with one billion dollars or more in revenue showed a lack of internal systems to monitor sustainability impacts and mitigate inaction risks. They note emergence of some key trends and action steps that are presented here:

1. Build same transparency and rigor as financial reporting into sustainability reporting.
2. Engage CFOs in measures to monitor and report on environmental and sustainability.
3. Engage employees as key stakeholders to embed sustainability into the corporate culture.
4. Recognize that greenhouse gas disclosure, with growing interest in water, has value outside of the regulatory arena for both internal and external stakeholders.
5. Integrate plans to manage risks in access to key resources in sustainability reporting.
6. Understand the value of sustainability reporting to ranking and ratings organizations, particularly those of interest to investors.

Traditional investors focus on tangible financial measures, but these effects are already reflected in the stock price. Sustainability performance indicators are typically intangible, and investors have difficulty valuing them. Current reporting frameworks do not provide adequate information, and greater transparency is required, especially in presenting the risks and interdependencies of financial and nonfinancial Environmental (E), Social (S) and Governance (G) information. The biggest challenge is *materiality*. The impact of E, S, and G factors is not uniformly relevant to all firms, making a global common reporting standard difficult. Integrated reporting is a step forward and could show the quantified impact of financial and nonfinancial performance of a firm on each other, using established frameworks.

1.3.3.4 Organizational Process to Enhance Corporate Social Responsibility Reporting

1. Advising operations on how to integrate sustainability measures into management actions to deliver results helps identify appropriate data to collect and the type of analysis required. It also helps build credible results.

2. Training sustainability professionals with financial skills helps them understand why it is important to find monetary values for sustainability impacts.
3. Training finance professionals in sustainability streamlines data analysis by providing a better understanding of the link between sustainability data and business value.
4. Building cross-functional teams between sustainability and finance builds credibility to the sustainability data and helps in decision making.

1.3.4 SUSTAINABLE DEVELOPMENT AND RESPONSIBLE INVESTING

The financial sector focused on socially responsible investment is growing fast. It has grown from \$2.7 trillion in 2007 to \$21.4 trillion in 2014.[146] Investors in this sector actively prefer to invest in corporations that have been vetted by and are high on the dominant sustainability indexes.[147] Also, per Ocean Tomo,[148] today intangible assets represent more than 80% of the market value of S&P 500 companies.

1.3.4.1 Evolution of Socially Responsible Investing

The earliest known socially responsible investors were faith-or values-based. For example, Quakers in the 1500s and churches in the 1920s used a negative screening and deliberately opted out of investing in gambling, tobacco, and alcohol. Half a century later, the Vietnam War drove some investors to opt out of nuclear and military weapons production, and the Global Sullivan Principles for social justice motivated others to selectively divest from South Africa to dissent apartheid.

This year, in response to the rising awareness and demand by victims of gun related violence, large money managers like BlackRock offer its clients the ability to opt out of investing in gun manufacturers.

In the 1990s, driven by the Brundtland Commission's sustainability, CSR took into account social and environmental behavior. At this phase CSR driven socially responsible investing continued on the path of exclusionary investment approach, maximizing financial return with a social alignment by negative screening of unacceptable social and environmental conduct. These CSR guided negative screening socially responsible investors sought to build portfolios of assets that exclude companies deemed irresponsible or ones that are contrary to the mission or values of the investors.

A further shift toward incorporating environmental and social factors in investment decisions occurred, but these investors explicitly sought financial returns as well, using a risk and return approach that includes non-traditional criteria (such as policies). The mantra was to do good for society but not do harm to financial returns. The key shift was the inclusion of positive screening for best-in-class sustainability performance and growth in active ownership or shareholder activism. This allowed socially responsible investors to aggregate the *triple bottom-line* economic, environmental, and social performance of organizations. For clarity, these investors are termed CSR guided positive screening for best-in-class triple bottom-line investors and investments.

In the early 2000s, increasing emphasis on governance, in addition to economic, environmental, and social factors emerged with the passage of Sarbanes-Oxley in 2002. Institutional investors, also known as *Universal owners* of private enterprise, generally have investments that are diversified across asset classes, sectors, and geographies with long-time horizons and closer ties to the markets and economies as a whole. *Universal owners* and other mission-driven high-net-worth individuals and foundations sought greater insight into the risk and opportunities in the nonfinancial performance of organizations. These investors engaged actively as shareholders with the corporations they invest in, rather than just mandate negative screening, and incorporated ESG factors into their investment process. While early faith-based investors were driven by *inside-out* value of the investor, today's responsible investors incorporate *outside in* external realities. Faith-based and corporate social responsibility guided investment evolved into ESG integrated investment, one that uses ESG factors in a best-in-class approach. Responsible or ESG investment strategy considers ESG criteria to achieve competitive and long-term financial return while making a positive societal impact. Capturing the upside needs appropriate, often industry disruptive innovation strategy that in turn requires better understanding of the ESG advantage and leveraging the information arbitrage. Today the focus is on which ESG factors are *material*.

In 2007 the Rockefeller Foundation coined the term *impact investing*, "an umbrella term to describe investments that create positive social impact beyond financial returns."[149] In contrast to the CSR guided negative screening investors with exclusionary strategy, impact investors focus on inclusion that is positive screening for best-in-class. It is emerging as a separate asset class that could be structured to serve different program or mission (e.g., agriculture, health) areas and use different legal entities (e.g., benefit corporations and community interest companies). The impact investment industry is estimated to grow to US$500 billion by 2020.[150]

1.3.4.2 Sustainability Operations and Investment Principles

One effective way to ensure companies and governments conduct their activities responsibly, is to build in a set of principles to guide investment decisions. While there are numerous sets under various stages of development, some of the dominant ones that cover large investments and investors, ranging from governmental development projects and multinational enterprise expansions, to private equities and mission-driven charities, are highlighted here. These investment principles include OECD, Equator, PRI, and UNGC. Several other niche groups focus on a narrower range of investments and/or objectives. For instance, the Global Sustainable Investment Alliance,[151] with a vision to integrate sustainable investment into financial systems, CDC,[152] the development fund arm of UK, with a focus on Africa and South Asia, Insead's Global Private Equity Initiative for assimilating ESG in Private Equity, and IRIS, an initiative of the Global Impact Investing Network,[153] with a goal to increase the scale and effectiveness of impact investing.

1.3.4.2.1 *Guidelines for Multinational Enterprises (Organization for Economic Cooperation and Development Principles)*[154]

One of the earliest sets of principles to emerge as socially responsible investment guidelines was the Organization for Economic Cooperation and Development (OECD) Guidelines for Multinational Enterprises (MNE),[155] adopted in 1976. OECD is an intergovernmental economic organization with 35-member countries. These guidelines establish legally non-binding principles and standards for responsible business conduct for multinational corporations covering such areas as human rights, disclosure of information, anticorruption, taxation, labor relations, environment, competition, and consumer protection. Select components are highlighted as follows:

1. *Policies*: Consider country policies and other stakeholder views and contribute to economic, social, and environmental progress for sustainable development. The policies should respect human rights, and promote human capital formation, capacity building, and good governance.
2. *Disclosure*: Ensure timely, regular, reliable, and relevant information is disclosed regarding activities, structure, financial situation, and performance. The disclosures should be of high quality and cover financial and required nonfinancial information, including environmental and social performance.
3. *Employment and industrial relations*: Respect, within the framework of applicable law, employees' right to form trade unions; abolish child labor and any forced labor; avoid discrimination in employment based on race, color, sex, religion, political opinion, national extraction, or social origin; and take adequate steps to ensure occupational health and safety.
4. *Environment*: Protect the environment, public health and safety, and operate in a manner that contributes to the wider goal of sustainable development.
   a. Collect and evaluate adequate and timely information on the environmental, health, and safety impacts of enterprise activities, and verify progress toward measurable goals.
   b. Engage in timely communication and consultation with the public and employees directly affected by the environmental, health and safety (EHS) policies, and activities of the enterprise.
   c. Incorporate, in decision-making, the foreseeable environmental, health, and safety-related impacts associated with the processes, goods and services, and, when needed, prepare an appropriate environmental impact assessment.
   d. Maintain contingency plans for preventing, mitigating, and controlling serious events, and not use scientific uncertainty to postpone cost-effective measures to mitigate damage.
   e. Improve environmental performance by the adoption of technologies and development of products or services with better EHS performance.
   f. Provide adequate education and training to employees in the safe handling of hazardous materials and the prevention of accidents.
   g. Help develop environmentally meaningful and economically efficient public policy.
5. Other guidelines address issues such as combating bribery, protecting consumer interest, building local science and technology capacity, fair competition, and the timely payment of appropriate amount of taxes.

1.3.4.2.2   Equator Principles

At the dawn of this millennia, because of the growing social expectations associated with the move from shareholder to stakeholder primacy, the financial investment sector came under pressure to commit to sustainability, which calls for measuring environmental and social impacts, continuous improvement of portfolios, proactively fostering sustainability, building capacity, and linking performance. In 2002, the Collevecchio Declaration on Financial Institutions was a move by more than 100 NGOs to advocate environmentally responsible behavior in the financial sector[156] and served as a precursor to the Equator Principles.

The first principle is sustainability, calling for measurements of environmental and social impacts, continuous improvement of portfolios, and proactive fostering of sustainability, building capacity, and performance. The second principle is to do no harm, which requires the creation of sustainability procedures and the adoption of international standards. The next three principles involve taking full responsibility for impacts, accountability for public consultation and stakeholder rights, and transparency through corporate sustainability reporting and information disclosure. The final principle is sustainable markets/governance, which cover public policies and regulations that recognize the government's role to discourage unethical use of tax havens and currency speculation.

Around the same time, the World Bank and its project financing arm, International Finance Corporation (IFC), were sued by impacted parties and NGOs for not ensuring that their borrowers operate their projects responsibly. In 2003, this lawsuit led to the development of the industry group voluntary initiative called Equator Principles,[157] designed to manage environmental and social risk in project financing. Although it was led by the IFC, later signatories include Goldman Sachs and Citigroup.

Equator Principles (2003) include conducting environmental and social impact assessments (ESIA), compliance with all applicable social and environmental standards, covenants in financial documentation, public consultation and disclosure, grievance mechanisms, and independent review, monitoring, and reporting. Furthermore, the public consultation and disclosure process requires consultation with all stakeholders for the development of the ESIA, disclosure of ESIA results to public and ongoing consultations during construction and operation. These consultations and engagements must be conducted in local languages, showing respect for local traditions and ensuring that the groups consulted are representative.

1.3.4.2.3   Principles for Responsible Investing[158]

Following a finding that environmental, social, and governance (ESG) issues affect long-term shareholder value, which in some cases could be profound, the UN launched the Principles of Responsible Investing (PRI) in 2006. PRI is not associated with any government; and while supported by, it is not part of the UN. PRI is specifically designed for institutional investors and the financial sector and reflects the core values of large investors whose investment horizon is long, and portfolios are diversified. PRI has grown to more than 1700 signatories and US$62 trillion associated assets under management (AUM).

The six Principles for Responsible Investment for incorporating ESG factors into investment practice are:

Principle 1: Incorporate ESG issues into investment analysis and decision-making processes.

Principle 2: Be active owners and incorporate ESG issues into ownership policies and practices.

Principle 3: Seek appropriate disclosure on ESG issues by the entities invested (or investing) in.

Principles 4–6: Promote acceptance, enhance effectiveness, and report implementation progress on the principles.

1.3.4.2.4   The UN Global Compact[159]

The UN Global Compact's ten principles are derived from the Universal Declaration of Human Rights, the International Labor Organization's Declaration on Fundamental Principles and Rights at Work, the Rio Declaration on Environment and Development, and the UN Convention against Corruption. These ten principles cover the areas of human rights, labor, environment, and anticorruption, and the signatories are required to provide annual communication on their progress. Failure to do so can result in expulsion.

Principle 1: Support and respect the protection of internationally proclaimed human rights.

Principle 2: Make sure that they are not, unwittingly or otherwise, complicit in human rights abuses.

Principle 3: Uphold the freedom of association and the effective recognition of the right to collective bargaining.

Principle 4: Eliminate all forms of forced and compulsory labor.

Principle 5: Support the effective abolition of child labor.

Principle 6: Eliminate discrimination in respect of employment and occupation.

Principle 7: Support a precautionary approach to environmental challenges.

Principle 8: Undertake initiatives to promote greater environmental responsibility.

Principle 9: Encourage the development and diffusion of environmentally friendly technologies.

Principle 10: Work against corruption in all its forms, including extortion and bribery.

1.3.4.3   Sustainable or Sustainability Investing

Sustainability issues affect the various sectors of finance and financial approaches, and integrating sustainability principles and practices into finance can be used to help business become more efficient and effective, reduce risks, create opportunities, and develop competitive advantage. Socially responsible investing covers a broad range of investments, faith- or values-based, CSR guided negative screening, CSR guided best-in-class triple bottom line, ESG integrated, and program- or mission-related impact investing.

Sustainable investing includes all the socially responsible investments that enhance one or more of the sustainability components or objectives, without significantly harming the other. For instance, a mission or program-related investment aligned with sustainability goals may focus on chemicals that harm unborn children.

1.3.4.4   Impact Investing

Impact investing is an umbrella term to describe investments that create positive social impact beyond financial returns.[160] It was coined by the Rockefeller Foundation in 2007, and is estimated grow to US$500 billion by 2020.[161,162] Impact investors invest with an intent to generate a measurable social and environmental impact while making financial returns. The idea is to align profit making with generating a positive social impact. Also, known as social investing, they could be broadly categorized, based on primary motives such as the following:

1. Impact first to primarily maximize impact
2. Investment first to primarily get financial returns
3. Catalyst first to seed funds to collaborators to initiate or strengthen impacts

They could be structured to serve different programs or missions (e.g., agriculture, health) areas. Impact investing includes program-related investments (PRIs), which have been around since the 1970s, and mission-related investments (MRIs),[163] a term coined in the last decade. PRI is below market rate investment by foundations, deeply focused on impact, and count toward endowment payout requirements for foundations. MRI is a market rate investment by private foundation endowments that use the tools of social investing, sometimes including shareholder advocacy and positive and negative screening.[164]

Social enterprise is an impact investing business that reinvests profits directly to serve social needs. Unlike nonprofit entities, social enterprise does not seek support from government or philanthropists. Also, it is distinct from a socially responsible business that engages in CSR. One example is an energy savings mission supported by energy conservation consulting services.

Impact investing could use different forms of hybrid organizations (e.g., community interest companies in the UK). Examples of such legal entities include the low-profit limited-liability company, Benefit Corporation, and B corporation to meet the investors' specific legal, tax, and mission needs and achieve financial returns while prioritizing social benefit objectives.

1.3.4.4.1   Low-Profit Limited Liability Company

A low-profit limited liability company is a hybrid of for-profit and nonprofit. It limits liabilities and protects officers from shareholder lawsuits that question business choices that prioritize social or environmental returns over profits. It can also attract charitable donations or funds that accept below market return.

1.3.4.4.2   Benefit Corporation

A benefit corporation is a for-profit company that creates a material positive impact or public benefit. They cannot seek charitable contributions and must produce benefits report to rigorous standards with third party independent assessment, that adhere to high transparency and accountability. Officers are not liable for damages if the public benefit is not achieved. However, they are required to consider a broad array of stakeholders.

1.3.4.4.3 B Corporation

B corporations are organizations that are certified by B Lab, a nonprofit third-party entity, much like the Underwriters Lab, to ensure that the B corporation meets social and environmental transparency, accountability, and performance standards. Unlike a benefit corporation, a B corporation must be certified. Some examples of public benefits that B corporations provide are buying from low-income communities and making donations to other nonprofit organizations.

1.3.4.5 Philanthropic Capitalism

Historically, commercial and social capitals have been clearly separated. Traditionally, the approach was to get rich using the commercial capital and then indulge in philanthropy. Rockefeller, Carnegie, and, today, Bill and Melinda Gates and Warren Buffett are icons of business philanthropy. Over the years, corporate philanthropy became more professionalized, but philanthropic capitalism—the business effort to do well by doing good—could not yield a superior model of capitalism. Former US president Bill Clinton calls the Clinton Global initiative a laboratory for testing philanthropic-capitalism ideas. He says "the twenty-first century has given people with wealth unprecedented opportunities … to advance public good … our interdependent world is too unequal, unstable, and because of climate change, unsustainable. The business effort to do well by doing good failed to turn around our global environmental, social, and ethical trends, and it may in fact be distracting us from true systemic sustainability and responsibility."[165]

A more recent phenomenon is venture philanthropy—the idea that corporate foundations can improve effectiveness through monitoring where they invest, providing management support and staying long enough until those ventures become self-supporting. Other emerging models include traditional foundations practicing high-engagement grant-making, organizations funded by high-net-worth individuals but with all engagements done through professionals, and a partnership model where both the partner and individuals donate the financial capital and engage with the grantees.

ENDNOTES

[11] Report of the World Commission on Environment and Development: Our Common Future, UN Brundtland Commission (United Nations, 1987).

[12] Solow, R.M., On the intergenerational allocation of natural resources, *Scandinavian Journal of Economics*, 1986, 88, 141–149.

[13] Ramanan, R. and H. Taback, Environmental ethics and corporate social responsibility. In Dhiman, S. (Ed.), *Spirituality and Sustainability: New Horizons and Exemplary Approaches* (New York: Springer, 2016).

[14] Elkington, J., Cannibals with forks: The triple bottom line of 21st century business, Capstone, 1997, Accessed March 2017 and available at http://www.economist.com/node/14301663.

[15] Kaplan, R.S. and D.P. Norton, The balanced scorecard—Measures that drive performance, *Harvard Business Review*, January–February 1992.

[16] Ramanan, R., Environmental, safety & health costs and value tracking, *Townley Global Management Center for Environment, Health and Safety* (Manhattan, NY: The Conference Board, April 1, 1998).

[17] Ibid.

[18] Mercer 2007, as cited in DB climate change advisors, *Sustainable Investing*, Accessed March 2016 and available at https://www.db.com/cr/en/docs/Sustainable_Investing_2012.pdf.

[19] Waite, M., SURF framework for a sustainable economy, *Journal of Management and Sustainability*, 2013, 3, 25.

[20] McKeown, C., Interpreting the quadruple bottom line, Accessed June 16, 2015 and available at http://futureconsiderations.com/2013/05/quadruple-bottom-line/; also see Pope Francis.

[21] Lawler III, E.E., The quadruple bottom line: Its time has come, Accessed June 16, 2015 and available at http://www.forbes.com/sites/edwardlawler/2014/05/07/the-quadruple-bottom-line-its-time-has-come/.

[22] Ramanan, R., Ethics-The 4th bottom-line of sustainability, Volume 10 of Compendium—Spirituality for Corporate Social Responsibility, Good Governance and Sustainable Development, ISOL (Integrating Spirituality & Organizational Leadership) Foundation, under publication by Bloomsbury to be released at the 5th International Conference on Integrating Spirituality & Organizational Leadership at Chicago Art Institute, September 10, 2015.

[23] Ramanan, R. and H. Taback, Environmental ethics and corporate social responsibility. In Dhiman, S. (Ed.), *Spirituality and Sustainability: New Horizons and Exemplary Approaches* (New York: Springer, 2016).

[24] Taback, H. and R. Ramanan, *Environmental Ethics and Sustainability* (Boca Raton, FL, CRC Press, 2013).

[25] Friedman, M., The social responsibility of business is to increase its profit, *The New York Times Magazine*, September 13, 1970, Accessed October 2017 and available at https://www.colorado.edu/studentgroups/libertarians/issues/friedman-soc-resp-business.html.

[26] Koehler, D.A. and E.J. Henspenide, Drivers of long-term business value, *Deloitte*, 2012.

[27] Investopedia, Accessed March 2017 and available at http://www.investopedia.com/terms/c/corporate-profits.asp#ixzz4caLZSu6e.

[28] United Nations Sustainable Development Goals, Accessed February 2016 and http://www.un.org/sustainabledevelopment/sustainable-development-goals/.

[29] Global Reporting Initiative, G4 sustainability reporting guidelines, Accessed March 2017 and available at https://www.globalreporting.org/resourcelibrary/GRIG4-Part1-Reporting-Principles-and-Standard-Disclosures.pdf.

[30] Singh, G., Mahatma Gandhi—A sustainable development pioneer, *Eco Localizer*, Accessed October 14, 2008 and available at http://ecoworldly.com/2008/10/14/mahatma-gandhi-who-first-envisioned-the-concept-of-sustainable-development/ in http://www.mkgandhi.org/articles/environment1.htm.

[31] Carson, R., *Silent Spring* (Boston, MA: Mariner Books, 2002), Accessed December 2012 and available at http://www.goodreads.com/work/quotes/880193-silent-spring.

[32] Pope Francis encyclical on climate change, On care for our common home, Accessed June 24, 2015 and available at http://w2.vatican.va/content/francesco/en/encyclicals/documents/papa-francesco_20150524_enciclica-laudato-si.html.

[33] United Nations Sustainable Development Goals, Accessed February 2016 and available at http://www.un.org/sustainabledevelopment/sustainable-development-goals/.

[34] Global Reporting Initiative, G4 sustainability reporting guidelines, Accessed March 2017 and available at https://www.globalreporting.org/resourcelibrary/GRIG4-Part1-Reporting-Principles-and-Standard-Disclosures.pdf.

35 Einstein, A., Speech at the California Institute of Technology, Pasadena, CA, February 16, 1931, as reported in *The New York Times*, February 17, 1931, p. 6.

36 United Nations Sustainable Development Goals, Accessed February 2016 and available at http://www.un.org/sustainabledevelopment/sustainable-development-goals/.

37 Global Reporting Initiative, G4 sustainability reporting guidelines, Accessed March 2017 and available at https://www.globalreporting.org/resourcelibrary/GRIG4-Part1-Reporting-Principles-and-Standard-Disclosures.pdf.

38 Taback, H. and R. Ramanan, *Environmental Ethics and Sustainability* (Boca Raton, FL, CRC Press, 2013).

39 Ramanan, R. and H. Taback, Environmental ethics and corporate social responsibility. In Dhiman, S. (Ed.), *Spirituality and Sustainability: New Horizons and Exemplary Approaches* (New York: Springer, 2016).

40 Global Reporting Initiative, G-4 56-58 ethics and integrity within Governance metrics of Global Reporting Initiative, https://g4.globalreporting.org/general-standard-disclosures/governance-and-ethics/ethics-and-integrity/Pages/default.aspx.

41 Ibid.

42 SASB, Materiality assessment, Accessed February 2016 and available at http://www.sasb.org/materiality/materiality-assessment/.

43 KPMG, Building business value in a changing world, Accessed December, 2012 and available at http://www.kpmg.com/Global/en/IssuesAndInsights/ArticlesPublications/Documents/building-business-value.pdf.

44 Oxfam, Oxfam says wealth of richest 1% equal to other 99%, Accessed March 2016 and available at http://www.bbc.com/news/business-35339475.

45 Oxfam, An economy for the 1%, Accessed February 2016 and available at https://www.oxfam.org/sites/www.oxfam.org/files/file_attachments/bp210-economy-one-percent-tax-havens-180116-en_0.pdf.

46 UN FAO, World agriculture towards 2030/2050, Accessed February 2016 and available at http://www.fao.org/docrep/016/ap106e/ap106e.pdf.

47 World Economic Forum, The global risk report 2016, Accessed March 2016 and available at http://www3.weforum.org/docs/Media/TheGlobalRisksReport2016.pdf.

48 Ramanan, R., How to build sustainability issues in corporate decision making, Net Impact Student Chapter Seminar, IIT Stuart School of Business, Chicago, IL, March 25, 2011.

49 Prahlad, C.K. and Hart, S.L., *The Fortune at the Bottom of the Pyramid* (Upper Saddle River, NJ: FT Press, 1998); Hamel, G. and C.K. Prahlad, *Competing for the Future* (Boston, MA: Harvard Business School Press, 1994).

50 Prahlad, C.K. and S.L. Hart, The fortune at the bottom of the pyramid, Strategy and Business Issue 26 First Quarter 2002, Accessed December 2012 and available at http://www.cs.berkeley.edu/~brewer/ict4b/Fortune-BoP.pdf.

51 Karnani, A., Eradicating poverty through enterprise, Presented at the University of Michigan, November 2007, Accessed December 2012 and available at http://www.un.org/esa/coordination/Eradicating%20Poverty%20through%20Enterprise.Karnani.ppt; Karnani, A., Fighting poverty together: Rethinking strategies for business, governments, and civil society to reduce poverty (London, UK: Palgrave Macmillan, 2011, p. 44).

52 Sen, A., Nobel Laureate Economist quoted by Aneel Karnani at the United Nations General Meeting in New York, 2007, Growth strategies, Accessed December 2012 and available at http://www.un.org/esa/coordination/Eradicating%20Poverty%20through%20Enterprise.Karnani.ppt.

53 Karnani, A., *Fighting Poverty Together: Rethinking Strategies for Business, Governments, and Civil Society to Reduce Poverty* (London, UK: Palgrave Macmillan, 2011, p. 232).

54 Singh, G., Mahatma Gandhi—A sustainable development pioneer, *Eco Localizer*, Accessed October 14, 2008 and available at http://ecoworldly.com/2008/10/14/mahatma-gandhi-who-first-envisioned-the-concept-of-sustainable-development/ in http://www.mkgandhi.org/articles/environment1.htm.

55 Bowen, H., *Social Responsibilities of the Businessman* (New York: Harper & Brothers, 1953).

56 Elkington, J., Cannibals with the forks: The triple bottom line of 21st century business, *Capstone*, 1999, Accessed March 2016 and available at http://www.johnelkington.com/archive/TBL-elkington-chapter.pdf.

57 Ramanan, R., Environmental, safety & health costs and value tracking, *Townley Global Management Center for Environment, Health and Safety* (Manhattan, NY: The Conference Board, April 1, 1998).

58 Jensen, M., Value maximization, stakeholder theory and the corporate objective function, Accessed March 2016 and available at http://www.facstaff.bucknell.edu/jcomas/readings/jensen2002.pdf.

59 Pope Francis encyclical on climate change, On care for our common home, Accessed June 24, 2015 and available at http://w2.vatican.va/content/francesco/en/encyclicals/documents/papa-francesco_20150524_enciclica-laudato-si.html.

60 Porter, M.E. and M.R. Kramer, Strategy and society, *Harvard Business Review*, 2006, pp. 78–88.

61 Franklin, D. (Ed.), *The Economist's World in 2009 Yearbook*; Visser, W., Ages and stages of CSR. In *The Age of Responsibility* (Hoboken, NJ: John Wiley & Sons, 2011), p. 94.

62 Karnani, A., Case against CSR, *Wall Street Journal*, August 23, 2010.

63 ABC news, Missouri Jury awards $72 million in Johnson and Johnson cancer suit, Accessed February 2016 and available at http://abcnews.go.com/US/wireStory/st-louis-jury-awards-72m-johnson-johnson-cancer-37142765s.

64 Ramanan, R. and H. Taback, Environmental ethics and corporate social responsibility. In Dhiman, S. (Ed.), *Spirituality and Sustainability: New Horizons and Exemplary Approaches* (New York: Springer, 2016.)

65 Porter, M.E. and M.R. Kramer, Strategy and society. *Harvard Business Review*, December 2006, pp. 78–88.

66 Karnani, A., Case against CSR, *Wall Street Journal*, August 23, 2010.

67 Visser, W., Ages & stages of CSR. In *The Age of Responsibility*, 1st ed. (Hoboken, NJ: John Wiley & Sons, 2011), p. 131.

68 Taback, H. and R. Ramanan, *Environmental Ethics and Sustainability* (Boca Raton, FL, CRC Press, 2013).

69 Porter, M.E. and M.R. Kramer, Creating shared value, *Harvard Business Review*, January–February 2011, pp. 62–77.

70 Ibid; Porter, 2011.

71 Darcy, K.T., The last decade, Presented at Business Roundtable, Institute for Corporate Ethics, Charlottesville, VA, November, 2009.

72 Currell, D., Weathering the integrity recession, Presented at Business Roundtable, Institute for Corporate Ethics, Charlottesville, VA, November, 2009.

73 Carson, R., *Silent Spring* (Boston, MA: Mariner Books, 2002), Accessed December 2012 and http://www.goodreads.com/work/quotes/880193-silent-spring.

[74] Bruner, R.F., The economic climate's impact on corporate culture and ethics, Presented at Business Roundtable, Institute for Corporate Ethics, Charlottesville, VA, November, 2009.

[75] CBS News, UBS to pay $1.5 billion in fines, Accessed February 2016 and available at http://www.cbsnews.com/news/ubs-to-pay-15b-in-fines-for-libor-rate-manipulation/.

[76] 2009 Compliance and ethics forum summary report: Leading thoughts and practices, Business Roundtable, Institute for Corporate Ethics, November 2009.

[77] Taback, H. and R. Ramanan, *Environmental Ethics and Sustainability* (Boca Raton, FL, CRC Press, 2013).

[78] Visser, W., Ages and stages of CSR. In *The Age of Responsibility*, 1st ed. (Hoboken, NJ: John Wiley & Sons, 2011).

[79] Friedman, M., The social responsibility of business is to increase its profit, *The New York Times Magazine*, September 13, 1970, Accessed October 2017 and available at https://www.colorado.edu/studentgroups/libertarians/issues/friedman-soc-resp-business.html.

[80] Koehler, D.A. and E.J. Henspenide, Drivers of long-term business value, *Deloitte*, 2012.

[81] Freeman, E., *Strategic Management: A Stakeholder Approach* (Boston, MA: Pitman, 1984).

[82] Taback, H. and R. Ramanan, The new social contract. In *Environmental Ethics and Sustainability* (Boca Raton, FL, CRC Press, 2013).

[83] Gordon Gekko's speech in Movie, *Wall Street*, 1987, Accessed April 1, 2018 and available at https://m.imdb.com.

[84] Carroll, A., The pyramid of corporate social responsibility, *Business Horizons*, 1991, 42, pp. 39–48.

[85] Taback, H. and R. Ramanan, The new social contract. In *Environmental Ethics and Sustainability* (Boca Raton, FL, CRC Press, 2013).

[86] Ibid.

[87] Moss Kanter, R., It's time to take full responsibility, *Harvard Business Review*, October 2010, p. 1.

[88] European Commission initiative for Mandatory Environmental, Social and Governance Disclosure in the European Union.

[89] Frankel, C. et al., Redefining value: The new metrics of sustainable business, Accessed February 2016 and available at http://e.sustainablebrands.com/REPORTNewMetricsofSustainableBusiness2013.html.

[90] SAS, BusinessWeek Research Services, White paper on emerging green intelligence, Business Analytics and Corporate Sustainability, 2009.

[91] Insead, Global private equity initiative, Accessed March 2017 and available at https://centres.insead.edu/global-private-equity-initiative/research-publications/documents/ESG-in-private-equity.pdf.

[92] Deloitte.com, Analytics for the sustainable business, Accessed April 5, 2014 and available at http://www.deloitte.com/assets/Dcom-UnitedStates/Local%20Assets/Documents/IMOs/Corporate%2Responsibility%20and%20Sustainability/usdsccfbusinessanalytics_011711.pdf.

[93] EY, Using data analytics to improve EHS and sustainability performance, Accessed May 2017 and available at http://www.ey.com/Publication/vwLUAssets/ey-using-data-analytics-to-improve-ehs-and-sustainability-performance/$FILE/ey-using-data-analytics-to-improve-ehs-and-sustainability-performance.pdf.

[94] SAP.com, SAP Sustainability Solutions, Accessed April 1, 2014 and available at http://www.sap.com/solution/lob/sustainability/software/ehs-management-overview/index.htm.

[95] Oracle.com, Oracle Sustainability Solutions, Walmart Drives Sustainability with Oracle RightNow, Accessed April 2, 2014 and available at http://www.oracle.com/us/solutions/green/it-infrastructure/index.html.

[96] Mercer 2007, as cited in DB climate change advisors, Sustainable Investing, Accessed March 2016 and available at https://www.db.com/cr/en/docs/Sustainable_Investing_2012.pdf.

[97] United Nations, Sustainable development knowledge platform, Accessed February 2016 and available at https://sustainabledevelopment.un.org/majorgroups.

[98] United Nations Sustainable Development Goals, Accessed February 2016 and available at http://www.un.org/sustainabledevelopment/sustainable-development-goals/.

[99] United Nations, The Millennium Development Goals Report 2012, 2012. UN Millennium Goals Indicators website - http://mdgs.un.org/unsd/mdg/Default.aspx.

[100] United Nations, Accessed February 2016 and available at https://www.un.org/pga/wp-content/uploads/sites/3/2015/08/120815_outcome-document-of-Summit-for-adoption-of-the-post-2015-development-agenda.pdf.

[101] United Nations Sustainable Development Goals, Accessed February 2016 and available at http://www.un.org/sustainabledevelopment/sustainable-development-goals/.

[102] Baumol, W.J., On taxation and the control of externalities, *American Economic Review*, 1972, 62, 307–322.

[103] Hardin, G., The tragedy of the commons, *Science*, 1968, 162, 1243–1248.

[104] Coase, R.H., The nature of the firm, *Economica*, 4, 386. doi:10.1111/j.1468-0335.1937.tb00002.x; Coase, R.H., The problem of social cost, *Journal of Law and Economics*, 3, 1–44.

[105] Baumol, W.J., On taxation and the control of externalities, *The American Economic Review* 1972, 62, 307–322.

[106] Miceli, T.J., Chapter 6, The economics of property law: Fundamentals. *The Economic Approach to Law*, 2nd ed. (Palo Alto, CA: Stanford University Press, 2008), Accessed December and available at http://www.sup.org/economiclaw/?d=Key%20Points&f=Chapter%206.htm.

[107] Report of the World Commission on Environment and Development: Our Common Future, UN Brundtland Commission (United Nations, 1987).

[108] P. R. Koutstaal, Tradeable CO_2 emission permits in Europe: A study on the design and consequences of a system of tradeable permits for reducing CO_2 emissions in the European Union (PhD Diss., University of Groningen, 1996), p. 17, available at http://www.unicreditanduniversities.eu/uploads/assets/CEE_BTA/Dora_Fazekas.pdf.

[109] Coase, 1937, *Economica*.

[110] Taback, H. and R. Ramanan, *Environmental Ethics and Sustainability* (Boca Raton, FL, CRC Press, 2013), p. 46.

[111] Ibid.

[112] Ibid, p. 53.

[113] European CEO, Accessed March 2017 and available at http://www.europeanceo.com/business-and-management/unilever-ceo-paul-polman-is-redefining-sustainable-business/.

[114] Investopedia, Public private partnerships, Accessed March 2017 and available at http://www.investopedia.com/terms/p/public-private-partnerships.asp.

[115] EY, Tomorrow's investment rules: Global survey of institutional investors on nonfinancial performance, 2014.

[116] Columbia University, Accessed March 2017 and available at http://spm.ei.columbia.edu/files/2015/06/SPM_Metrics_WhitePaper_2.pdf.

[117] International Federation of Accountants, Accessed April 2017 and available at https://www.ifac.org/publications-resources/international-standard-assurance-engagements-isae-3000-revised-assurance-enga.

[118] ISO, Accessed April 2017 and available at https://www.iso.org/standard/66455.html.

[119] Institut RSE Management, The Grenelle II Act in France: A milestone towards integrated reporting, 2012.

[120] International Federation of Accountants, Accessed April 2017 and available at https://www.ifac.org/publications-resources/international-standard-assurance-engagements- isae-3000-revised-assurance-enga.

[121] Sustainable Stock Exchanges Initiative, Notice on strengthening listed companies'assumption of social responsibility, Accessed March 23, 2017 and available at http://www.sseinitiative.org/fact-sheet/sse/.

[122] Shenzhen Stock Exchange Social Responsibility Instructions to Listed Companies, Accessed March 2017 and available at http://www.szse.cn/main/en/rulseandregulations/sserules/2007060410636.shtml.

[123] BSD Consulting, Accessed March 23, 2017 and available at http://www.bsdconsulting.com/insights/article/sustainability-reporting-standards-in-china.

[124] Hong Kong Exchange, ESG reporting guide, Accessed March 23, 2017 and available at http://www.hkex.com.hk/eng/rulesreg/listrules/listsptop/esg/guide_faq.htm.

[125] Business for Social Responsibility (BSR), Accessed March 23, 2017 and available at https://www.bsr.org/en/our-insights/blog-view/india-companies-act-2013-five-key-points-about-indias-csr-mandate.

[126] European Commission initiative for Mandatory Environmental, Social, and Governance Disclosure in the European Union.

[127] U.S. Securities and Exchange Commission, Fact sheet: Disclosing the use of conflict minerals, 2014.

[128] Taback, H. and R. Ramanan, *Environmental Ethics and Sustainability* (Boca Raton, FL, CRC Press, 2013), p. 56

[129] Global Reporting Initiative, G4 sustainability reporting guidelines, Accessed March 2017 and available at https://www.globalreporting.org/information/sustainability-reporting.

[130] Personal conversation with Jeanne Chi Yun Ng, Director, Group Environmental Affairs, CLP, March 2017.

[131] EY, Accessed March 2017 and available at http://www.ey.com/Publication/vwLU-Assets/EY_Sustainability_reporting_-_the_time_is_now/$FILE/EY-Sustainability-reporting-the-time-is-now.pdf.

[132] Oster, S., What do we make of CSR reporting? Accessed December 2012 and available at http://www.forbes.com/sites/csr/2010/05/11/what-do-we-make-of-csr-reporting/.

[133] Bloomberg LP, The sustainability edge: Sustainability update 2011.

[134] KPMG, Corporate sustainability a progress report, in cooperation with the economist intelligence unit, Accessed December, 2012 and available at http://www.kpmg.com/Global/en/IssuesAndInsights/ArticlesPublications/Documents/corporate-sustainability-v2.pdf.

[135] KPMG, Accessed March 2017 and available at https://assets.kpmg.com/content/dam/kpmg/pdf/2016/05/carrots-and-sticks-may-2016.pdf.

[136] Ernst and Young, Accessed March 2017 and available http://www.ey.com/Publication/vwLUAssets/EY_Sustainability_reporting_-_the_time_is_now/$FILE/EY-Sustainability-reporting-the-time-is-now.pdf.

[137] MIT Sloan Management Review and the Boston consulting Group. (2012). Sustainability nears a tipping point. MIT Sloan Management Review Research Report Winter.

[138] Dhaliwal, D.S., O.Z. Li, A. Tsang, Y.G. Yang, Voluntary non-financial disclosure and the cost of equity capital: The case of corporate social responsibility reporting, February 15, 2009, Available at SSRN: http://ssrn.com/abstract=1343453 or http://dx.doi.org/10.2139/ssrn.1343453.

[139] KPMG, 2012, State of corporate responsibility reporting.

[140] SAP, Impact through innovation SAP integrated annual report 2016, Accessed May 2017 and available at https://www.sap.com/docs/download/investors/2016/sap-2016-integrated-report.pdf.

[141] LeBlanc, B., B. Miller, and J. Osborn, Driving value by combining financial and non-financial information into a single, investor grade document, Ernst and Young, Accessed December 2012 and available at http://www.ey.com/Publication/vwLUAssets/Integrated_reporting:_driving_value/$FILE/Integrated_reporting-driving_value.pdf.

[142] Rogers, J., CEO SASB, Integrated reporting: Solution to global reporting challenge, Accessed March 2016 and available at http://www.huffingtonpost.com/jean-rogers/integrated-reporting-solu_b_9373994.html.

[143] Integrated Reporting, Accessed April 2017 and available at http://integratedreporting.org/resource/ir-training/.

[144] Ibid.

[145] Ernst and Young, Six growing trends in corporate sustainability, In cooperation with GreenBiz Group, 2012, Accessed December 2012 and available at http://www.ey.com/US/en/Services/Specialty-Services/Climate-Change-and-Sustainability-Services/Six-growing-trends-in-corporate-sustainability_overview; http://www.ey.com/US/en/Services/Specialty-Services/Climate-Change-and-Sustainability-Services/Six-growing-trends-in-corporate-sustainability_Trend-7.

[146] Global Sustainable Investment Alliance, Accessed March 2017 and available at http://www.gsi-alliance.org/wp-content/uploads/2015/02/GSIA_Review_download.pdf.

[147] Voorhes, M. et al., Executive summary—Fig. B: Growth of SRI $2.7 trillion in 2007 to $3.0 trillion in 2010, in 2010 report on socially responsible investing trends in the United States, Social Investment Forum Foundation, Accessed December 2012 and available at http://ussif.org/resources/research/ documents/2010TrendsES.pdf.

[148] Ocean Tomo, Accessed March 2017 and available at http://www.oceantomo.com/blog/2015/03-05-ocean-tomo-2015-intangible-asset-market-value/.

[149] Griffin, M.H., Impact investing a guide for philanthropist and social investors, *Northern Trust*, October 2013.

[150] Monitor Institute, 2009.

[151] Global Sustainable Investment Alliance, Accessed March 22, 2017 and available at http://www.gsi-alliance.org/.

[152] CDC Investment Works, UK's Development Finance Institution (DFI) and wholly owned by the UK Government, Accessed March 21, 2017 and available at http://www.cdcgroup.com/Who-we-are/Key-Facts/.

[153] Global Impact Investing Network (GIIN), Accessed March 22, 2017 and available at https://iris.thegiin.org/about-iris.

[154] Organisation for Economic Co-Operation and Development, Accessed March 21, 2017 and available at http://www.oecd.org/corporate/mne/1922428.pdf.

[155] Organisation for Economic Co-Operation and Development, Accessed March 20, 2017 and available at http://www.oecd.org/investment/mne/38783873.pdf.

[156] Collevecchio Declaration, BankTrack (Amsterdam, the Netherlands: BankTrack, January 2003), Accessed December 2012 and available at http://www.banktrack.org/download/collevechio_declaration/030401_collevecchio_declaration_with_signatories.pdf.

[157] Equator Principles, Accessed October 2017 and available at http://www.equator-principles.com/index.php/about.

[158] UN Principles of Responsible Investment, Accessed March 21, 2017 and available at https://www.unpri.org/about.

[159] UN Global Compact, Accessed March 22, 2017 and available at https://www.unglobal-compact.org/what-is-gc/mission/principles.

[160] Griffin, M.H., Impact investing a guide for philanthropist and social investors, *Northern Trust*, October 2013.

[161] Monitor Institute, The future of impact investing, Accessed March 2016 and available at http://monitorinstitute.com/downloads/what-we-think/impact-investing/Impact_Investing.pdf.

[162] Morgan, J.P., Impact investments—An emerging asset class, Accessed April 2017 and available at https://www.jpmorganchase.com/corporate/socialfinance/document/impact_investments_nov2010.pdf.

[163] Rockefeller Philanthropy Advisors, Mission related investing—A policy and implementation guide for foundation trustees, Accessed March 2016 and available at http://rockpa.org/document.doc?id=16.

[164] Monitor Institute, The future of impact investing, Accessed March 2016 and available at http://monitorinstitute.com/downloads/what-we-think/impact-investing/Impact_Investing.pdf.

[165] Bill Clinton in his Foreword in Mathew Bishop and Michael Green, *Philanthrocapitalism—How Giving Can Save the World* (New York: Bloomsbury Press, 2008).

2 Sustainability Analytics and Decision Acumen

2.1 SUSTAINABILITY METRICS, MATERIALITY, AND INDEXES

2.1.1 Chapter Overview

This chapter identifies sustainability indicators that are useful and relevant for each of the quadruple bottom lines. A review of the dominant sustainability frameworks, metrics or indicators, materiality, and indexes across some sectors is provided. For select dominant indexes, the aggregation of metrics and ranking methodology is presented. Also, examples of measuring sustainability of select entities, such as, corporations, colleges, cities, and countries are presented.

Metrics or indicators are simply measures that describe the current progress level or state of a sustainability aspect, within a category, of a value creating or producing entity or activity (for instance, an operation's energy consumption or greenhouse gas emission). Metrics or indicators are the basic units that go into reporting frameworks as well as indexes.

Indexes are aggregates of metrics or indicators, designed and defined by the provider of that index. An index is an aggregate of multiple metrics, commonly used by investors. For instance, Dow Jones Sustainability Index (DJSI)[166] tracks financial performance of sustainability-driven companies and provides an integrated economic, social, and environmental assessment to serve as benchmarks for sustainability portfolio asset managers.

A framework is a disclosure of a structured comprehensive set of metrics or indicators stated individually about the sustainability performance and impact of an operation, with a focus on the quadruple bottom line, namely prosperity (economic or profit), planet (environmental), people (social), and purpose (ethics or governance). Thus, each element that goes into a framework requires a metric. For instance, the Global Reporting Initiative (GRI)[167] G4 is a voluntary framework for sustainability reporting on economic, environmental, and social dimensions of products, activities, and services. The recent inclusion of ethics and integrity aspect within the governance category addresses the fourth of the quadruple bottom line.

Figure 2.1 is a visual representation of the sustainability element information flow starting with resource input through to the integrated financial and sustainability report.

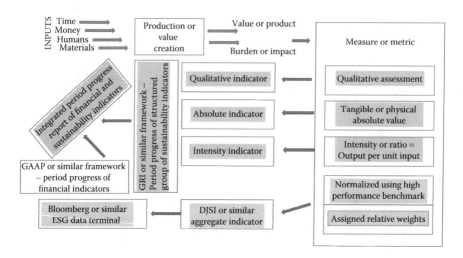

FIGURE 2.1   Sustainability element information flow.

The sustainability category and aspect, beginning as inputs, enter into the *production or value creation* crucible. Inputs include time, money, human resources, and materials and come out as product, value, impact, or burden. The sustainability aspects of the inputs and outputs are measured by a simple metric. The metric could provide a qualitative assessment, a tangible or physical absolute value, or a ratio or intensity value. For instance, the basis for selecting stakeholders for engagement is a qualitative statement that becomes a qualitative indicator. Reporting the total number and total volume of a significant spill is an example of a tangible or physical absolute value and is an absolute indicator. Disclosing the greenhouse gas (GHG) emissions intensity ratio is a common example of ratio or intensity metric or indicator. Typically, intensity indicators are a ratio of an output over some significantly related input (tons of raw material) or output (number of cars or units of energy produced).

These indicators form the basis for input into the structured framework calling for various metrics or indicators (for instance, the GRI G4 reporting framework). This becomes the sustainability disclosure or corporate social responsibility/citizenship report. Some pioneering leaders in the sustainability area take it a step beyond by producing the integrated report that combines financial performance and nonfinancial sustainability progress indicators. For instance, they may utilize the Integrated Reporting Framework.

Indexes are sets of indicators aggregated into a single composite measure. There is a process for converting indicators to indexes. First, a group of indicators, meaningful and relevant to the index, are identified. The metrics for these indicators have to be normalized for aggregation. This process involves transforming original value of all the select group of indicators into a more generic dimensionless value. The final step involves assigning relative weights against each other, and a weighted average value yields aggregate indexes such as the Dow Jones Sustainability Index. This information is then accessed by investors through information channels that provide environmental, social, and governance (ESG) data to investors, usually as a subscription service such as the Bloomberg terminal where ESG performance data is housed alongside more traditional financial information.

Sustainability metrics or indicators can transform a vast amount of complex data into concise decision analysis and usable information. The need to quantify sustainability categories and aspects into metrics and indexes has been frequently observed in literature.[168] According to Sudhir Anand and Nobel laureate Amartya Sen,[169] the sustainability framework is a preferred form of disclosure because all information is presented and not hidden behind aggregated data. Structured frameworks can be used to isolate indicators, do not require aggregation, and are hence more revealing and accurate. Unlike frameworks, indexes use metrics and aggregate toward a composite index, or score rating, which is easier to use and to understand. Indexes or aggregate metrics may be more useful in sustainability analytics for prioritization decision-making or ranking within or across investment project portfolios.

2.1.2 SUSTAINABILITY METRICS OR INDICATORS

Metrics or indicators are the basic units that go into reporting frameworks as well as indexes. Sustainability metrics or indicators must meet the requirement for established common measures to accumulate, or universal methods to aggregate, the seemingly intangible capitals and their performances.[170] For instance, the unit used for the metric or indicator is a critical factor. The Sustainability Accounting Standards Board (SASB),[171] at the level of accounting metrics, defines the following as the criteria the sustainability metric must meet (or provide) for organizations:

1. Be a fair representation of performance, both adequate and accurate
2. Provide useful information to manage operational performance and financial analysis
3. Be applicable to most in the relevant industry sector
4. Provide comparable quantitative and qualitative data for peer-to-peer benchmarking
5. Provide complete information to understand and interpret aspect performance
6. Be verifiable to support internal controls for data assurance
7. Be aligned with other similar metrics already in use
8. Be neutral and free from bias and value judgment to yield objective disclosure
9. Be distributive to yield a discernable range of data to differentiate performance

An extensive study of sustainability metrics or indicators from organizational perspectives at Columbia University[172] found close to 600 sustainability metrics, 200 in each category of environmental, social, and governance. Within environmental metrics, the highest number of metrics observed were clearly around energy. Energy is reported variously as total consumption in gigawatt hours to tons of oil equivalent by some and in terms of energy efficiency by others. Closely behind energy was GHG emissions. It was captured in tons of carbon equivalent by one group and as carbon intensity by another. The most common environmental metrics reported were energy, emissions, water, materials, effluents and waste, and biodiversity. Social metrics, also nearly 200, were focused on equality, justice, and other social impacts. The aspects covered most commonly included human rights, safety and health, education, employee training, and community activities. Interestingly, there were equal numbers of governance metrics. These were, as one would expect, transparency, equality and fairness, corruption, and diversity. Ethics or purpose, the fourth bottom line, is incorporated within the governance metric.

From a government or nation perspective, the 2007 version of the United Nations Commission for Sustainable Development (CSD)[173] Indicators of Sustainable Development has 96 indicators, with 50 core ones. CSD indicators are not categorized into the four pillars of sustainable development but placed in a framework of 14 themes, which are poverty; governance; health; education; demographics; natural hazards; atmosphere; land; oceans, seas, and coasts; freshwater; biodiversity; economic development; global economic partnership; and consumption and production patterns. Many of these metrics are beyond the control of an organization's reach and require government intervention. Finally, metrics that relate to special applications include, charitable organizations, making decisions on impact investing such as Acumen, and activists protecting indigenous people may need building of newer metrics.

2.1.3 SUSTAINABILITY MATERIALITY[174,175]

There are numerous ESG issues that can affect a firm's financial performance and therefore be highly material to investors. ESG issues are clearly material to company performance—the question is which ones.[176] The materiality of any issue varies from one industry to the other. Also, "ESG is just one ingredient and is not the full recipe."[177]

The Public Company Accounting Oversight Board (PCAOB)[178] refers to the US Supreme Court interpretation[179] of securities laws in its materiality guidance. The Securities and Exchange Commission (SEC) follows the PCAOB guidance and the SASB uses this SEC definition of materiality. Like the PCAOB, SASB defines material information as information that represents a substantial likelihood that its disclosure will be viewed by the reasonable investor as significantly altering the total mix of information made available.[180] Sustainability investments that affect financial performance for example by affecting customer satisfaction, loyalty, employee engagement, and regulatory risk may be classified as material. The SASB,[181] in order to protect the investors from risk and to help organizations, especially publicly listed corporations, meet the needs of the *new social contract*, has identified a universe of sustainability issues to gauge material ESG matters within an industry for select sectors and organized them in five broad dimensions; these are included in Appendix A.

SASB has published materiality guidance for health care, financials, technology and communications, non-renewables, transportation, and services sectors. Materiality maps for 88 industries in ten sectors are underway. Each map prioritizes 43 ESG issues, ranking their materiality for a given industry on a scale from 0.5 to 5, with 5 being most material with the greatest probable impact on a firm's financial performance.

SASB might classify an issue that may be important for other stakeholders as immaterial from an investor standpoint. The investor focus of SASB for materiality is narrower compared to other organizations such as the GRI, which has a multi-stakeholder focus. The GRI defines materiality as the threshold at which *aspects* that reflect the organization's significant economic, environmental, and social impacts or substantively influence the assessments and decisions of stakeholders become material and important enough to be reported.

SASB evaluates evidence of interest by different types of stakeholders and evidence of economic impact of that issue. It conducts a three-component evidence of materiality test comprising of evidence of interest, evidence of financial impact, and forward impact adjustment. Evidence of interest is gathered from Form 10-Ks, Corporate Social Responsibility (CSR) reports, shareholder resolutions, and media. The results reveal the intensity with which issues arise in each industry. The evidence of economic or financial impact on revenue, assets, liabilities, or cost of capital, uses a value framework developed by McKinsey. Forward-looking impact, although rarely used, acknowledges an emerging issue that is not yet reflected in these evidence-based tests but may generate externalities in the future and still be relevant to investors. For instance, BlackRock's funds for investors that want to screen out gun manufacturers - given the sentiment around gun violence. If SASB assessments for their industry are not yet available, firms can use a similar approach to find out which ESG issues are most material to their investors.

2.1.4 SUSTAINABILITY INDEXES AND DATA ANALYTICS

As noted earlier, metrics or indicators are the basic units that go into reporting frameworks as well as indexes. Indexes are aggregates of multiple metrics or indicators, designed and defined by the provider of that index, and are commonly used by investors. For instance, the DJSI[182] tracks the financial performance of sustainability driven companies and provides an integrated economic, social, and environmental assessment to serve as benchmarks for sustainability portfolio asset managers. Bloomberg is another example of a financial data and analytics provider that collects ESG data from published company material and integrates it into the Equities and Bloomberg Intelligence platforms that are accessed by financial researchers. The number of financial organizations that provide sustainability indexes is growing; so is the family of indexes. Not all motivation for voluntary sustainability rating participation is dangling carrots—there are sticks too! Delisting from a sustainability index has serious consequences or drawbacks.[183]

Besides ESG indexes for private sector companies and investors, there are indexes for scoring stock exchanges, governments, and nations.

2.1.4.1 Environmental, Social, and Governance Ratings for Investors

Many business drivers, such as ESG,contribute to financial performance and investment returns but are not included in a company's financial report. A few specialized ESG rating agencies provide information to investors about the level of socially responsible behavior by firms. However, there is a lack of uniformity and transparency in the methods. Research[184] finds that a level of subjectivity is inevitable in ESG ratings and the need for uniformity may inhibit innovation. However, increased transparency of the rating methods will make them more robust.

The ESG index field is very dynamic at this stage with ongoing innovations, as sustainability and the linkage with financial performance is being understood. Typically, an in-house grown, acquired, or reputed sustainability data analytics research group tracks sustainability data for thousands of companies based on public data or surveys. This data is then sub-aggregated, and ratings are developed and made available to the index provider. This dynamic rating then powers the indexes. For instance, the Thomson Reuters[185–187] Corporate Responsibility Indexes utilize the ASSET4 ESG database, which rates the ESG practices of 4,600 companies globally in 226 key indicators of ESG performance. The indexes employ the data and apply different levels of weightage depending on the industry, country, and regional focus of company. In the spirit of transparency, information is available on the Web on the rating and aggregating methodology.

A generic hypothetical data structure for ESG rating, generally based on FTSE Russell, is presented in Figure 2.2. The data is structured around the three pillars of ESG (environmental, social, and governance). Each pillar has four themes as shown. The process starts with hundreds of indicators of the various ESG aspects, which serve as the foundation. Thematically, these measures are sub-aggregated into scores and exposures per theme. The score is the measure of quality of management of the theme, issue, or the collective pillar, and exposure is the measure of relevance of the pillar. Thematic scores and exposures are aggregated into relevant pillars, and the aggregated pillar scores are compounded with appropriate weights to get the composite ESG rating.

2.1.4.1.1 Bloomberg[188]

Founded in 1981, Bloomberg has more than 325,000 global subscriptioners to its data, analytics, and information-delivery service. Bloomberg ESG Disclosure Scores[189] rate companies based on their disclosure of quantitative and policy-related ESG information and provide data on more than 120 indicators for more than 10,000 publicly listed companies globally from company-sourced filings and third-party information, covering virtually the entire publicly investable universe. This information enables all market participants, investors, company management, and others to understand ESG-related risks and opportunities, including intangible and reputational value factors.

Bloomberg evaluates companies based on the extent and robustness of their ESG disclosure and related corporate policies. The company evaluations using 120 indicators and rating is based on a proprietary computer model. Indicator is a measure of an issue, for instance for the climate change issue, a product could use the amount of greenhouse gas emitted as the indictor. The coverage is global—all companies that are in a major index or those that disclose quantitative environmental or social data.

2.1.4.1.2 Dow Jones Sustainability Index[190]

Launched in 1999 as the first global sustainability benchmark, the family of the DJSI[191] tracks the stock performance of the world's companies that are leaders in terms of economic, environmental, and social criteria. Created jointly by the S&P and Dow Jones Indexes, an established index provider, and RobecoSAM, a specialist in sustainability investing to select the most sustainable companies from across

ESG Rating = Indicator of overall quality of a company's management of ESG issues		
Environmental	Social	Governance
Score = Measure of quality of management of pillar issue and Exposure = Measure of relevance of the pillar issue		
Score = Measure of quality of management of theme issue and Exposure = Measure of relevance of the theme issue		
Biodiversity	**Labor** *Freedom of association and collective bargaining, Labor management relations, Occupational health and safety, Diversity and equal opportunity, Equal remuneration for women and men*	**Assessment-internal and supplier** *Human rights, Environmental, Labor practices, Impacts on society*
Climate change	**Human rights** *Non-discrimination, Indigenous, Supplier, No child or forced labor, Security practices*	**Compliance** *Environmental, Social, Product and customer responsibility*
Resources *Energy Water Materials*	**Communities** *Local communities, Fair taxes, Employment, Training and education, Supplier assessment for impacts on society, Labor practices, and Environmental*	**Grievance mechanisms** *Human rights, Environmental, Labor practices, Impacts on society*
Pollution *Emissions Effluents Waste*	**Product and customer responsibility** *Customer privacy, Marketing communications, Customer health and safety, Product and service labeling, Products and services-fair pricing and access*	**Ethics and integrity** *Public policy, Anti-competitive behavior, Anti-corruption*
Hundreds of indicators distributed over the themes		

FIGURE 2.2 Hypothetical ESG index rating data structure (3 pillars, 12 themes, 100s of indicators).

60 industries. RobecoSAM[192] provides the research backbone for DJSI. In 2009, Dow Jones exited the STOXX[193] joint venture. STOXX sustainability index families continue to provide access to companies that are leaders in terms of environmental, social and governance criteria for global and regional markets. Dow Jones acquired Trucost in 2016.

Total sustainability scores of companies coming from the RobecoSAM Corporate Sustainability Assessment (CSA)[194] are powered through Dow Jones. RobecoSAM invites the world's largest 2,500 publicly traded companies to participate in RobecoSAM CSA. They are placed into 60 Global Industry Classification Standards (GICS) categories for analysis using industry-specific questionnaires. No industries are excluded from this process. Company evaluation is based on financially relevant sustainability—economic, environmental, and social criteria. The CSA Resource Center[195] provides a complete overview of the criteria weights for each of the 60 industry categories. Only the top-ranked companies within each industry are

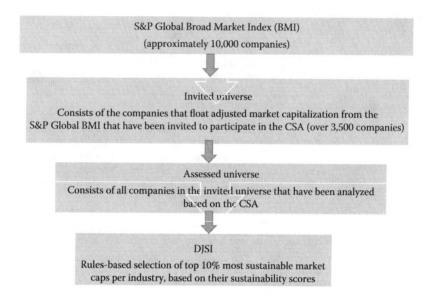

FIGURE 2.3   The Dow Jones sustainability index—rules-based component selection.[196]

selected for inclusion in the Dow Jones Sustainability Index family. The DJSI applies a rules-based component selection process, as shown in Figure 2.3.

The DJSI family is comprised of global, regional, and country benchmarks and includes the world, North America, Europe, Asia Pacific, Australia, and emerging markets. DJSI also offers indices with exclusion criteria such as armaments and firearms, alcohol, tobacco, gambling, and adult entertainment. All DJSI indexes are calculated in both price and total return versions and are disseminated in real time.

2.1.4.1.3   FTSE Russell[197]

FTSE Russell brings more than 15 years of ESG experience and provides data analytics, ratings, and indexes covering thousands of companies worldwide. Using an ESG index rating data structure, similar to the one shown in Figure 2.2, the FTSE Russell ESG ratings are comprised of an overall rating system that is built up from the exposures and scores of the three pillars of ESG and a dozen thematic issues. The score is the measure of quality of management of the theme issue or the collective pillar, and exposure is the measure of relevance of the theme issue or pillar.

These pillars and themes are built on a foundation of 300 individual indicator assessments of the company's sustainability elements or aspects. This approach, using a separate measure for exposure, allows users to identify or specify which ESG issues are most relevant for a given company. Furthermore, using an exposure-weighted average to calculate the ratings allows the most material ESG issues to be given the most weight when determining a company's scores.

2.1.4.1.4 Morgan Stanley Capital International[198] Sustainability Index Family[199]

Launched in 1972, Morgan Stanley Capital International (MSCI) indexes were some of the first indexes to arrive. The MSCI Global Sustainability Indexes (GSI) are constructed by applying a best-in-class selection process to companies in a global equity index consisting of developed and emerging market countries. MSCI GSI target the highest ESG-rated companies making up 50% of the adjusted market capitalization in each sector for investors seeking exposure to companies with strong sustainability profiles. The selection universe for the GSI is the constituents of the MSCI Global Investable Market Indexes. In addition, companies showing involvement in alcohol, gambling, tobacco, nuclear power, and weapons are excluded from the indexes.

MSCI Global ESG Environmental Indexes allows institutional investors to consider climate change risks and opportunities and to manage carbon exposure in their portfolios. These indexes support low carbon investment strategies and include Low Carbon and Climate Indexes. The MSCI index family also includes the MSCI KLD 400 Social Index, which was launched in May 1990 and is one of the first Socially Responsible Investment (SRI) indexes. KLD Innovest[200] is a research and analytics firm.

2.1.4.1.5 Thomson Reuters Corporate Responsibility Indexes and Ratings

Thomson Reuters acquired ASSET4 in 2009 and launched a new family of ESG indexes in 2013.[201] Their first generation of ESG indexes targeted the very best performers under the presumption that they would provide better returns. However, that resulted in concentrating on certain sectors and regions, and consequently excessive potential risk. These new families of indexes mirror the performance of US and international markets, yet only stocks with better than average ESG ratings are included. The current Thomson Reuters Corporate Responsibility Indexes utilize the ASSET4[202] ESG database, which rates the ESG practices of 4,600 companies globally in 226 key performance indicators of ESG performance. The indexes employ these data and apply different levels of weightage depending on the industry, country, and regional focus of a company. Individual members of the family of the Thomson Reuters Corporate Responsibility Indexes targets ESG in US, developed (non-US), and emerging markets.

2.1.4.1.6 Consolidations and Collaboration of Environmental, Social, and Governance Data Providers

An increasing number of financial data providers for research and analysis are incorporating ESG data within their offerings. In recent years, Bloomberg[203] and Thomson Reuters[204] have started providing environmental, social, and notably governance (includes ethics) metrics on their terminals to serve the rapidly growing group of socially responsible investors for assessment of non-financial factors. For instance, in 2014, access to Sustainalytics ESG information was made available on Bloomberg[205] platforms, an indication of the mainstreaming of sustainability in capital markets. Per Jantzi,[206] "Having corporate ESG performance data housed alongside more traditional financial information allows investors to more easily integrate ESG factors into their fundamental analysis."

Other major ESG research and integrity data provider consolidations include: ASSET4 (acquired in 2009 and now a Thomson Reuters business); Riskmetric (which bought Innovest and KLD and was eventually acquired by Morgan Stanley); Jantzi Research, which merged with Sustainalytics and are now associated with Bloomberg; and Trucost, which was acquired in 2016 by Dow Jones.[207]

In September 2016, RobecoSAM became "the first company to offer a complete ESG company ranking dataset to mainstream investors through the Bloomberg Professional service."[208] The implication offers an avalanche of new DJSI ESG data points on ESG issues delivered to more than 300,000 Bloomberg screens. This is a game changer in terms of the transparency of a company's ESG performance to investors. So far only the names and aggregate scores of the leading companies in each industry have ever been disclosed. But now, one can view the percentile ranking for twenty-plus sustainability themes of more than 2,000 individual companies. Non-responders to the DJSI sustainability rating ESG surveys can also be benchmarked on the basis of public information. Exposure of a company's ESG performance will go mainstream, way beyond the socially responsible investors.

2.1.4.2 Sustainable Development Rating for Governments and Nations

Engelbrecht[209] classifies sustainable development indexes into two groups. The first group is associated with the capital approach to development. The second group of sustainability indexes are composite indexes that go beyond economic sustainability. Engelbrecht's[210] research makes some very insightful observation from a study of 26 Organization for Economic Co-operation and Development (OECD) country data. Firstly, if sustainability is defined as non-declining natural capital per capita, then the group of countries as a whole is probably on a non-sustainable path, whereas in terms of economic sustainability by other capital approach indexes it is on a sustainable path.

Secondly, the low correlation coefficients between the change in total wealth per capita versus any one of the three, tangible wealth per capita, adjusted net savings, and adjusted net savings per capita, is intriguing. These three are currently the most widely used economic sustainability indexes. Is that appropriate, or is it misleading? It would be interesting to explore which intangible capital, like human and social capital, accounts for most of the difference.

Nobel laureate Amartya Sen is critical of composite indexes. In the context of sustainable development indexes for nations, Stiglitz et al.,[211] expresses concern about the normative implications of the weightings used to aggregate the components are rarely made explicit or justified, and such indexes arguably lack a well-defined meaning of sustainability. Likewise, in the context of corporations, in Anand et al.,[212] a preference is expressed for frameworks to disclose sustainability because all information is presented and not hidden behind aggregated data. Indicators can be isolated from structured framework because of non-aggregation and making the information more revealing and accurate. Unlike frameworks, indexes aggregate metrics toward a composite index or score.

2.1.4.2.1 Capital Approach Sustainable Development Indexes

1. Adjusted net savings is "gross savings minus consumption of fixed capital, plus education expenditure, minus depletion of natural resources, minus pollution damages."[213] Adjusted net savings per capita equals adjusted net savings divided by the related population.
2. Change in total wealth per capita is the sum of "change in tangible (i.e., partial) wealth per capita" and intangible wealth (human capital and social capital) per capita. The sum is a more comprehensive measure of economic sustainability. It supports the view, "Whether levels of well-being can be sustained over time depends on whether stocks of capital that matter for our lives (natural, physical, human, social) are *all* passed on to future generations."[214]
3. Change in natural capital per capita considers development to be sustainable if natural capital per capita does not decline between two points in time. This is a valid measure of sustainability if most of the concern is about depletion and destruction of the natural environment—that is, environmental sustainability.

2.1.4.2.2 Beyond Economic Composite Sustainable Development Indexes

1. *Happy planet index*[215] is an ecological efficiency measure—that is, it aims to capture the degree to which happy life years are achieved per unit of environmental impact. It measures "what truly matters to us—our well-being in terms of long, happy, and meaningful lives—and what matters to the planet—our rate of resource consumption." This measure of sustainability explicitly includes life outcomes and scales them by a measure of resource use.
2. *Environmentally responsible happy nation index*,[216] modifies "happy planet index by accounting for external costs of environmental disruption negatively."
3. *Environmental efficiency of well-being*, a large positive value over one indicates a country with high well-being relative to their environmental consumption (and vice versa) index that takes social well-being and the ecological footprint into account.
4. *Environmental Performance Index (EPI)*,[217] and its predecessor *Environmental Sustainability Index (ESI)* are both designed to better measure actual policy performance with respect to (1) reducing environmental stresses on human health and (2) promoting ecosystem vitality and sound natural resource management. This is done by measuring country performance against absolute targets. The EPI is based on aggregating only 16 indicators, while ESI aggregated 21 indicators.

5. *Ecological Footprint*[218] "measures humanity's demand on the biosphere in terms of the area of biologically productive land and sea required to provide the resources we use and to absorb our waste." The ratio of required to available resources is a measure of ecological sustainability, that is, if it is greater than one, the current consumption level and lifestyle is unsustainable.

2.1.4.2.3   United Nations Commission for Sustainable Development Indicators of Sustainable Development—National Indicators

The CSD[219] Indicators of Sustainable Development serve as reference for countries to develop or revise national indicators of sustainable development (Please see Section 2.1.2).

2.1.4.2.4   Customized Rating for Nations

2.1.4.2.4.1   The Natural Resource Protection and Child Health Indicators   The World Economic Forum[220] has developed, in collaboration with Yale and Columbia universities, national scale sustainability metrics for nations to be compared on their environmental performance. It is a depository of environmental data. The Natural Resource Protection and Child Health Indicators, 2014 Release, (formerly Natural Resource Management Index) was produced in support of the US Millennium Challenge Corporation as selection criteria for funding eligibility.

2.1.4.2.4.2   The Gross National Happiness Index   The government of Bhutan surveys its citizens using a GNH Index[221] in addition to the traditional gross domestic product (GDP). It surveys its citizens on nine key aspects of happiness: psychological well-being, health, time-use (time or work-life balance), cultural diversity and resilience, good governance, community vitality, ecological diversity and resilience (biodiversity), living standard (material well-being), and education (Figure 2.4).

The Happiness Alliance, a Seattle-based organization inspired by the government of Bhutan's GNH index, and endorsed by Seattle City Council, has developed objective indicators to measure happiness. The indicators measure poverty rates, air emissions, voter turnout, graduation rates, volunteer rates, rates of domestic violence and other crime, life expectancy, length of commute, work hours, and so on. The findings allow policymakers to make more effective decisions when serving their communities.

2.1.5 Measuring Sustainability: Select Significant Sectors

2.1.5.1 Measuring Sustainability: Stock Exchanges

Capital Knights[223] has been tracking sustainability disclosures and rating stock exchanges since 2012. This approach tracks the disclosure of the seven *first-generation* sustainability indicators, namely employee turnover, energy, GHG emissions, injury rate, payroll, waste, and water. The analysis of individual stock exchanges is based on disclosure rates according to Bloomberg, growth in disclosure rates on a trailing

FIGURE 2.4   Bhutan 2010 GNH index part 1.[222]

five-year basis, and disclosure timeliness. In 2016, the most recent year for which data is available, 45 stock exchanges, each with more than 10 listed companies with a total listing of 4,281 companies were covered.

Data from Bloomberg's ESG database on the seven indicators—(1) energy use (energy); (2) carbon emissions (GHGs)—scope 1 (direct from owned or controlled sources) and Scope 2 (indirect from consumption of purchased energy e.g. electricity, heat or steam); (3) water use (water); (4) waste generation (waste); (5) rate of employee injury (injury); (6) rate of employee turnover (employee turnover); and (7) personnel costs (payroll)—are used for the analysis.

The ranking model uses a weighted average of the three measures of performance:

1. The Disclosure Score (50% weight) measures the proportion of the large listings of the exchange that disclosed the seven indicators. Disclosure score of an exchange is a simple average of the seven percentage-rank scores. The indicators are equally weighted in terms of their contribution to the disclosure score.
2. The Disclosure Growth Score (20% weight) measures the growth rate in that disclosure score proportion over the five-year period. First, the annualized compound growth rate in the disclosure of a given indicator is calculated for each exchange. Second, the resulting annualized compound growth rates are percentage-rank scored, using the high percentage benchmark method, with the highest percentage receiving the highest score. This is repeated for each of the remaining six indicators. Disclosure growth score is a simple average of the seven percentage-rank scores.

3. The Disclosure Timeline Score (30% weight) measures how quickly after the end of the fiscal year, large companies listed in that stock exchange disclose sustainability data to the market.'

2.1.5.2   Measuring Sustainability: Process Industry[224]

In 2002, the Institution of Chemical Engineers (IChemE) UK launched the sustainability metrics for the process industry. The reporting format is consistent with that of the GRI. The metrics are presented in the three groups: economic, environmental, and social, which reflect the three components of sustainable development. Economic progress, a key element of sustainability, is the success of industry in creating wealth. The economic indicators go beyond the conventional reporting of financial progress in describing the creation of wealth or value, as well as its distribution and reinvestment for future growth. Both human and financial capital are considered.

The environmental metrics provide a balanced view of the environmental impact of inputs—resource usage and outputs—emissions, effluents, and waste and the products and services produced. Good social performance is important in ensuring a company's license to operate over the longer term. The social indicators reflect company treatment of its employees, suppliers, contractors, and customers and its impact on society at large. The preferred unit of product or service value is the value added. However, the value added can sometimes be difficult to estimate accurately, so surrogate measures such as net sales, profit, or even mass of product may be used. Alternatively, a measure of value might be the worth of the service provided, such as the value of personal mobility, the value of improved hygiene, health, or comfort.

The IChemE UK indicators can be used to measure the sustainability performance of an operating unit, which could be a process plant, a group of plants, part of a supply chain, a whole supply chain, a utility, or other process systems. Most metrics are calculated as appropriate ratios or intensity; this provides a measure of impact and benefit or value, independent of the scale of operation that allows comparison between different operations. For example, in the environmental area, the unit of environmental impact per unit of product or service value is a good measure of eco-efficiency or impact intensity. Because of the ratio or intensity approach used, they can be scaled or aggregated to present a view of a larger operation on a company, industry, or regional basis.

This engineer-developed set of metrics provides absolute, normalized, and qualitative measures. These are representative of industry sector specific metrics for sustainability reporting covering two of the largest and highest sustainability impact categories - integrated petroleum and chemical industry. Most process industries' products pass through the chain resource extraction: transport, manufacture, distribution, sale, utilization, disposal, recycling, and final disposal. Suppliers, customers, and contractors all contribute to this chain, so boundaries have to be clear in reporting the metrics.

Finally, a unique feature of the UK Sustainability Metrics is the Environmental Burden for Emissions to Air (Appendix A) and Environmental Burdens for Emissions to Water (Appendix B). Think of this as the *boson*—invaluable in integrated conceptual modeling for sustainability investment prioritization and decisions. An averted or reduced environmental burden offers a relative measure of benefits.

2.1.5.3 Measuring Sustainability: Consumer Goods and Walmart Suppliers

The Sustainability Consortium (TSC) is a multi-stakeholder organization focused on developing a science-based, sustainability measurement, and reporting system for the consumer goods industry. TSC's key performance indicators (KPIs) are questions that help assess manufacturers' performance against the identified sustainability hotspots; hotspots are activities that create materially significant social or environmental impacts within a single life-cycle stage of a product category. An extract of the TSC methodology[225] for the development of the product sustainability toolkit and design principles of KPIs is presented below:

1. Define a unique product category and decide types of products that are included, components, materials and ingredients considered, and major processes that occur in the product life cycle—based on industry norms, the similarity of supply chains, and stakeholder feedback. Prioritize categories with significant environmental and social impact.
2. Review scientific sources that describe the sustainability impacts of the entire product category life cycle such as life-cycle assessments (LCAs).
3. Research and identify up to 15 actionable hotspots, which are activities within a single life-cycle stage of a product category that creates materially significant social or environmental impacts and has actionable improvement opportunity.
4. Research and identify at least one with a maximum of three actionable hotspot improvement opportunities, which are specific actions that manufacturers can take to address the hotspots. Actionable means manufacturer visibility into the supply chain to effect a change.
5. Evaluate the evidence and determine the materiality of the hotspots by considering either the number and quality of the sources, or the largest impacts identified by life-cycle assessment.
6. Design KPIs questions to quickly assess manufacturer performance against the identified sustainability hotspots that are designed to be answered by brand manufacturers either at the request of a customer or for self-assessment. The KPIs have to be traceable back to the original scientific research. They have to be objective and clear, actionable by a brand manufacturer, measurable, differentiating, strategic, and have consistent, forms and metrics for the same hotspots.
7. Conduct review by multi-stakeholders from business, civil society, government, and academia who collaboratively develop the product sustainability toolkits.
8. Publish the toolkit comprising of the product category sustainability profile and short descriptions of the hotspots improvement opportunities for manufacturer to customer sustainability reporting.
9. Update the toolkit, balancing the need to improve usability with the value of reporting consistency over time.

Walmart's Sustainability Index Program[226] is a tool that helps understand, monitor, and enhance the sustainability of Walmart products and supply chain. The program allows Walmart to integrate sustainability into their core business and to increase customer's trust in Walmart and the brands they carry. The Walmart Sustainability Index uses questionnaires that are based on KPIs developed by TSC.[227]

2.1.5.4 Measuring Sustainability: Colleges and Universities

Sustainability Tracking, Assessment, and Rating System[228] (STARS) of the Association for the Advancement of Sustainability in Higher Education (AASHE) is a transparent, self-reporting framework for colleges and universities to measure their sustainability performance. Sierra Club[229] collaborates with the AASHE and uses their STARS reporting tool to gather raw data for their rankings. This tool gives colleges and universities a method for tracking and assessing their sustainability programs; it uses a self-reporting survey method where responses are not audited for accuracy. Cool Schools Rankings evaluate about 200 four-year, degree-granting undergraduate colleges and universities in the United States.

Now suspended, the Green Report Card (2007–2011) of the Sustainable Endowments Institute[230] examined sustainability practices at 322 colleges and universities in the United States and Canada, with combined holdings of more than \$325 billion in endowment assets, which represents more than 95% of all university endowments. A school's overall grade was calculated from scores in nine equally weighted categories that comprised of a total of 52 indicators used to evaluate performance within the categories. These categories are administration, climate change and energy, food and recycling, green building, student involvement, transportation, endowment transparency, investment priorities, and shareholder engagement. Indicators do not include teaching, research, or other academic aspects concerning sustainability.

Besides the physical operations of the facilities, universities teach, conduct research, and develop our next generation thought leaders. In that context, the International Alliance of Research Universities (IARU), a collaboration of ten of the world's leading research-intensive universities, jointly addresses grand challenges facing humanity and provides a *Green Guide* for universities.[231] A comprehensive course-based evaluation of sustainability content in engineering education and research in US universities has been evaluated by Murphy et al.[232] Ramanan et al.[233] present a status of higher education in sustainability from a North American and East Asian perspective.

2.1.5.5 Measuring Sustainability: Cities

Stakeholders of a city need a variety of data to make a wide range of decisions. One driver may be the need to evaluate outcomes of different scenarios for policy development to prioritize rank resource deployment, including continuation or termination of an ongoing program. Another could be the prioritization issues such as zoning of living, business, and industrial regions; benefit options such as a public park versus a sports stadium; and support to different underserved people, regions, and services.

Cities, in addition to measuring their total sustainability, need to evaluate the impact of specific programs and services. Distinction between input, output, and impact KPIs become relevant. KPIs often reflect measures useful to program managers but ignore the goals of the funders or program recipients. For instance, funding a remedial program may track the number of additional staff or failing students served, but it may not consider how much the program alters the students' well-being and prospects. Very rarely are these KPIs available; often the program may not change the KPI or achieve the mission funders goal.

Impacts are hard to determine, may be subjective, and may not be evident until well after the end of the program. Input KPIs measure the resources that go into a service. Output KPIs are direct measures of the results. Impact KPIs reflect long-term effects and purpose of the activity. Many impacts cannot be measured within program timeframes.

Price Waterhouse Cooper[234] identifies *the 10 cities of opportunity indicators*. These progress factors and the predominance of infrastructure across them is shown in Figure 2.5.

1. *Economic clout*: Ability to finance infrastructure and other investments for prosperity
2. *Sustainability and natural environment*: Available green space and waste recycling programs in place or within reach for environmental quality
3. *Technology readiness*: Connectivity in terms of internet and telecommunications for social and economic bottom lines
4. *Intellectual capital and innovation*: Quality of schools, universities, research, and libraries for innovation and human capital
5. *Health, safety, and security*: Availability and access to quality hospitals, hospices, police and fire stations, and courts for social and health aspects

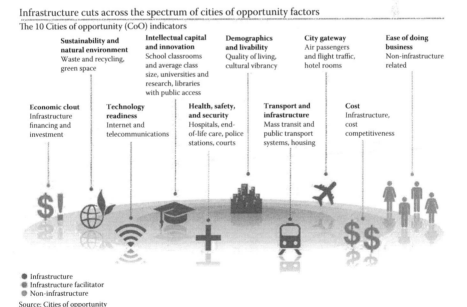

FIGURE 2.5 The 10 cities of opportunities indicators.[235]

6. *Demographics and livability*: Cultural vibrancy, access to good quality of living with reasonable means, for social aspects

7. *Transport and infrastructure*: Public transport systems including mass transit and affordable quality housing

8. *City gateway*: Access to airports and train stations, air passengers and flight traffic, affordable quality hotel rooms

9. *Cost infrastructure*: Cost competitiveness for infrastructure and public projects, credit rating

10. *Ease of doing business*: Non-infrastructure related, zoning, and other regulatory constraints; availability of educated and trained workforce for prosperity

Each factor is relatable to one or more of the quadruple sustainable development indicators. A recent report by KPMG[236] provides a review of the existing guidelines for measuring cities.

The UN Global Compact Cities Circles of Sustainability method, used by numerous cities (such as Delhi for sustainability profiles), is based on economic, ecological, political, and cultural *domain* scores. Each of the four domains is divided into seven sub-domains: ecology, materials and energy, water and air, flora and fauna, habitat and settlements, built-form (Human made environment) and transport, embodiment, and food emission and waste. This approach considers a range of indicators, and allows understanding the impact of actions across diverse social settings (Figure 2.6).

Green City Index[239] is a tool for cities to compare their environmental performance. The index uses approximately 30 indicators, including air quality, buildings, CO_2 emissions, energy, environmental governance, land use, transport, water and sanitation, and waste management. A mix of quantitative and qualitative indicators measures not only each city's current environmental performance, but also identifies potentials to become greener. Cities can learn policies and strategies from each other to minimize their environmental footprint, accommodate population growth, and promote economic prosperity today and in the future (Figure 2.7).

2.1.5.6 Measuring Sustainability: Green Buildings

The term green building encompasses planning, design, construction, operations, and ultimately end-of-life recycling or renewal of structures. Green building pursues solutions that represent a healthy and dynamic balance between environmental, social, and economic benefits.[241]

The top four global systems for measuring sustainability of green building are LEED, Green Globes, Green Star, and BREEAM.[242] Leadership in Energy and Environmental Design (LEED) continues to grow through the Green Building Councils in countries like South Africa, New Zealand, Germany, and the US. However, Building Research Establishment Environmental Assessment Method (BREEAM) still leads in total buildings certified. Green Globes has experienced success as well with its unique system adopted by many, including the US Department of Veterans Affairs. The Green Star, developed by the Green Building Council of Australia as a national, voluntary appraisal tool, continues to be a solution for many

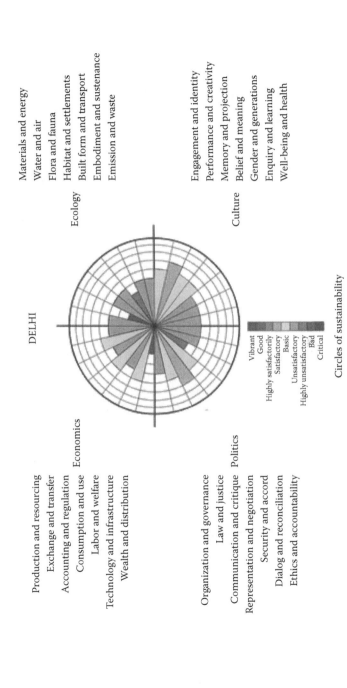

FIGURE 2.6 The circles of sustainability method indicators (e.g., Delhi, 2012). (From Circles of Sustainability.[237] Urban Profile Process per United Nations Global Compact Cities Program.[238] http://www.circlesofsustainability.org/tools/urban-profile-process/.)

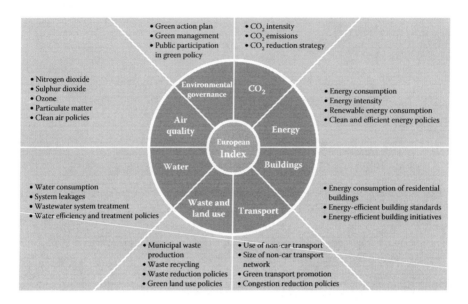

FIGURE 2.7   The Green City Index. (From Siemens, A.G., The Green City Index.[240])

in the Southern Hemisphere. BREEAM, LEED, and Green Globes are currently available for buildings outside of their home countries.

2.1.5.6.1   *Building Research Establishment Environmental Assessment Method*

Established in 1990 in the UK, BREEAM focuses on eight categories of environmental impacts: health and well-being, energy, transport, water, materials, waste, land use and ecology, and pollution. Each category gets credits and is scored, weightings applied, and then weighted scores from each category are added together to produce an overall score. The method covers a dozen rating schemes applicable to communities, courts, education, health care, homes, industrial, international, multi-residential, offices, prisons, retail, and other. There are five certification levels: pass, good, very good, excellent, and outstanding. As of 2016, more than 250,000 buildings have BREEAM certification.

2.1.5.6.2   *The Leadership in Energy and Environmental Design*

The LEED rating system was developed by the US Green Building Council (USGBC) in 1998. The first LEED rating system developed was for new construction, but today LEED has been expanded to include commercial interiors, core and shell, existing buildings, health care, homes, neighborhood development, new construction, retail, and schools. LEED focuses on nine categories of impacts: effects of the building on the ecosystem; location of the building; water and energy consumption; sustainable use and transportation of materials; indoor air and lighting quality; utilization of technology innovation; regional issues and priorities; awareness and education; and innovation and design. There are more than 100,000 LEED certified buildings.

Step 1 for LEED certification is registering the building with the Green Building Certification Institute (GBCI), which is responsible for all certifications. The US

Green Building Council develops and manages the LEED rating systems and provides checklists for the prerequisites and credits for each rating system. The checklist helps determine which credits are feasible and what certification level should be sought for the building. Certification levels are certified (40–49 points), silver (50–59 points), gold (60–79 points), and platinum (80-plus points). Cost of achieving a credit requires attention; costs for LEED registration can be found at www.gbci.org.

2.1.5.6.3   Green Globes

Green Globes started in 2000, and today there are about 3,000 certified buildings. Green Globes focuses on six categories of environmental impacts: energy reduction, environmental purchasing, development area, water performance, low-impact systems and materials, and air emissions and occupancy comfort. The system is heavily weighted toward energy reduction and integration of energy-efficient systems. The Green Globes tool includes a life-cycle assessment to evaluate the impact of building materials over the lifetime of the building. There are two certification schemes—existing buildings and new construction. Ratings of certification are from basic practice compliance (level 1), 70%–79% (level 2), 80%–89% (level 3), and 90%–99% (level 4).

The phases involve project initiation, site analysis, programming, concept design, design development, construction documents, contracting and construction, and commissioning. The system is composed of 1,000 points. Each category gets a score and the building gets an overall score, and that dictates globes level eligibility. A third-party verifier must examine the building and supporting documentation for certification.

2.2   SUSTAINABILITY ACCOUNTING, REPORTING, AND STAKEHOLDER ENGAGEMENT

2.2.1   Chapter Overview

The chapter helps practitioners identify environmental and social aspects of an organization's products and services, set sustainability goals, communicate progress toward goals effectively, and engage with and communicate the environmental and social impacts and benefits of their business to a wide array of stakeholders. This chapter begins with an introduction of concepts and practices in national and corporate sustainability accounting, reporting and stakeholder engagement. It discusses economic approaches to understanding and managing pollution and natural resources in the context of the UN System of Environmental Economic Accounting (SEEA) in the section on national standards. In the next section on standards for corporations, it presents fundamentals of traditional accounting and examines the potential pitfalls of the standard measures of growth, risk and return. The final section offers comprehensive coverage and discussion on dominant sustainability frameworks such as GRI G4, SASB, and International Organization for Standardization (ISO) 26000. It goes on to introduce the sustainability performance disclosure standards for private equity as well as impact investors.

2.2.2 Total Impact Accounting and Integrated Reporting

2.2.2.1 Total Impact Accounting

Sustainability stewards, both government and corporate, must be familiar with the terminology, practices, and consequences of accounting, finance, and operational and sustainability performance measurements as we move to an era of *green* economy and the new social contract. Our current economic and accounting systems have been essential for enabling markets, companies, investors, governments, and consumers to function. They are, however, not perfect and have presented stakeholders at all levels with inadequate or even misleading information that is relied upon to determine financial risk and strategic opportunity values.

"Green accounting provides a foundation of historic and future financial and nonfinancial measurement practices."[243] The primary indicator that affects the allocation of natural, built, human, and social capital resources is the economic bottom line. It drives the strategic decisions of government and business leaders and shapes the consumption decisions of individuals. It helps develop the skills to analyze integrated environmental and social bottom line measures to improve and enhance economic bottom line. Improved accounting and value measurement practices are mandatory to achieve a transformation to green economy and advance a long-term sustainable society.

2.2.2.2 International Integrated Reporting Framework

The International Integrated Reporting Community (IIRC) is a global alliance of regulators, investors, corporations, accounting professionals, and NGOs. The IIRC's vision is to align capital allocation and corporate behavior to financial stability and sustainable development through the integrated reporting and thinking. IIRC published an International Integrated Reporting (IR) Framework[244] in 2014, which offers guiding principles and content elements governing an integrated report to explain to providers of financial capital how an organization creates value over time and benefits all stakeholders, including employees, customers, suppliers, business partners, local communities, legislators, regulators, and policymakers.[245]

IIRC defines an integrated report as, "An integrated report is a concise communication about how an organization's strategy, governance, performance and prospects, in the context of its external environment, lead to the creation of value over the short, medium and long term."[246] The IIRC Framework identifies information to be included for use in assessing the organization's ability to create value and is written primarily in the context of private sector, for-profit companies.

Resources and relationships used and affected by an organization are identified as *capitals*, whose value increases, decreases, or transforms through the activities and outputs of the organization. These capitals are categorized in this IIRC Framework as financial, manufactured, intellectual, human, social and relationship, and natural capital. The report attempts to explain how the organization interacts with the external environment and the capitals to create value over the short, medium and long term. When the value the organization creates for stakeholders and society at large are material to the organization's ability to create value for itself, they are included in the integrated report.

1. *Guiding principles*:
   a. Provide insight into strategic focus and future orientation linkage to create value.
   b. Provide a holistic picture showing connectivity of information to create value.
   c. Provide insight into nature and quality of key stakeholder relationships
   d. Disclose material information that could substantively affect ability to create value.
   e. Make report concise.
   f. Provide all material matters without material error for reliability and completeness.
   g. Make report consistent over time and comparable to other relevant organizations.
2. *Content elements*:
   a. What is the organizational and external environment it operates within?
   b. What is the governance structure and how it supports value creation?
   c. What is the business model?
   d. What are the specific risks and opportunities and what are the response plans?
   e. What is the strategic goal and what is the resource allocation plan to get there?
   f. What level of strategic goals have been achieved and what is the impact on the capitals?
   g. What are the challenges pursuing its strategy and what are the potential implications?
   h. How are matters to be included, identified, and evaluated?

2.2.3 SUSTAINABILITY ACCOUNTING AND REPORTING FRAMEWORKS: NATIONS

2.2.3.1 US Government Accounting

Businesses and governments operate in different modes. Governments are funded through the mandatory payment of taxes, and do not face threat of liquidation, except for geopolitical situations, such as coups, sanctions, systemic corruption, or currency manipulations. Governments do not have returns expecting equity owners; instead, they are accountable to citizens for the use of resources. Governments need financial reports and accounting standards, which are different from the publicly listed corporations.

In the US, the Government Accounting Standards Board (GASB)[247] was established in 1984 to provide standards for state and local governments. Like the Financial Accounting Standards Board (FASB), the GASB is under the auspices of the Financial Accounting Foundation (FAF). In 1999, the AICPA recognized the Federal Accounting Standards Advisory Board (FASAB), which was set up in 1990, as the board that sets generally accepted accounting principles (GAAP) for federal entities.

2.2.3.2 United Nations Guidelines for National Accounting

The UN System of National Accounts (SNA, 2008),[248] an international statistical standard for national accounts, was adopted to create a macroeconomic database for analyzing and evaluating the performance of an economy through key aggregates such as GDP and GDP per capita. The United Nations encouraged member countries to use the 2008 SNA as the framework for compiling and integrating economic and related statistics, as well as in the national and international reporting of national accounts statistics.

Under UN SNA, institutional sectors are categorized as total economy, nonfinancial corporations, financial corporations, general government, households, nonprofit institutions serving households, and the rest of the world. By definition, corporations are market producers and could be split between nonfinancial and financial corporations. Financial corporations are principally engaged in providing financial services, such as insurance and pension funding services.

The five conceptual elements of SNA accounting structure for the institutional sectors are as follows:

1. Integrated economic accounts that trace production and income onto wealth accumulation
2. Supply and use frameworks, which trace production, use as intermediate input, or final product
3. Sociodemographic and employment tables for per capita and productivity analysis
4. Three-dimensional accounts of financial transactions and stocks of financial assets showing the relationship between sectors
5. Functional accounts, such as expenditure by government (health, education, defense, etc.), households (housing, food, health, etc.), and corporations (intermediate use and investment)

The *Monetary and Financial Statistics Manual and Compilation Guide* (MFSMCG)[249] focuses on monetary and financial statistics as a building block for policy purposes. The UN Handbook of National Accounting: Financial Production, Flows and Stocks in the System of National Accounts (UN HNA 2015)[250] focuses on the financial corporation sector in relation to the other sectors of an economy and the rest of the world. The SNA, 2008, UN HNA, and MFSMCG are all complementary to one another.

2.2.3.3 The System of Environmental Economic Accounting for Nations

The System of Environmental-Economic Accounting 2012—Central Framework (SEEA Framework)[251]—is a multipurpose conceptual framework that describes the interactions between the economy and the environment, and the stocks and changes in stocks of environmental assets. The UN SEEA Framework guides the compilation of consistent and comparable statistics and indicators for policymaking and potentially sustainability disclosure. It has been released under the auspices of the United Nations, the European Commission, the International Monetary Fund, and the World Bank Group. The effect of human activity on the environment has become a significant policy issue. Concern is the growing effect of burgeoning economic activity upon the local and global environment. Also, it is becoming increasingly obvious that continuing economic growth and human welfare depends on natural resources and the environment. Environmental-economic accounting comprises the compilation of physical supply and use tables, environmental protection expenditure accounts and natural resource asset accounts. It also reflects the evolving needs of its users, new developments in environmental-economic accounting and advances in methodological research.

SEEA Framework is designed with the capacity to organize and integrate physical and monetary data that have common scope, definitions, and classifications into combined presentations; for instance, it could measure water, energy, air emissions, or forest products. It takes a systems approach to organize the stocks and flows of environmental and economic assets. Another key feature is that it allows for the integration of environmental information (often measured in physical terms) with economic information (often measured in monetary terms) in a single framework, because of its capacity to present information in both physical and monetary terms coherently. This integration in a single measurement system provides an information base on water, minerals, energy, timber, fish, soil, land and ecosystems, pollution and waste, production, consumption, and accumulation for the development of models and for detailed analysis of interactions between the economy and the environment. That in turn allows the development of aggregates, indicators, and trends across a range of environmental and economic aspects, such as natural resources, emissions, and discharges to the environment from economic activity, and even the amount of economic activity to protect the environment.

Core to the measurement of physical flows are the tracking of the movement of natural inputs, products, and residuals. The measurement boundary for physical and monetary flows aligns to the production periphery and the economic territory of a country and includes all intra-enterprise flows. Only assets that have an economic value per SNA, including natural resources and land are included. Also, production of environmental goods and services, for environmental protection and for resource management, such as the production of energy (e.g., cogeneration turbines) for internal consumption are to be included. However, only the activities whose primary purpose is to reduce or eliminate an environmental impact or to make more efficient use of natural resources are to be tracked. Other economic activities such as natural resource use and minimization of natural hazards are not considered environmental activities.

Natural resources depletion and environmental assets valuation are two significant issues that are not fully developed yet. The following highlights some key aspects of the challenge. The SEEA Framework recognizes the change in the value of natural resources that can be attributed to depletion. The term *natural resources* is used to cover natural and biologically cultivated resources (e.g., timber and aquatic ecosystem resources), mineral and energy resources, water resources, and land. The area of land does not change over the reporting timeframe, whereas the capacity of soil resources, and all other natural resources to deliver benefits can diminish over time. Likewise, natural biological resources such as timber resources and aquatic resources could also deplete over time, depending on protections or cultivations in place.

Depletion is both a physical and a monetary concept, and without physical depletion of a natural resource there can be no monetary depletion. In physical terms, depletion is the decrease in the quantity of the stock of a natural resource over an accounting period because extraction occurs at a level greater than that of regeneration.

Measures of depletion in physical terms can be valued to estimate the cost of using up natural resources due to economic activity. In prior frameworks like SNA, the value of depletion was lost alongside flows such as catastrophic losses or eminent domain under compensated takeovers. This clarifies the treatment of value of depletion as a cost, to be recorded against income, especially relevant to the extractive industry.

The UN SEEA Framework adopts the same market price valuation principles as the UN SNA. However, since observable market prices are usually not available for environmental assets, the UN SEEA Framework provides only an extensive discussion of the techniques that may be applied in the valuation of these assets and discount rates. This valuation is, as discussed in the cost-benefit section, highly relevant to compute net present value. The UN SEEA Central Framework does not provide any guidance on valuation of renewable and non-renewable natural resources and land beyond that within the SNA. And this remains an outstanding issue.

2.2.4 Sustainability Accounting and Reporting Frameworks: Corporations

Financial accounting and reporting standards have been around for nearly a half a century. The two dominant globally practiced ones, namely US GAAP and International Financial Reporting Standards (IFRS), are presented here to address the economic or prosperity bottom line.

Sustainability accounting and reporting is still in a nascent stage. There is a need for meaningful disclosure, to define consistent format and appropriate and relevant content, and an integrated measure to monitor and track progress of sustainability. A plethora of formats and industry-sector protocols are at various stages of development. Ramanan[252] classifies these developing reporting standards as across the sector, industry specific, and national scale frameworks. For instance, the GRI[253] G4 is a voluntary sustainability reporting framework across multiple sectors. Oil and gas industry guidance on sustainability reporting[254] creates an industry specific framework. SASB has developed several sector specific materiality guidance and standards.

The Institution of Chemical Engineers (IChE) UK Sustainability Metrics are based on a relative burden or intensity approach and provide absolute, normalized, and qualitative measures. These metrics can be used to scale or aggregate to present a view of a larger operation, of a company, industry or region. This process industry sector specific metrics for sustainability reporting covers two of the largest and highest sustainability impact categories, petroleum, and chemical.

These sector specific standards help understand the relevance of the ESG metrics and ensure reliable comparison (Lydenberg 2010),[255] especially for best-in-class screening of investment opportunity. World Business Council for Sustainable Development

(WBSCD) eco-efficiency framework combines economic and environmental indicators relevant to the business sector to assess sustainable development. World Economic Forum[256] has developed national scale sustainability metrics for nations to be compared on their environmental performance, which is a repository of environmental data.

Despite the transient stage of their development, slowly but surely dominant leading protocols are emerging. The GRI,[257] the SASB,[258] and the Integrated Reporting (IR) Framework of the IIRC are the top three emerging global frameworks for sustainability reporting. SASB is US-focused and industry specific; its mission is to help disclose all material nonfinancial information. IIRC collaborates with the de facto global leader GRI to make reported information better aligned with investor need. Significant issues related to each of these select dominant frameworks have been addressed extensively in the relevant quadruple bottom line.

2.2.4.1 US Generally-Acceptable Accounting Principles

The Great Depression in 1929, a financial tsunami, is primarily attributed to faulty and manipulative reporting practices among businesses[259]; a complete collapse of the then unrecognized fourth bottom line, ethics (or governance or purpose), was the root cause. In response, the federal government intervened and, along with professional accounting groups, set out to create standards for the ethical and accurate reporting of financial information. Notably, following the 1929 stock market crash, the American Institute of Accountants developed standard accounting principles, leading to the generally accepted accounting principle (GAAP).[260] rules and standards was mandated for the creation of uniform financial reports by publicly traded companies in the United States.

Federal endorsement of GAAP began with legislation like the Securities Act of 1933[261] and the Securities Exchange Act of 1934,[262] and the laws were enforced by the US SEC that regulate public companies. While the federal government requires public companies to file financial reports in compliance with GAAP, they are not responsible for its creation or maintenance. The FASB was formed as an independent board in 1973 to take over the task of the creation and maintenance of the standards and to continually monitor and update GAAP. The GASB was established in 1984 as a policy board charged with creating GAAP for state and local governments. The Financial Accounting Foundation (FAF)[263] formed in 1972 as an administrative corporation to oversee the FASB[264] and the GASB.[265]

Public companies in the United States have to follow GAAP standards when filing financial statements; private companies can choose their standards system. Although private US businesses are not required to follow GAAP, many do. The method used for preparing financial reports impact earnings per share (EPS). GAAP and non-GAAP methods can lead to different valuations. For investors, it is important to check how reports were prepared. GAAP requires accrual accounting, while non-GAAP methods could follow cash accounting. Accrual accounting under GAAP strives to more closely reflect a company's financial position, regardless of cash-flow. Accrual accounting differs from cash accounting in that, the related revenues and expenses are reported together at the time of transaction, rather than when the cash is actually exchanged. Non-GAAP methods for calculating EPS often pursue ways to more closely reflect cash flow, so EPS values prepared under non-GAAP methods, also called *adjusted* earnings, tend to be higher than those under GAAP. According to Investopedia,[266] many businesses assert that non-GAAP earnings more accurately reflect their positions.

There are 10 general principles of the main mission and direction of the GAAP standards. These provide guidelines to apply to the measurement and disclosure of financial information.

1. *Principle of Regularity*: Regularly adhere to GAAP rules and regulations as a standard
2. *Principle of Consistency*: Consistently apply the same standards throughout the process to prevent discrepancies, and to fully disclose and explain the reasons for change
3. *Principle of Sincerity*: Honestly try to accurately depict the financial situation
4. *Principle of Permanence of Methods*: Use consistent methods in financial reporting
5. *Principle of Non-compensation*: Not expect debt compensating and fully report both negatives and positives transparently
6. *Principle of Prudence*: Emphasize prudent fact-based non-speculative financial data
7. *Principle of Continuity*: Assume continuity of business while valuing assets
8. *Principle of Periodicity*: Distribute financial entries across the appropriate periods
9. *Principle of Materiality/Good Faith*: Make a good faith effort for full disclosure
10. *Principle of Utmost Good Faith*: Maintain implicit honesty in engagements. Originating from the Latin phrase *uberrimae fidei*, it implies honesty in transactions between the parties, analogical to the Hippocratic oath.

2.2.4.1.1   Gaps in Generally Accepted Accounting Principles for Sustainability Reporting

US GAAP has a major gap (pun unintended) in sustainability reporting. It does not have the mechanisms to incorporate sustainability indicators nor the methods to measure resource use, pollution, or social assets and activities. While the SEC has now called for including climate change and other material risks in the financial reports, very little real direction is available in GAAP. In that context, SASB, discussed later in this section, is establishing a framework for sustainability reporting and also developing sector-specific materiality guidance. Furthermore, IIRC, also discussed later in this section, has established a framework for integrated reporting of financial and nonfinancial information, such as ESG indicators in one integrated document. This concept of sustainability indicator disclosure in one integrated report is aligned with the UN SEEA Framework's design with the capacity to organize and integrate physical and monetary data that have common scope, definitions, and classifications into one combined presentation of monetary and physical flows for national asset accounting.

2.2.4.1.2 *Conventional Balance Sheet, Income, and Cash-Flow Statements*

Listed public companies, as well as some other entities, commonly prepare four financial statements: the balance sheet (also called the statement of financial position), income statement, cash flow statement, and statement of changes in equity.

1. A balance sheet is a statement or *snapshot* of the financial position of the assets, liabilities, and shareholder equity of the entity itself, not its owners at a specific point in time. Assets are items that could provide future economic benefits, liabilities are obligations of the firm against those assets, and owners' equity—defined by what remains after subtracting liabilities from assets. The key accounting equation is: Assets = Liabilities + Owners' Equity, or A = L + OE.

   Assets are current or long-term, and listed in order of decreasing liquidity. Current assets are expected to be used during the normal business period, usually one year, and are comprised of cash and cash equivalents, short-term investments, account receivables, inventory, and prepaid expense. Noncurrent assets typically include long-term investments such as property, plant, and equipment and other intangible assets such as goodwill (brand) and intellectual property (such as patents). The accounting practice of allocating the cost of a fixed asset over its useful life is termed depreciation; accounts for decline in useful value of a fixed asset due to wear and tear from use and passage of time. Liabilities are, current or non-current and listed in order of expected payment. Current liabilities are obligations expected to be satisfied within the same normal business period, usually one year and comprise of accounts payable, tax payable, accrued expenses, advances and deposit. Noncurrent liabilities include bonds payable and other forms of long-term capital. The structure of the owners' equity section depends on whether the entity is an individual, a partnership or a corporation. Examples include paid in capital, common stock, and retained earnings.

   Balance sheet data can provide key sustainability economic bottom lines or indicators on financial structure, and its ability to meet its obligations, such as working capital, current ratio, quick ratio, debt-equity ratio, and debt-to-capital ratio. For additional details, please see some standard accounting textbooks.[267]

2. The income statement (also known as the profit and loss statement or P&L) reflects total revenues and total expenses, and the profitability of a business. A fundamental accounting principle is the assumption of business continuity. The P&L is always for a specific period of time, and the periodic income statements are used by managers and investors to show if the company made or lost money over that period of time.

   The income statement is broken into income from continuing operations, net income, and unusual items such as closed operations, one-term charge for major restructuring, a change in accounting methodology, and other comprehensive income. Income from continuing operations, the heart of the P&L, includes sales revenue, cost of goods sold, operating expenses, gains and losses, other revenue and expense items that are unusual or infrequent but not both, and income tax expense. Gross margin, usually presented as a percentage of the revenue, is the difference between the revenue from product sales and the cost of goods sold. Cost of goods sold is what it cost the company to buy or create the products it sold to customers.

Several key sustainability economic indicators, or *bottom line,* come from an income statement. It informs the computation of the key profitability ratios of gross margin, operating margin, and pretax margin. Earnings per share is an economic indicator that helps investors assess the ability of the company to generate income from its activities. Earnings per share is presented on basic shares and if additional shares were issued in the period, on a diluted basis.

3. The statement of changes in owners' equity (OE) reconciles the various components of OE on the balance sheet at the start and the end of the P&L period.

4. The statement of cash flows (or sources and uses statement) describes the sources and uses of cash, including the investing and financing activities during the period. Like the income statement, the statement of cash flows is always for some specific period of time. Different from the financial statements that seek to record matching economic events regardless of when cash is actually received or used, cash flow statement provides the actual cash received and disbursed during the period. The direct method of cash flow statement shows the items that affected cash flow, such as cash collected from customers, interest received, cash paid to suppliers, and so on. The indirect method adjusts net income for any revenue and expense item that did not result from a cash transaction.

The cash flow statement typically contains: (1) net cash flow from operating activities, such as sales, inventories, rent, and insurance; (2) cash flow from investing activities, such as buying and selling plants and equipment; (3) cash flow from financing activities, such as selling common stock, paying off long-term debt; (4) exchange rate impact; (5) net increase (or decrease) in cash; (6) cash and equivalents at the start and end of the period; and (7) any non-cash financing and investing activities, such as the conversion of bonds.

2.2.4.2   Global Accounting and Reporting Standards

The IFRS[268] is a set of accounting standards that provides general guidance for the preparation of financial statements, rather than setting rules for industry-specific reporting. The IFRS standards are developed and set by an independent, not-for-profit organization called the International Accounting Standards Board (IASB).[269] The IFRS,[270] currently followed by more than 100 countries, has been required for countries in the European Union since 2006 and are replacing the national GAAPs of many countries, including Australia, Canada, and Japan. A single set of world-wide standards will simplify accounting procedures by allowing use of one reporting language and provide investors and auditors with a cohesive view of finances. Efforts are underway by the IASB[271] and the FASB to set standards that apply domestically and internationally.

However, GAAP and IFRS reporting standards have some major differences, such as in the rules for balance sheet intangible assets. GAAP recognizes intangible assets at fair value, but IFRS examines intangible assets, only if they can be associated with a future benefit. Also, IFRS are considered more principles-based than US GAAP. Principles-based systems offer broader guidelines in accounting treatment with exercise of best judgment; rules based systems are more prescriptive and specific. The difference in emphasis reflect the differing business and legal cultures. There is concern that a principles-based system will offer opportunity to tilt the numbers, but then GAAP did not prevent financial scandals either.

2.2.4.3 Sustainability Accountability Standards Board

The SASB is an independent nonprofit organization with a mission to develop and disseminate sustainability accounting standards to help publicly listed corporations comply with SEC requirements to disclose material ESG factors (including sustainability and climate change risks). For instance, in 2010 the US SEC created guidelines for companies to report on climate risks. In response, SASB standards were developed using a multi-stakeholder process. Accreditation by American National Standards Institute (ANSI) provides credibility to the SASB sustainability accounting standards as one that meets ANSI requirements for openness, balance, consensus and due process.

SASB has identified a universe of 30 sustainability issues and organized them under five broad dimensions: environmental capital, social capital, human capital, business model and innovation, and leadership and governance to gauge material ESG issues within an industry for select sectors. See Appendix A for a list of SASB sustainability components within each of the five SASB dimensions (Figure 2.8).

The SASB dimensions are categorized in the context of the quadruple bottom line as follows:

The traditional financial and physical capital dimensions, as well as intangibles such as brand and intellectual capital, are included in the long-term viability of the economic or the prosperity dimension. Natural capital covers the planet dimension, and the social dimension is addressed in the role of business in society. Purpose includes leadership and business ethics.

2.2.4.3.1 Prosperity (Economic or Profit)

1. Business model and innovation dimension incorporate environmental and social factor impacts on
   a. Innovation and business models
   b. Value chain integration and resource efficiency
   c. Production process or product innovation
   d. Product design, use, and disposal responsibility
   e. Owned or managed tangible and financial assets
2. Human capital dimension weighs the management of human resource as a key asset:
   a. To deliver long-term value
   b. To affect productivity through engagement, diversity, incentives, and compensation

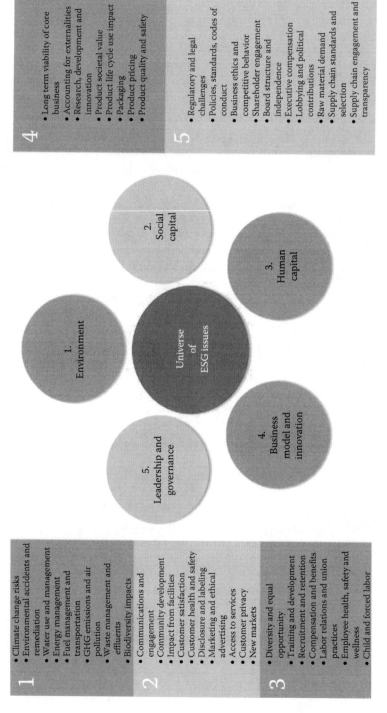

FIGURE 2.8   SASB universe of sustainability issues and five broad dimensions.[272]

   c. To attract and retain employees in highly competitive or constrained markets

   d. To ensure employee well-being by managing labor relations and pension liability

   e. To protect employee health and safety through safety culture

2.2.4.3.2 *Planet (Environment)*

Environmental dimension measures impact based on the concept of natural capital:

1. Eliminating or minimizing use of nonrenewable natural resources (e.g., fossil fuel, water, minerals, ecosystems, and biodiversity)
2. Minimizing impact of harmful releases into the environment, such as air and water pollution, waste disposal, and greenhouse gas (GHG) emissions

2.2.4.3.3 *People (Social)*

Social capital dimension relates to the perceived role of business in society

1. To meet the expectation of contribution to society in return for its social license to operate
2. To manage relationships with key external stakeholders, such as customers, local communities, the public, and the government
3. To secure access to products, services, and affordability
4. To ensure responsible marketing and customer privacy practices

2.2.4.3.4 *Purpose (Governance or Ethics)*

The leadership and governance dimension involves the management of potential conflict with broader stakeholder group interest, which may threaten the license to operate

1. Through inclusive stakeholder groups—government, community, customers, and employees
2. Through regulatory compliance, appropriate lobbying, and political contributions
3. Through management of risk, safety, supply chain, and resource
4. Through eliminating risk of conflict of interest, anticompetitive behavior, corruption and bribery, and business complicity with human rights violations

2.2.4.4 Global Reporting Initiative

Launched in 2000, the GRI[273] is a long-term multi-stakeholder, multinational process to develop and disseminate global guidelines. Into its fourth generation, GRI G4 is a voluntary guideline for reporting on economic, environmental, social, and governance dimensions of products, activities, and services. The guidelines offer reporting principles, general and specific standard disclosures, and an implementation manual for the preparation of sustainability reports regardless of size, sector, or location. A composite table of categories and aspects covered in the specific standard disclosures of the GRI G4 guidelines is included in Figure 2.9, GRI G4 Guidelines Categories and Aspects.[274] This table, from the GRI G4 Specific Standard Disclosure, addresses economic, environmental, and social categories of indicators and related aspects. The governance indictor category is addressed in the specific standards disclosure. Governance has been expanded to incorporate within, an element for ethics and integrity (Figure 2.9).

The GRI G4 Framework General Standard Disclosure, besides the indicators referenced earlier, calls for a company profile, a process used to define boundaries and to identify material aspects, dedication to external commitments, and the engagement of stakeholders. Forward-looking statement on sustainability trends, risks, and opportunities, as well as a report card on sustainability targets are sought in the strategy and analysis section. General disclosure calls for

1. A statement of the overall relevance of sustainability to the organization, its vision and strategic priorities
2. A report card on the status of achievements and failures with respect to targets
3. One concise narrative on the impact of organization on sustainability and key stakeholders
4. Another concise narrative on the impact of sustainability trends, risks, and opportunities on the organization's long-term prospects and financial performance

Another important component is the disclosure of management approach, where an organization has an opportunity to discuss how a material aspect impact is being or will be managed. An aspect refers to a subject covered by the guidelines; aspect boundary may vary with the aspect. Material aspect is one that either has significant potential to cause economic, environmental, or social impact or could substantively influence stakeholder assessments or decisions.

Select parts, namely governance (G4-34 and G4-45) and ethics and integrity (G4-56-58) are highlighted below for its significance in addressing the fourth bottom line—ethics or purpose. This recent inclusion of ethics and integrity aspect within the governance category in GRI G4[275] addresses the fourth of the quadruple bottom line.

G4-34: Reports the governance structure of the organization, including committees of the highest governance body. Identifies any committees responsible for decision-making on economic, environmental, and social impacts.
G4-45:
1. Reports the highest governance body's role in the identification and management of economic, environmental, and social impacts, risks, and opportunities. Includes the highest governing body's role in the implementation of due diligence processes.
2. Reports whether stakeholder consultation is used to support the highest governance body's identification and management of economic, environmental, and social impacts, risks, and opportunities.
G4-56: Describes the organization's values, principles, standards, and norms of behavior such as codes of conduct and code of ethics.

Category	Economic		Environmental	
Aspects	Economic performance	Materials	Emissions	Transport
	Market presence	Energy	Effluents and waste	Overall
	Indirect economic impacts	Water	Products and services	Supplier environmental assessment
	Procurement practices	Biodiversity	Compliance	Environmental grievance mechanisms
Category	Social	Social	Social	Social
Subcategories	Labor practices and decent work	Human rights	Society	Product responsibility
Aspects	Employment	Investment	Local communities	Customer health and safety
	Labor/management relations	Non discrimination	Anti corruption	Product and service labeling
	Occupational health and safety	Freedom of association and collective bargaining	Public policy	Marketing communications
	Training and education	Child labor	Anti competitive behavior	Customer privacy
	Diversity and equal opportunity	Forced or compulsory labor	Compliance	Compliance
	Equal remuneration for women and men	Security practices	Supplier assessment for impacts on society	
	Supplier assessment for labor practices	Indigenous rights	Grievance mechanisms for impacts on society	
	Labor practices grievance mechanisms	Supplier human rights assessment		
		Human rights grievance Mechanisms		
		Assessment		

FIGURE 2.9 Global reporting initiative G4 guidelines categories and aspects.[276]

G4-57: Report the internal and external mechanisms for seeking advice on ethical and lawful behavior and matters related to organizational integrity, such as help lines or advice lines.

G4-58: Reports the internal and external mechanisms for reporting concerns about unethical or unlawful behavior and matters related to organizational integrity, such as escalation through line management, whistleblowing mechanisms, or hotlines.

2.2.4.5 Sectoral and Special Purpose Sustainability Reporting Frameworks

2.2.4.5.1 Oil and Gas Sector Sustainability Reporting Guidelines—IPIECA, API, and OGP

Launched in 2005, the oil and gas sectors have developed and follow the oil and gas industry guidance on sustainability reporting.[277] The current version, released in 2015, provides a set of performance indicators appropriate to industry specific sustainability issues, categorized as environmental, health and safety, and social and economic. Environmental subcategories include climate change and energy, biodiversity and ecosystem, water, and local environmental impacts. Health and safety is categorized into workforce protection, product environmental, health and safety, and process safety and asset integrity. The social and economic issue is subdivided into community and society, local content, human rights, business and transparency, and labor practices.

Each category or subcategory has multiple indicators, each indicator becoming a reporting element. For instance, under the environmental category and the climate change and energy subcategory, there are four indicators—greenhouse gas emissions, energy use, alternative energy use, and flared gas. Each indicator provides a choice of reporting elements depending on the depth or accuracy required. Sectoral specificity makes these attractive for the reporters as well as the broad range of users who find comparison easier and meaningful.

2.2.4.5.2 Operations Sector Environmental and Social Responsibility Guidance

International Organization for Standardization (ISO) has a membership of 160 national standards institutes from a wide range of countries. ISO 14000 addresses environmental aspects, and ISO 26000 focuses on social responsibility.

ISO 14000 family of environmental management standards facilitate the fusion of business and environmental goals by encouraging the inclusion of environmental aspects in product design. Most of the guidance standards help reliably assess the environmental aspects associated with the process, product, services, or supply chain. Communication on the environmental aspects of products and services is an important way to use market forces to influence environmental improvement. Truthful and accurate information provides the basis on which consumers can make informed purchasing decisions. ISO 14000 family of guidance standards, including some relevant forthcoming ones are listed to help capture awareness of useful future resources. The brief purpose of each standard is also indicated.

1. ISO 14001 is a framework for environmental management systems (EMS) that addresses the environmental aspects of processes, products and services. ISO 14004, provides additional guidance on EMS, and ISO 19011 guides EMS audits.
2. ISO 14031 provides guidance on how to evaluate environmental performance as well as how to select suitable performance indicators.
3. The ISO 14020 series of standards addresses environmental labels, eco-labels (seals of approval), and quantified environmental information about products and services.
4. The ISO 14040 standards provide guidelines on the principles and conduct of Life-cycle assessment (LCA) studies. LCA is a tool for identifying and evaluating the environmental aspects of products and services from the cradle to the grave.
5. ISO 14064 is an international GHG accounting and verification standard for GHG emission assessment and reduction projects. ISO 14065 provides accreditation of GHG validation or verification bodies. ISO 14066 will specify competency requirements for greenhouse gas validators and verifiers.
6. ISO 14063 provides environmental communication guidelines to link to external stakeholders.
7. ISO 14045 will provide guidance on the assessment of eco-efficiency, which relates environmental performance to the value created.
8. ISO 14051 will provide guidelines for material flow cost accounting (MFCA), a management tool to promote effective resource utilization.
9. ISO 14067 and ISO 14069 will provide requirements for the quantification and communication of carbon footprint of product, services, and supply chains to measure GHG intensity.
10. ISO 14005 will provide guidelines on EMS and environmental performance evaluation for small- and medium-sized enterprises.
11. ISO 14006 will provide guidelines on eco-design.
12. ISO 14033 will provide guidelines for compiling and communicating quantitative environmental information.

ISO 26000 provides guidance to businesses and organizations on how to operate in a socially responsible way. This means acting in an ethical and transparent way that contributes to the health and welfare of society."[278] ISO 26000:2010 is a guideline; it cannot be certified to. It helps businesses and organizations understand and translate principles into effective actions and share best practices relating to social responsibility, globally. It is aimed at all organizations regardless of their activity, size, or location. Per ISO 26000, by definition, *social responsibility* is not decided in a vacuum; it always involves reference to the guiding principles, and awareness of impact on others.

ISO 26000 goes on to define the seven principles and presents seven core aspects of social responsibility.

The seven principles of ISO 26000 are:

1. *Accountability*: Involves top decision makers and the complete chain of command taking responsibility for decisions and policies and being answerable to all potentially impacted stakeholders.
2. *Transparency*: Involves openness and willingness to communicate in a clear, timely, honest, and complete manner.
3. *Ethical behavior*: Involves deciding the right course of action in accordance with accepted principles of right or good conduct in the context of a particular situation.
4. *Respect for stakeholder interests involves identifying groups of stakeholders*: Those who are affected by your decisions and actions, and responding to their concerns, but does not imply letting them make the decision for the organization. In particular, identifying the vulnerable populations among its stakeholders, and to work to ensure their fair treatment is an important element of respect for stakeholders.
5. *Respect for the rule of law*: Means compliance with all applicable laws and regulations even if they are not adequately enforced. Where the law or its implementation does not adequately protect environmental or social safeguards, respect international norms of behavior as a minimum.
6. *Respect for international norms of behavior*: Involves following generally accepted principles of international law, or intergovernmental agreements that are largely universally recognized and sourced from organizations such as the United Nations, International Labor Organization (ILO).
7. *Respect for human rights*: Involves taking steps to protect respect for human rights, and to avoid taking advantage of situations where human rights are not protected.

The United Nations adopted the Universal Declaration of Human Rights in 1948. The standard recommends avoiding or terminating business relationships with the client, supplier, or subcontractor who violates human rights because that is complicity and infringes human rights laws. Workplace human rights issues include the freedom to associate and collectively bargain, elimination of forced labor and child labor, and to follow equal opportunities and nondiscrimination. The ILO, an international governing body for labor practices and standards recommends adopting and internalizing the good practices in employee hiring, working conditions, health and safety, and training and education Please also see The UN Global Compact's ten principles in Section 1.3.4.2.4.

Figure 2.10 shows the seven core subjects of ISO 26000. Organizational governance requires structuring strategic decision-making processes in a manner that enables application of social responsibility principles. The environmental core aspect deals with taking responsibility for the environmental impacts and improving performance across the value chain, as well as recommends following the precautionary principle, considering long-term cost effectiveness, internalizing the cost of pollution, and preventing pollution to avert the need to mitigate impacts.

FIGURE 2.10   ISO 26000 social responsibility guidance—seven core aspects. (From International Standards Organization 26000 Training.[279])

The core aspect of fair operating practices focuses on the fourth bottom line—purpose or ethics—and seeks ethical conduct in all dealings with stakeholders throughout the value chain. It recommends support of anti-corruption, anti-bribery, fair trade, and fair-wage policies; as well as respect for physical and intellectual property rights, responsible competitive behavior, and responsible political involvement across the workforce and the value chain.

Consumer issues deal with taking responsibility for minimizing health and safety risks (especially to vulnerable consumers) in the use of a product or service, for informing consumers so they can make educated choices, based on social, economic, and environmental impacts of the product or service, and for promoting responsible consumption principles across the entire value chain. The marketing information must not be misleading, unfair, or unclear, minimizing information asymmetry between the transacting parties. Finally, ISO 26000 states that "Community involvement and community development are both integral parts of sustainable development." It recommends contribution through diversifying economic activities and creating more employment, social investments, supporting skills development programs, and promoting culture and arts.

2.2.4.5.3   Private Equity Environmental, Social, and Governance Disclosure Framework: American Investment Council Private Equity Growth Capital Council

The American Investment Council (AIC), formerly the Private Equity Growth Capital Council (PEGCC), is an organization launched by a consortium of private equity firms in February 2007. Driven by the "conviction that companies that address ESG issues can achieve better growth, cost savings, and profitability, while strengthening stakeholder relations and improving their brand and reputation, General Partners (GP), Limited Partners (LP), associations, and the private equity industry at large have an aligned interest in communicating how the management of ESG factors contributes positively to risk-adjusted returns."[280]

Because of the diverse nature of the private equity sector transactions, discussions between a GP and its LPs is the best means of ESG disclosure. The two distinct phases of a private equity transaction are the fund raising and life of a fund.

1. During fundraising, GP should disclose adequate information to enable LPs to assess GPs:
   a. ESG-related policy and investment belief alignment
   b. Methodology to identify and manage ESG-related value drivers, material risks, and future development opportunities
   c. Influence and support of its portfolio companies' management of ESG driven risks and opportunities
   d. Support to LP to monitor and ensure that the GP is performing per agreed-upon ESG-related policies and practices
   e. Approach to managing and disclosing material incidents at the GP and its portfolio companies
2. During the life of a fund, GP should disclose adequate information to enable LPs to assess GPs:
   a. Performance is per agreed-upon ESG-related policies and practices
   b. Positive and negative ESG-related developments that may impact portfolio companies in the fund are understood
   c. Responses to GP and portfolio company incidents and reporting are consistent with investment terms and the fund's policies

Insead Global Private Equity Initiative (GPEI) references PwC's Total Impact Measurement and Management, a PwC Framework for the Private Equity Industry ESG measurement, and valuation techniques to guide decision-making.[281]

2.2.4.5.4   Impact Investing Reporting and Investment Standards and B Impact Assessment

World Economic Forum defines impact investing as "an investment approach that intentionally seeks to create both financial return and positive social or environmental impact that is actively measured."[282] Another study[283] suggests that impact measurement efforts could be classified by measurement objectives as: (1) Estimating impact—for due diligence prior to investment, (2) Planning impact—selecting metrics and data collection methods to monitor impact, (3) Monitoring impact—to improve program, (4) Evaluating impact—to prove social value. The same study identifies four impact measurement methods: (1) Expected return, one that takes into account the anticipated social benefits of an investment against its costs, discounted to today's value, (2) Logic model, a tool used to map a theory of change by outlining the linkage from input, to activities, to output, to outcomes, and ultimately to impact, (3) Mission alignment method measure the social value criteria and scorecards to monitor and manage key performance metrics, and (4) Experimental and quasi-experimental methods, typically, after-the-fact evaluations of the impact of the intervention compared to the status quo.

1. *Impact reporting and investment standards (IRIS)*: Acumen Fund, B Lab and the Rockefeller Foundation founded the Impact Reporting and Investment Standards (IRIS) to create a common framework for defining and reporting impact capital performance. The IRIS[284] is a catalog of metrics for impact investors that measure the performance of an organization. The key metrics include:
   a. Financial performance, including standard financial reporting metrics
   b. Operational performance, including metrics to assess investees' governance policies, employment practices, and the social and environmental impact of their business activities
   c. Product performance, including metrics that describe and quantify the social and environmental benefits of the products, services, and processes offered by investees
   d. Sector performance, including metrics that describe and quantify impact in particular social and environmental sectors, including agriculture, financial services, and health care
   e. Social and environmental objective performance, including metrics that quantify progress toward specific objectives such as employment generation or sustainable land use
2. *B impact assessment*: B Impact Assessment (BIA)[285] is another tool to assess a company's overall social and environmental performance. The impact of a business on all stakeholders is assessed through an online, easy to use platform. The BIA is a free, confidential service administered by the nonprofit organization B Lab. The BIA uses IRIS metrics in conjunction with additional criteria to come up with an overall company or fund-level rating, as well as targeted sub-ratings in the categories of governance, workers, community, environment, and socially and environmentally focused business models.

2.2.4.5.5 *Task Force on Climate-Related Financial Disclosures*

An essential function of financial markets is to price risk to support informed, efficient capital-allocation decisions. Accurate and timely disclosure of current and past operating and financial results, must be supplemented by the governance and risk-management context. Without the right information, investors and others may incorrectly price or value assets, leading to a misallocation of capital. "Increasing transparency makes markets more efficient and economies more stable and resilient"[286]—Michael R. Bloomberg.

Regarding financial implications of climate change especially, the exact timing and severity of physical effects are difficult to quantify. The large-scale and long-term nature of the problem makes it uniquely challenging, especially in the context of economic decision-making. While the implications of climate change may be long term they are relevant to decisions made today; transition to a lower-carbon economy, coupled with decreasing costs and increasing deployment of clean energy, could have significant near-term impact on the sectors extracting, producing, and using coal, oil, and natural gas.

A 2015 study[287] estimates the value of the total global stock of manageable assets at risk to range from \$4.2 trillion to \$43 trillion between now and the end of the century. However, they also create significant opportunities—estimated to require around \$1 trillion of investments a year for climate change mitigation and adaptation solutions. Moving out of certain assets may not be adequate protection against risk, because the impact will come through weaker growth and lower asset returns across the board, compounded by the fact that present valuations do not adequately factor in climate-related risks because of insufficient information. This also has major implications for the global financial system, especially in terms of avoiding financial dislocations and sudden losses in asset values.

The Task Force on Climate-Related Financial Disclosures (TCFD) was tasked by G-20 countries to develop guidelines for voluntary, consistent climate-related financial disclosures useful to investors, lenders, and insurance underwriters in understanding material risks. The TCFD has produced the guidance documents in three parts:

1. Recommendations of the Task Force on Climate-Related Financial Disclosures:[288]

    Key features of the recommendations include: (1) Adoptable by all organizations, (2) Included in financial filings, (3) Designed to solicit decision-useful, forward-looking information on financial impacts, and (4) Having a strong focus on risks and opportunities related to transition to lower-carbon economy. Core elements include governance, strategy, risk management, metrics, and targets.

    The task force considered existing voluntary and mandatory climate-related reporting frameworks (including CDP, CDSB, GRI, IIRC, and SASB) in developing its recommendations and provides alignment of its recommendations and guidance to those frameworks in the annex. The disclosures related to the Strategy and Metrics and Targets recommendations are subject to an assessment of materiality for climate-related issues consistent with how they determine the materiality of other information included in their financial filings.

2. Implementing the Recommendations of the Task Force on Climate-Related Financial Disclosures[289]

    a. TCFD does not specify a timeframe for implementing the recommendations.

    b. TCFD recommends that organizations provide climate-related financial disclosures in their mainstream (i.e., public) annual financial filings.

    c. Asset managers and asset owners should use their existing means of financial reporting.

    d. Organizations with over US \$1 billion equivalent (USDE) in annual revenue should disclose information related to the Strategy and Metrics and Targets, but not deemed material, in other official company reports.

e. Established a threshold annual revenue of greater than US$1 billion to consider conducting more robust scenario analysis to assess the resilience of their strategies.

f. Streamlined the supplemental guidance for specific nonfinancial groups to reduce redundancy and focus on key metrics.

g. Changed the carbon foot-printing metric for asset owners and asset managers from GHG emissions normalized for every million dollars invested to a weighted average carbon intensity metric.

3. The Use of Scenario Analysis in Disclosure of Climate-Related Risks and Opportunities[290]

Scenario analysis is a tool for companies to consider, in a structured way, potential scenarios that are different from business-as-usual and to evaluate how their strategies might perform under those circumstances. One key recommended disclosure focuses on the resilience of an organization's strategy, under different climate-related scenarios, including a 2°C or lower scenario. In December 2015, nearly 200 governments agreed to work toward "holding the increase in the global average temperature to well below 2°C above pre-industrial levels and to pursue efforts to limit the temperature increase to 1.5°C above pre-industrial levels," referred to as the Paris Agreement. Impact of the recent withdrawal of the US from the Paris Climate Change Agreement by President Trump is unknown. As a result, a 2°C scenario is a common reference.

2.2.5 ETHICAL STAKEHOLDER ENGAGEMENT

Through sustainability reporting and communications, a company conveys its sustainability strategy and how its ESG policies, practices, and processes are helping the company to comply with regulations, manage risk, and create value through efficiencies and innovation. While the US government does regulate green marketing through the Federal Trade Commission (FTC), sustainability disclosure is voluntary.

Despite a lack of consensus on allocation of responsibility between the public and private sectors, most stakeholders, such as shareholders, employees, customers, suppliers, communities, governments, and regulators, are demanding that companies, not only boost the bottom line, but also recognize a broader role for themselves in addressing some of the world's greatest sustainability challenges in the context of the quadruple bottom line—economic, environmental, social, and governance issues. Responsible corporations are progressively engaging stakeholders, and doing so ethically, to understand their concerns and perspectives, and integrate them in to the company's strategic decision-making processes.

2.2.5.1 Stakeholder Engagement Drivers

2.2.5.1.1 Recognition of Contribution to Long-Term Corporate Value

Companies are becoming cognizant of the long-term sustainability and shareholder value associated with stakeholder engagement, comprising

1. Enhanced business intelligence through stakeholder interactions and consequent inputs that enable corporate leadership make better informed decisions to avoid or mitigate business risks
2. Enlarged business opportunities through a more inclusive and comprehensive process of identifying ways to expand brand value and reputation
3. Accelerated innovation by bringing together diverse perspectives

2.2.5.1.2   Response to Rising Shareholder Activism

Shareholder activism on ESG issues has been on the rise ever since the eruption of the financial and environmental tsunamis, driving the need for more ethical and transparent engagement between companies and their shareholders and other stakeholders. Also, there is an upward trend in shareholder support for these resolutions. Topping the list are requests to address disclosure of political contributions and lobbying, followed by transparency in sustainability disclosure on environmental issues such as financial risks of climate change, energy efficiency and water constraint issues. Benefits from effective stakeholder engagement are:

1. Proactive discussion and in-time action on related stakeholder concerns
2. Averting a potentially expensive, disruptive and prolonged proxy fight
3. Better nurturing of relationships between companies and their stakeholders

2.2.5.1.3   Attracting and Retaining Growing Responsible Investors

Both mainstream and responsible investors are increasingly evaluating nonfinancial ESG factors in analyzing investment options. Institutional investors, also known as universal owners of the private enterprise, are increasingly integrating ESG factors in their investment criteria. For instance, the California Public Employees' Retirement System (CalPERS) total fund approach to investment integrates ESG factors— corporate governance, such as executive compensation; climate change, such as GHG emissions and risk of water stress; and human capital, such as of health and safety, and diversity.[291] The financial sector focused on socially responsible investment is growing fast. It has grown from US$2.7 trillion in 2007 to US$21.4 trillion in 2014.[292] Investors in this sector actively prefer to invest in corporations that have been vetted by and are high on the dominant sustainability indexes.[293]

One of the earliest set of principles to emerge as socially responsible investment guideline was the OECD Guidelines for Multinational Enterprises (MNE),[294] adopted in 1976. These guidelines establish legally nonbinding principles and standards for responsible business conduct for multinational corporations covering such areas as human rights, disclosure of information, anticorruption, taxation, labor relations, environment, competition and consumer protection. Implicit in these is the need for ethical stakeholder engagement. Global treaties and the potential liability of legal action by the people impacted by funding of major development projects are additional drivers for consideration of ESG factors. For instance, the World Bank faced some major lawsuits prior to the new phase of imposed mandatory environmental and social impact assessments, stakeholder engagements, and defined statements.

2.2.5.2   Stakeholder Engagement and Corporate Social Responsibility

CSR and stakeholder engagement are like monozygotic twins, both having emerged from one zygote—the need for human well-being; stakeholder engagement is an intrinsic part of the CSR report; and CSR contains information on a broad spectrum of issues that matter to the stakeholders. "A primary objective of corporate stakeholder engagement is to build relationships with stakeholders to better understand their perspectives and concerns on key issues (including CSR issues) and to integrate those perspectives and concerns (when and where feasible and prudent) into the company's corporate strategy."[295] While stakeholder engagement may be broader in scope, CSR certainly is a key component of stakeholder strategy. Stakeholder engagement comprises both formal and informal connectivity of a company to its stakeholders.

The CSR report provides an invaluable opportunity to communicate CSR efforts to the corporation stakeholders. It is also a medium for transparency and may be used as an effective outreach tool for shareholder relations campaign to deter shareholder activism and threat of litigations. CSR should focus on the important areas of interaction between the company and its key stakeholders and address value creation actions as part of the company's strategy. In addition, the CSR report provides investors with ESG information to assist in evaluating investment decisions.

2.2.5.3   Stakeholder Engagement and Communication

Stakeholder engagement should follow a streamlined process to identify and continually update the list of relevant stakeholders, the knowledgeable corporate executives trained in public communication, appropriate forums and frequency, legal review and follow up and grievance process. An effective engagement strategy is to provide periodic updates from the CEO or the board on the companie's positions and actions regarding CSR issues and how they impact or may impact stakeholders. However, a legal review of CSR disclosures and communications must be built in to avoid release of material nonpublic information, ensure consistency of contents with disclosures in other public documents, such as securities and other regulatory filings; and assure alignment with other company policies and practices.

1. The CSR report typically provides disclosures on key stakeholder concerns: (1) shareholders and leadership, such as business structure and models, corporate governance, and the role of the board in sustainability risk management; (2) employees, such as diversity, health, and safety, training and mentoring, employee relations, and wages and benefits; (3) customers, such as product safety, customer service, and privacy; (4) suppliers such as labor standards and CSR programs; (5) communities, such as corporate philanthropy and charitable contributions, community investment and partnerships, volunteerism, and the local environmental and social impact of operations; and (6) governments and regulators, such as addressing lobbying, public policy, and the effects of and compliance with environmental and tax regulations. Each one is important enough to engage the relevant stakeholder(s) for the long-term sustainability of the organization.

2. Most CSR reporting guidelines recommend disclosures on the following factors: (1) Economic bottom lines, such as impact on the economic conditions and on local, national and global economic systems; (2) environmental bottom lines such as operations' impacts on natural systems—such as land, air, water, and ecosystems, inputs—such as energy and water, outputs—such as emissions, effluents and waste, as well as environmental compliance performance and expenditures; (3) social bottom lines, such as impacts on the social systems, human rights, society, and product responsibility; (4) ethics and integrity, such as company's values, principles and standards, internal and external mechanisms for seeking advice on matters of integrity and ethical behavior and reporting concerns; and (5) stakeholder engagement, during the reporting period and not limiting it to that conducted for preparing the CSR report.
3. Disclosures may include, in addition to CSR reports, supplemental reports (e.g., carbon disclosure), regulatory filings, the annual shareholders' meeting and direct interactions with stakeholders such as community, employees and supplier forums.
4. Stakeholder engagement, post report disclosure is an important element and helps avoid misinterpretations and frustrations. This is a year-round public affairs function.
5. Some useful tactical measures include: (1) focusing on material issues most significant to the company and its stakeholders; (2) use of simple language, balancing detail versus brevity; and (3) adding third-party evaluations and awards and recognitions received, where available, to verify the reliability and credibility.

2.2.5.4 Principles of Effective Disclosure—What Makes a Good Corporate Social Responsibility Report?

1. *Responsiveness*: The CSR report is like the user interface in a software application, and care should be taken to ensure it is responsive to stakeholder concerns and needs.
2. *Balance*: A balanced view and set of contents should be presented to retain credibility.
3. *Comparability*: Allows stakeholders, activists, and investors to make better informed choices when they compare peers' reports.
4. *Accuracy*: Presence of any significant error greatly affects trust and credibility.
5. *Timeliness*: Availability of report on time with current information is seen as a reasonable stakeholder expectation.
6. *Clarity*: Because of stakeholder diversity, with varying levels of education, simple and clear communication is mandatory. Clarity must be balanced against brevity.
7. *Reliability*: Stakeholders expect the report to be reliable and a true reflection of the intent and actions of the corporation. There should be no misleading disclosure.

8. *Boundary setting*: The scope of the report and boundary of the operations and impacts covered should be clearly defined.
9. *Assurance and auditing*: Both internal and third-party validations add significant credibility.

2.3 ETHICAL ASSESSMENT AND DECISION-MAKING PROCESS

2.3.1 CHAPTER OVERVIEW

This chapter focuses on the ethical dimension of sustainable development and eco-system management. The chapter highlights the significance of ethics as the fourth bottom line of sustainability and expands the triple bottom line context of people, planet and profit with a fourth component—purpose—to emphasize the power of ethics as a balancing force to preempt the disastrous pitfalls of economics without ethics.[296] The ethics bottom line of sustainability issues and their import in decision-making are also addressed.

2.3.2 DEFINITION OF ETHICS

Ethics is a mandatory dimension—the fourth bottom line to complement the triple bottom line of sustainability. Ramanan and Taback, defines ethics as "the difference between what a person has the right to do and the right thing to do!" A person's right to act is often defined by the law of the land. The right thing to do is the action taken in response to a situation that will result in the greatest benefit and the least harm to all the stakeholders. Arjoon[297] observes that ethics are more akin to the spirit of the law and inspires excellence. This definition is consistent with Paine's[298] elegant distinction that ethical decisions go beyond the legal tenet of limits to stay within. Paine elegantly presents the differences in the legal and ethical approaches to decision-making.[299] While the legal tenet is to define a set of limits that must be met with a goal to prevent unlawful action, ethics provide a set of principles to guide choices to act responsibly.

2.3.3 REASONING TO MAKE ETHICAL DECISIONS

Making any decision involves reasoning, ranging from a whim followed by rationalization to meticulous systematic logical consideration of all the impacts, their intensity and likelihood. The first of two aspects of ethics involve the ability to distinguish right from wrong, good from bad, and propriety from impropriety. Incorporating the ethical factor involves recognizing why unethical behavior occurs, what qualities are essential, and what pitfalls or barriers must be overcome. A journey through the evolution of the framework development leads to the process of building ethical filters in the sustainability decisions to minimize or mitigate ethical bias.

The second is a commitment to do what is right and proper. Ethics is an action concept, not simply something to think and rationalize about. As US president Lyndon Johnson once said, "A president's hardest task is not to do what is right but to know what is right."[300] Understanding underlying ethical implications and commitment to the right action requires character and courage to meet the challenge when doing the right thing costs more than the stakeholder wants to pay.

2.3.3.1   Unethical Behavior—Why Does It Occur in Organizations?

Sims[301] offers a list of twelve unethical encounters commonly seen in corporations based on sixty *Wall Street Journal* articles. These are stealing, lying, fraud, influence buying, hiding or failing to protect data, cheating, working below par, personal or organizational abuse, not complying with legal requirements or noticing an act of noncompliance and not reporting it, and choosing one option with an inherent self-centered bias in the case of an ethical dilemma. The following list provides a sustainability context for some of these common unethical situations that a policy or corporate steward may face.

1. *Lying*: Misrepresenting statements to stay under the applicability of a more stringent regulatory requirement or pollution control equipment
2. *Fraud/deceit*: Misleading consumers using green-washing commercials
3. *Influence buying*: Paying bribes to expedite the clearance of environmental or social permits
4. *Hiding or failing to disclose/protect data*: For example, hiding the effects of tobacco on personal health or not disclosing material risks - for example the failure of Facebook to protect private data from potential alleged misuse by Cambridge Analytica.
5. *Noncompliance*: Using creative interpretations to avoid a regulatory requirement
6. *Accessory*: Failing to blow the whistle when a client or a superior refuse to stop toxic releases

2.3.3.2   Ethical Decsion-Making—Characteristics

Ethical decision-making is affected by three qualities: (1) competence in identifying issues and evaluating consequences; (2) self-confidence in seeking different opinions and deciding what is right; and (3) willingness to make decisions when the issue has no clear solution. The development of these qualities in individuals depends on their intrinsic personality and their stage of moral development at the point of decision. People could be fatalistic or believe that it is choice and not chance that leads to consequences. Another significant personality trait is Machiavellianism—the willingness to do whatever it takes to achieve a goal. People with this trait are comfortable manipulating others and, unlike Gandhi, believe their desired end justifies the means. Finally, people's cognitive stage of moral development is affected by their group interactions and personal growth. The early self-stage is shaped by reward and punishment. The focus in this stage is on the expectation of others in authority, including parents, peers and society. At the mature level, people begin to recognize and accept diversity in value systems, and become confident and often fixed in their own choices. The social contract practiced by the society defines right and wrong based on principles of justice and fairness. Often individuals also become fixated on their self-selected absolutes for what is ethical or moral.

Other factors that contribute to individuals' ethical decision-making process in the workplace are the moral intensity of the consequence of the action; the individual's empathy, knowledge, and intellectual and emotional ability enabling one to recognize the potential impacts on stakeholders; and the influence of the business and job environment. For instance, sending a personal e-mail from an office computer may not be seen as unethical at all. Some individuals are more capable of understanding the broader impacts of an issue than others. For example, the natural attenuation remediation of a contaminated site may be better understood by someone with expertise in the area. They may also realize that this method of remediation is a good use of the community's resources and that it reduces health risks. However, this does not imply that such an individual is more ethical and will make more ethical decisions. For instance, one may be an environmental activist with a set agenda to make the company responsible for the contamination pay more for the remediation; or could be an environmental remediation services company who may choose the less ethical option of dig, move, and treat the dirt that provides them additional contract work and compensation.

Sims draws upon Reidenbach's moral development pyramid and finds similarity between the moral maturity of organizations and individuals.[302] Interestingly, Carroll's CSR pyramid is quite similar and has broadly equivalent criteria.[303] The ethical culture matures from an amoral focus on profitability and an obsession with shareholder primacy to following the letter of the law, in which the understanding is that if it is legal, it is ethical. The next level is termed responsive, and consists of seeking social responsibility as long as it is in the corporation's financial interest to do so. This is consistent with Porter's strategic CSR.[304] The next two stages up are called the emerging ethical and developed ethical, in which a balance between profits and ethics along with commitment to codes and ethical principles are in play. This more closely parallels Porter's Creating Shared Value (CSV), Visser's Systemic CSR, and Ramanan's Creating Social Impact (CSI), which is inclusive of pure philanthropy and mission driven impact investing.[305]

Regardless of the stage, corporate culture plays a role, and Sims provides a set of nine broad categories that have differing levels of impact on the ethical decision-making process carried out by individuals in the organization. These are driven by ego, benevolence, and principles. Personal self-interest, company profit, and operating efficiency are ego-driven climates. A climate of benevolence promotes individual friendships, team interests, and social responsibility. An organizational ethical climate, at the top of the pyramid based on principles, leads to personal morality/ethics, rules, standard procedures, laws, and professional codes.

2.3.3.3 Groupthink—A Dysfunctional Phenomenon

"Groupthink is a psychological phenomenon that occurs within groups of people, in which the desire for harmony in a decision-making group overrides a realistic appraisal of alternatives."[306] "[Groupthink] is an enormously powerful force in today's world of global harmony."[307] Haidt captures the problems with groupthink brilliantly with the idea of morality binds and blinds.[308]

Groupthink is described as a dysfunctional process by Janis,[309] and Sims provides a set of valuable symptoms to look for and to avoid separating the ethical decision process from the malaise of groupthink. Humans are strategically altruistic, 90% chimps and 10% selfless and team-spirited bees. However, when taking one for the team, assimilation and acceptance in groups are predicated on getting along and not rocking the boat, because dissent from the group is discouraged. The symptoms include a range of illusions, group morality, unanimity, and a false sense of invulnerability that lead to no reexamination of ethics. Again, because of peer pressure, which rejects dissent and promotes a selective bias against new information, members of a group indulge in rationalization and self-censorship, and leaders actively shield groups from any external negative feedback. This provides such a false sense of assurance that the correct path has been selected that no one investigates impact on stakeholders who may have been completely overlooked or the possibility of a better solution.

2.3.3.4   The Evolution of a Framework for Making Ethical Decisions

2.3.3.4.1   Early Foray—The Four-Way Test[310] (1932)

For thousands of years, religions have been pursuing the absolute in morals and ethics. Socrates called for first learning what is good before pursuing it. Plato assigned cardinal virtues to man by class: prudence or wisdom to the ruler, courage or fortitude to the warrior, temperance or moderation primarily to the worker; and justice to all regardless of class.[311]

One of the early forays into formal ethical decision-making by business associations was created in 1932 and adopted eleven years later by Rotary International as their *code of ethics*. It was both simple and elegant and has survived the test of time. Every decision becomes ethically robust as it goes through the following four-way test:

1. Is it the truth?
2. Is it fair to all concerned?
3. Will it build goodwill?
4. Will it be beneficial to all concerned?

2.3.3.4.2   Examining the Ethics of a Business Decision (1981)

Nash differentiates between the ethical expectations from man and the corporate ethos.[312] She places corporate ethos halfway between amoral and Platonic virtues; it fulfills the (then, at the time of her study) social contract that centers on avoiding social harm. She sets out a set of twelve very pragmatic questions for ethical enquiry that are easily comprehensible to the business executive.

1. Have you defined the problem accurately? The focus should be on factual neutrality and not an attempt to colorfully articulate group or individual morality. An example is presenting select ESG data to advocate stringent regulation.
2. How would you define the problem if you stood on the other side of the fence? Calling for articulating the other side (stakeholders) interrupts self-interest and brings in empathy. For a long time, the auto industry resisted any efforts to impose fuel efficiency. However, once they recognized the

younger consumer group's call for greener cars, enlightened self-interest kicked in. However, today the pendulum is swinging all the way back - serious efforts are underway to roll back the fuel efficiency rules in the US by President Trump's administration, citing need for deregulation to be essential to make America Great Again.

3. How did the situation occur in the first place? Investigating history often provides clues to the root cause rather than just the symptoms. Unethical acts often start inadvertently and are not recognized until they become a crisis. Recurring patterns of seemingly trivial events often point to the real issue. This has been most recently seen in the disastrous BP Deepwater Horizon Gulf oil spill, where the culture of relentless cost cutting at the expense of plant safety was ultimately discovered as the root cause.

4. To whom and to what do you give your loyalty as a person and a member of the corporation? Conflicts of loyalty are real and identifying them is just the beginning. All structural precautions, whether they are mandates or industry or corporate standards are inadequate unless people involved also have a strong enough sense of integrity to put upholding the principles above personal gain. Code of ethics, and so on, are not worth the paper they are printed on unless backed by organizational culture and leadership commitment. Both Enron and BP had highly publicized codes of ethics that were routinely overridden by the executives to meet their personal agenda.

5. What is your intention in making this decision? A good intent is not enough to justify making a decision in an unknown area that could result in more harm than good; relevant expertise is needed. An oil company learned the hard way that human interaction with the natives of an ultra-pristine area inadvertently led to an avoidable epidemic of the common cold virus, to which the local population had not developed any immunity. The intention was noble—to engage the native stakeholders directly. However, as soon as they recognized the onset of the epidemic, they did react responsibly and brought in all the emergency medical personnel and medication to address and resolve the epidemic.

6. How does this intention compare with the probable result? This question brings into focus two inherent dangers in the decision process: knowledge of the future is often impossible, and blinding overconfidence driven by good intent is a recipe for disaster. Banning DDT is an example. While it may have helped eliminate the human health impact of this chemical, it is claimed to have set off a resurgence of malaria and numerous deaths, in regions like India, because of the lack of alternate ways to contain the spread of disease-carrying mosquitoes.

7. Whom could your decision or action injure? This is an extremely significant question in today's context. Innovations and new products have significantly outpaced our ability to manage their production and contain their distribution or disposal. Businesses often only investigate potential impacts when hit by lawsuits—such as in the case of asbestos. Nanotechnology, genetics, and biological and viral antidotes are prime examples of the unknown.

8. Can you discuss the problem with the affected parties before you make your decision? This is an avant-garde effort. As we are aware today, initiatives such as Equator Principles and UN Global Compact mandate disclosure and the engagement of relevant stakeholders in the environmental and social impact assessment process prior to funding or permitting a major project.[313]

9. Are you confident your position, though it now seems valid, will continue to be valid over a long period of time? While executive compensation systems and, more importantly, speculator investor incentive systems are typically designed for the next quarter and at best one year. Long-range planning calls for a much longer timeframe. Over longer time frames, uncertainty in a business climate is inevitable; more so today when the globe is shrinking and manufacturing and money are moving at lightning speed. It becomes extremely significant in the case of addressing climate change—an issue that brings in an intergenerational time frame and conflict across nations over distributive justice.

10. Could you disclose your decision to all stakeholders as a whole without qualms? This is an absolute necessity in today's world of synchronous interactive connectivity. Today, one does not have to wait for radio or print-media press coverage. In addition, integrated reporting of nonfinancial metrics (e.g., social, environmental, and governance metrics), along with financial factors is becoming a global norm. As billionaire businessman Warren Buffett once said, it takes twenty years to build a reputation, and in today's context about 20 minutes to lose it; and only then do you discover you had a reputation to begin with. CSR reporting and disclosure of material risks in operations is increasingly becoming a mandate in most advanced economies.

11. What is the symbolic potential of your action if understood or misunderstood? The role of corporations in society is becoming increasingly symbiotic. As seen in the earlier sections, CSR has morphed and transformed into a new genre, in which making a social impact is becoming integral to the core business. This significantly increases the visibility of any corporate initiative in the social arena. Charity, philanthropy, or even the creation of shared value/impact through public–private partnerships is subjected to intense scrutiny. Also, perception is reality and matters, so every action has to be also evaluated in light of the perception it may create. The furor over donations by *Chick-fil-A* to a hardcore religious organization set off protests by the gay and lesbian alliance and impacted the food chain's expansion plans. For instance, the optics of renting a place from the spouse of a lobbyist, whose client's business is in the regulator's jurisdiction is certainly not good!

12. Under what conditions would you allow exceptions to your stance? This question sets the level of consistency with which the policy will be implemented. The waiver of conflict of interest or ethical code became a norm in Enron and eventually led to its demise.

2.3.3.4.3 Ethical Values Checklist for an Environmental Professional (1998)

Taback provides a practical values checklist that gives the environmental professional an opportunity to examine a situation from all sides.[314] The format could be used whenever a professional is faced with an environmental dilemma. It is based on the Six Pillars of Character: trustworthiness, respect, responsibility, justice and fairness, caring, and civic virtue and citizenship.[315] It suggests that decision makers create a table that lists the various sub-elements of the Six Pillars of Character as values on one side and evaluates the issues and related actions on the other side. After completing such a table, the professional should take the contemplations to colleagues, peers, and supervisors for validation and select the right action.

2.3.4 BUILDING ETHICAL FILTERS IN SUSTAINABILITY DECISIONS

Sustainable development and ecosystem management commonly involve tough sociopolitical choices and often face a conflict between economic benefits and environmental degradation. This calls for stewardship from both corporate and public policy leaders to protect and preserve our only planet for future generations. Sustainability stewards routinely encounter ethical dilemmas such as how much regulation to impose and how to balance human needs, how to incorporate equity and environmental justice, and how to preserve other living species and the natural world.

The simplest ethics filter to apply in everyday life may be the basic yet robust four-way test used since 1932 by the Rotary Club,[316] a charitable organization. Is it true and fair, and will it build goodwill and benefit all? Nearly half a century later, Nash[317] set out a dozen pragmatic queries for the business executives. Another quarter century later Paine[318] transformed those into eight business-specific ethics principles. Sucher[319] suggests evaluating each action from multiple ethical perspectives.

Paine's eight business-specific ethical principles[320] are fiduciary, dignity, property, transparency, reliability, fairness, citizenship, and responsiveness. Ramanan[321] suggests applying Paine's business-specific ethics principles to build ethics filters as well in the corporate decision process. Paine's business-specific ethics principles are presented here with expanded sustainability ethics examples to illustrate the building blocks of an ethics filter that must be applied in any decision-making process.

1. *Fiduciary duty*: The materiality of potential climate change risks, conducting environmental, social and governance audits of current operations, conducting due diligence audits of companies to be acquired, and setting environmental and ethical standards for operations in countries with less stringent standards are all areas to which the fiduciary duty principle applies extensively in both corporate and public policy sustainability management.

2. *Dignity of life*: Protecting public health and safety is one of the primary values of a public policy or corporate sustainability steward. Concern for human dignity, in particular for the vulnerable—setting air and water quality standards that protect the most vulnerable segment of the public, including children, sick people, and the elderly—is a given. Dignity often goes beyond public health and welfare to include nonhuman species as well.

3. *Property protection*: Protecting others' property, including nature's beauty, is inherent in all the decisions that sustainability professional makes. Responsible use of one's property is most evident through minimizing the impact of any activity on the people's property and leaving the activity site after its use in at least as pristine a condition as before. It also includes the natural inanimate world as something to be protected for generations to come and enjoy.

4. *Transaction transparency*: Disclosure of material information is built into every report or document to which the sustainability steward is connected. For instance, any findings on the potential health impacts of a new chemical or a newly discovered impact needs to be documented and reported to the appropriate authorities so they can take necessary protective action. Many infractions of regulatory requirements are managed through self-reporting by the corporation and its sustainability stewards.

5. *Expertise reliability*: The society depends on the reliable performance of the sustainability standards and professionals. In light of this, one must maintain skills and expertise, but also not make commitments or provide services in areas that are beyond one's expertise. For instance, a board-certified air quality specialist should not commit to work in wastewater or hazardous waste management areas. Also, professionals should make sure they are not over-committing, which could result in below-par work and could result in injury to society. Is there adequate access and availability of relevant expertise to make decisions?

6. *Fairness to most*: This principle applies to every aspect of the work the sustainability steward performs, especially in dealing with clients, colleagues and competitors. Fairness becomes particularly relevant when decisions related to sustainable development policy come up. Environmental justice, in which the lower economic strata are disproportionately impacted by lower environmental quality must be addressed in the context of the fairness principle. The central idea of the Brundtland Commission's definition of sustainable development is that of intergenerational equity. Climate change is a classic case for intergenerational equity and distributive justice. Is the decision fair to most?

7. *Respect for citizenship*: Respect for law and regulation is pervasive in a sustainability steward's job. Given the market externality of environmental issues, government intervention becomes mandatory. Sustainability stewards must also participate in advocacy, especially in areas in which they have expertise.

8. *Crisis responsiveness*: Addressing public complaints, participating in public hearings and crisis response planning, responding to any environmental, social, or governance crisis in which they could have an effective role are all relevant to this principle.

It may be noted from the previous examples that each ethical principle could have implications for the moral duties of the sustainability stewards. Decision makers have to ensure comfort with the consequences, respect for duties and rights of others, and compliance with community norms and company commitments.

Finally, the Ethics Resource Center[322] presents the PLUS method to test all corporate decisions for meeting the ethics threshold. The process uses the mnemonic PLUS to reflect **p**olicies and procedures of the corporation, **l**aws and regulations that apply, **u**niversal ethical principles that the corporation has adopted, and the **s**elf-defined values of the individual making the decision. They suggest surfacing ethical issues in defining any problem, assessing the ethical impacts in evaluating identified alternatives and resurfacing any remnant ethical issue in the selected option.

2.3.5   ETHICAL MANAGEMENT OF SUSTAINABILITY BIAS

"Earth provides enough to satisfy every man's needs, but not every man's greed."[323] Extreme greed, whether for money or nature's resources, has disastrous consequences. Today, given the unprecedented size and speed of global transactions, the world is poised for potential tsunamis in many areas because "the avenues to express greed have grown so enormously" and "man has acquired significant power to alter the nature of his world."[324]

People are inherently self-or inner-group serving. Without purpose as a moderator, one could easily skew the objective through the inherent bias of self- or inner-group interest. In the sustainability/environmental arena, unethical behavior most often manifests itself in the form of caring only for people who mimic us, protecting only parts of the ecosystem that overtly serve us, and, of course, generating profits only for a subsection of the stakeholders. Furthermore, Ramanan[325] suggests absent ethics as the fourth bottom line; in other words, sans purpose as a moderator, even congruence of the other three sustainability bottom lines may be skewed, and not incorporate issues such as distributive justice or intergenerational equity in the decision process.

Rich countries often indulge in irresponsible exploitation of natural resources, and businesses shift resources or shifting of high-pollution intensity industry to countries with less stringent regulations or enforcement. These decisions are frequently rationalized by calling them opportunity for economic development. Although signatories to OECD, United Nations-supported Principles for Responsible Investment (PRI), or Equator principles, led by International Finance Corporation, the financing arm of World Bank, have put in place requirements to conduct Environmental Impact Assessments and Social Impact Assessments to preempt irresponsible behavior; however, enforcement remains untenable. Similarly, dumping of toxic wastes in undeveloped regions of the world is an ongoing problem, despite numerous in-place global treaties. These examples highlight the need to manage ethical bias in sustainability decisions.

Similarly, there is a need to protect society at large from the avalanche of unethical green-washing. Demas[326] defines green-washing as the intersection of two behaviors: poor environmental performance and positive communication about environmental performance, and goes on to expand on it as the act of misleading consumers regarding the environmental practices of the company or the environmental performance of the product or service. At the product level, green-washing sins may involve claiming benefits with no evidence or where it involves a trade-off with some other negative impact, but that is not disclosed. Cigarettes is an age-old classic example of misleading consumers through *green-washing*. Routinely, people, especially younger adults, were targeted with misleading advertisements that showed healthy athletes smoking. There are several groups of drivers. The first group is market-based with pressure coming from competitors or from consumer or investor demand. Organizational culture, incentive structure, or internal pressures form the second group. Individual psychological drivers include individual bias or narrow framework. Finally, nonmarket group comprises lax regulatory environment and lack of monitoring by activists and media. Some valuable recommendations that could be critical from an environmental ethics perspective include transparency in environmental performance disclosure, whether voluntary or mandatory; awareness through sharing information about incidents of green-washing and reduction in regulatory uncertainty and most significantly ethical leadership and training for employees.[327]

The huge role that sustainability professionals play in integrated financial and sustainability reporting is related to the materiality and integrity of evaluating and reporting these highly technical and often complicated array of environmental factors; and hence ethical dilemmas are inevitable. Good comprehension of this emerging area, especially in the context of ethics, is crucial for sustainability stewards.

2.3.6   ETHICAL ISSUES IN DECISION SUPPORT MODELS

Electronic decision models are far from neutral. Decision support systems may raise ethical issues for both the builders and the users, and the best course is to anticipate potential ethical scenarios that could arise. Some example scenarios:

1. Is it all right to avoid a key metric because of the difficulty in capturing that data?
2. Is it all right to build the system, knowing fully that the input data quality is questionable?
3. Is it all right for users to have privileges of not using the business process monitoring system in the name of business exigency?
4. Is it all right to allow continued use of the knowledge-based system, knowing that the knowledge base is out of date?
5. Is it all right to allow data extracted from the system?

Policies must outline access rules, permissions to use system, and exceptions to preempt ethical lapses and poor data-use judgments.

Following is a select list of some common ethical issues affecting decision support systems per Power:[328]

1. Data quality assurance failure
2. Nontransparent data collection
3. Data error propagation
4. Obsolete system continuity
5. Privacy violation; misuse of customer information
6. Liability exposure from use or nonuse of system
7. Unauthorized data transfers
8. Poor policy or failure to enforce
9. Missing a key metric or non-validation of model

2.3.7 ETHICAL CONSIDERATIONS IN CLIMATE CHANGE

As choices arise in climate protection responsibility, in particular the ethical allocation of carbon share, to balance ecological effectiveness against economic efficiency and equity, positions taken on the following will be significant in shaping the outcome.

1. *Avoiding perceived vulnerability*: Climate change may endanger human health, wealth, and ultimately survival; in particular, the weaker sections of the world population. Many stakeholders call for action to protect the climate to protect humans. However, the primary objective is clearly the protection of man from perceived vulnerability, and not vice versa (protection of nature from man).

2. *Optimizing resource use*: Nature is considered a free and potentially infinite good. However, recognition of the value of resource use optimization results in egocentric ethics giving way to utilitarian ethics. The narrow pursuit of self-interest calls for collective rules instead and advocates the regulation of individual action in the name of the greater good of a greater number of people for a longer period of time. As a consequence, target selection is guided by aggregate benefits and costs rather than the individual actor's self-interest.

3. *Holding in trust for future*: Protecting the climate system for the benefit of present and future generations suggests considering the well-being of future generations as one of the factors to be considered for decision-making in the present. The approach shifts from posterity, seen only as future beneficiaries of progress, to a possible victim of it. "Justice across generations demands restraint today. The concept extends the principle of equity among the human community along the axis of time."[329] It is indeed a question of ensuring intergenerational equity. Being a beneficiary of the global commons today, therefore, also implies being their trustee.

4. *Beyond anthropocentric*: While ethicists like Peter Singer[330] value wildlife and wild animals with equal status and believe nonhuman beings have rights as well, people in general are anthropocentric, and give humans a strong preference over other species. Humans are not entitled to inflict climate change upon the communities of plants and animals, which—along with humans and inanimate matter—are not just instrumental but also have intrinsic value in the biosphere, for instance biodiversity. An associated driver could be the motivation of humans to rejoice in creation.

ENDNOTES

[166] Dow Jones Sustainability Indexes, Accessed December 2012 and available at http://www.sustainability-index.com/.

[167] Global Reporting Initiative, Accessed December 2012 and available at https://www.globalreporting.org (Author Dr. Ramanan served as a member of the GRI's G4 Academic Research Group).

[168] Székely, F. and M. Knirsch, Responsible leadership and corporate social responsibility: Metrics for sustainable performance, *European Management Journal*, 2005, 23, 628–647.

[169] Anand, S. and A.J. Sen, *Sustainable Human Development: Concepts and Priorities* (UNDP, 1994).

[170] Ramanan, R., Environmental performance reporting—State of the art, *Proceedings of the International Interdisciplinary Conference on Sustainable Technologies for Environmental Protection* (Coimbatore, India, January 2006).

[171] SASB Conceptual Framework, Accessed April 2017 and available at https://www.sasb.org/wp-content/uploads/2013/10/SASB-Conceptual-Framework-Final-Formatted-10-22-13.pdf.

[172] Cohen, B. et al. Columbia University, The growth of sustainability metrics, May 2014.

[173] United Nations, Accessed April 2017 and available at http://www.un.org/esa/sustdev/natlinfo/indicators/guidelines.pdf and http://www.un.org/esa/sustdev/natlinfo/indicators/factsheet.pdf.

[174] Soyka, P.A. and M.E. Bateman, Finding Common Ground on the Metrics that Matter, IRRC Institute, Accessed February 2012 and available at http://www.irrcinstitute.org/pdf/IRRC-Metrics-that-Matter-Report_Feb-2012.pdf.

[175] Governance and Accountability Institute, Sustainability—What matters? 2014, Accessed October 2017 and available at http://www.ga-institute.com/fileadmin/user_upload/Reports/G_A_sustainability_-_what_matters_-FULL_REPORT.pdf.

[176] Sustainability, Accessed April 2017 and available at http://sustainability.com/our-work/insights/what-do-esg-ratings-actually-tell-us/.

[177] WHEB Asset Management Group, Accessed April 2017 and available at http://www.whebgroup.com/just-what-is-the-point-of-esg/.

[178] The Public Company Accounting Oversight Board is a nonprofit corporation established by the U.S. Congress to oversee the audits of public companies in order to protect investors and the public interest by promoting informative, accurate, and independent audit reports. http://pcaobus.org/About/Pages/default.aspx.

[179] TSC Industries v. Northway, 426 U.S. 438, 449 (1976). See also Basic, v. Levinson, 485 U.S. 224 (1988).

[180] Khan, M., G. Serafeim, and A. Yoon, Corporate sustainability: First evidence on materiality, *The Accounting Review*, 2015. doi:10.2139/ssrn.2575912.

[181] SASB, Materiality assessment, Accessed February 2016 and available at http://www.sasb.org/materiality/materiality-assessment/.

[182] Dow Jones Sustainability Indexes, Accessed December 2012 and available at http://www.sustainability-index.com/.

[183] Russo, A. and M. Mariani, Drawbacks of a delisting from a sustainability index: An empirical analysis, *International Journal of Business Administration*, 2013, 4, 20.

[184] Stubbs, W. and P. Rogers, Lifting the veil on environment-social-governance rating methods, *Social Responsibility Journal*, 2013, 9, 622–640.

[185] Thomson Reuters, Accessed April 2017 and available at http://financial.thomsonreuters.com/content/dam/openweb/documents/pdf/financial/tr-corporate-responsibility-Indexes-overview.pdf.

[186] Ibid.

[187] Ibid.

[188] Bloomberg, Accessed April 2017 and available at https://www.bloomberg.com/professional/sustainable-finance/.

[189] Global Initiative for Sustainability Ratings, Accessed April 2017 and available at http://ratesustainability.org/hub/index.php/search/at-a-glance-product/24/99.

[190] Dow Jones Sustainability Indexes, Accessed April 2017 and available at http://www.sustainability-indices.com/index-family-overview/djsi-family-overview/index.jsp.

[191] Ibid.

192 http://www.sustainability-index.com/images/sam-csa-methodology-en_tcm1071-338252.pdf.

193 STOXX, Accessed April 2017 and available at https://www.stoxx.com/indices.

194 RobecoSAM, Measuring intangibles, Accessed April 2017 and available at http://www.robecosam.com/images/Measuring_Intangibles_CSA_methodology.pdf.

195 RobecoSAM CSA Resource Center, Accessed April 2017 and available at http://www.robecosam.com/en/sustainability-insights/about-sustainability/corporate-sustainability-assessment/resource-center.jsp.

196 RobecoSAM, Rules based component selection, Accessed April 2017 and available at http://www.sustainability-indices.com/index-family-overview/djsi-family-overview/index.jsp.

197 FTSE 4Good, Accessed April 2017 and available at http://www.ftse.com/products/downloads/ESG-ratings-overview.pdf.

198 Morgan Stanley ESG Index Family, Accessed April 2017 and available at https://www.msci.com/documents/10199/242721/MSCI_ESG_Indexes.pdf.

199 Morgan Stanley ESG Index Family, Accessed April 2017 and available at https://www.msci.com/esg-index-family.

200 Innovest was founded by Dr. Mathew Kiernen, a personal friend since mid-1990s, and a pioneer in the field of linking sustainability performance metrics and organizational value.

201 Thomson Reuters, 2017.

202 Thomson Reuters, Accessed April 2017 and available at http://www.trcri.com/index.php?page=asset4.

203 Park, A. and C. Ravenel, Integrating sustainability into capital markets: Bloomberg LP and ESG's quantitative legitimacy, *Journal of Applied Corporate Finance*, 2013, 25, 62–67.

204 Thomson Reuters (ASSET4), ESG/CSR content overview, Accessed December 2012 and available at http://thomsonreuters.com/products_services/financial/content_news/content_overview/content_az/content_esg/.

205 Bloomberg, Sustainability, Accessed December 2012 and available at http://www.bloomberg.com/bsustainable/.

206 Jantzi, M. CEO of Sustainalytics, B.V., An investment research firm that specializes in environmental, social, and governance (ESG) research and analysis. Accessed April 2017 and available at http://lgdata.s3-website-us-east-1.amazonaws.com/docs/1669/1300628/2014-06-02_Bloomberg_Guide.pdf.

207 Cision, P.R., Newswire, Accessed April 2017 and available at http://www.prnewswire.com/news-releases/sp-dow-jones-indices-acquires-trucost-300337852.html.

208 RobecoSAM, Accessed April 2017 and available at http://www.robecosam.com/images/160920-robecosam-bloomberg-release-en-vdef.pdf.pdf.

209 Engelbrecht, H.J., A comparison of sustainability indices: Mixed messages from OECD Countries, Paper presented to the 42nd Australian Conference of Economists, ACE 2013 (Perth, Western Australia, July 7–10, 2013). Accessed April 2017 and available at https://www.murdoch.edu.au/School-of-Business-and-Governance/_document/Australian-Conference-of-Economists/A-comparison-of-sustainability-indices.pdf.

210 Ibid.

211 Stiglitz, J., A. Sen, J.-P. Fitoussi, Report by the Commission on the Measurement of Economic Performance and Social Progress, 2009. www.stiglitz-sen-fitoussi.fr.

212 Anand, S. and A.K. Sen, *Sustainable Human Development: Concepts and Priorities* (UNDP, 1994).

213 Perman, R., Y. Ma, M. Common, D. Maddison, and J. McGilvray, *Natural Resource and Environmental Economics*, 4th ed. (Harlow, UK: Pearson Education, 2011), p. 662.

214 Stiglitz, J., A. Sen, J.-P. Fitoussi, Report by the commission on the measurement of economic performance and social progress, 2009. www.stiglitz-sen-fitoussi.fr, p. 11.

215 Murphy, M. (Ed.). The Happy Planet Index 2.0. (London, UK: New Economics Foundation, 2009). www.happyplanetindex.org, p. 1.

216 Ng, Y.-K., Environmentally responsible happy nation index: Towards an internationally acceptable national success indicator, *Social Indicators Research*, 2008, 85, 425–446.

217 Esty, D., Levy, M., Srebotnjak, T., and A. de Sherbinin, *Environmental Sustainability Index: Benchmarking National Environmental Stewardship* (New Haven, CT: Yale Center for Environmental Law & Policy, 2005).

218 World Wide Fund for Nature (WWF), Living planet report 2008 (Gland, Switzerland: WWF, 2008), p. 14.

219 United Nations, Accessed April 2017 and available at http://www.un.org/esa/sustdev/natlinfo/indicators/guidelines.pdf and http://www.un.org/esa/sustdev/natlinfo/indicators/factsheet.pdf.

220 World Economic Forum, Sustainability data and trends, Accessed December 2012 and available at http://sedac.ciesin.columbia.edu/theme/sustainability.

221 Happiness Alliance Gross National Happiness Index, Accessed April 2017 and available at http://www.grossnationalhappiness.com/articles/.

222 Ibid.

223 Measuring Sustainability Disclosure: Ranking the World's Stock Exchanges. Corporate Knights Capital, July 2016, Accessed April 2017 and available at www.corporateknights.com.

224 Institution of Chemical Engineers, UK, Sustainability metrics for the process industry, Accessed April 2017 and available at https://www.scribd.com/document/51848632/IChemE-Metrics-sustainability.

225 The Sustainability Consortium, Accessed April 2017 and available at https://www.sustainabilityconsortium.org/wp-content/uploads/2017/03/TSC-Toolkit-Methodology-Brief.pdf.

226 Wal-Mart, Accessed April 2017 and available at www.walmartsustainabilityhub.com.

227 The Sustainability Consortium, Accessed April 2017 and available at https://www.sustainabilityconsortium.org/product-sustainability/.

228 AASHE's STARS Program, Accessed April 2017 and available at https://stars.aashe.org/.

229 Sierra Club Cool School Rankings, Accessed April 2017 and available at http://www.sierraclub.org/sierra/coolschools-2016.

230 Sustainable Endowments Institute, The college sustainability report card. Accessed April 2017 and available at http://greenreportcard.org/report-card-2011/executive-summary.html.

231 IARU Green Guide for Universities, Accessed April 2017 and available at http://www.iaruni.org/images/stories/Sustainability/IARU_Green_Guide_for_Universities_2014.pdf.

232 Murphy, C., D.T. Allen, B. Allenby, J. Crittenden, C. Davidson, C. Hendrickson, and S. Matthews, Sustainability in engineering education and research at US universities, *Environmental Science and Technology*, 2009, 43, 5558–5564.

233 Ramanan, R., (Ed.–Special Issue–Post graduate environmental education) Environmental Manager, September 2012. Accessed April 2017 and available at http://pubs.awma.org/gsearch/em/2012/9/ramanan.pdf; Green MBA and integrating sustainability in business education, Accessed April 2017 and available at http://pubs.awma.org/gsearch/em/2012/9/ramanan.pdf.

234 Price Waterhouse Cooper, Cities of opportunity: Building the future, Accessed April 2017 and available at http://www.pwc.com/gx/en/industries/capital-projects-infrastructure/publications/cities-of-opportunity-building-the-future.html.

235 Ibid.

236 KPMG, The future of cities: Measuring sustainability, Accessed April 2017 and available at HOME.KPMG.COM/xx/en/home/insights/2016/04/the-future-of-cities-measuring-sustainability.html.

237 Circles of Sustainability Urban Profile Process, Accessed April 2017 and available at http://www.circlesofsustainability.org/tools/urban-profile-process/.

238 United Nations Global Compact Cities Programme, Accessed April 2017 and available at http://citiesprogramme.org/our-framework/.

239 Siemens, A.G., Green city index, Accessed April 2017 and available at https://www.siemens.com/entry/cc/features/greencityindex_international/all/en/pdf/gci_report_summary.pdf.

240 Ibid.

241 US Green Building Council, An introduction to LEED and green building, Accessed May 2017 and available at http://go.usgbc.org/Intro-to-LEED.html.

242 International Facility Management Association—Sustainability How-to-Series, Accessed April 2017 and available at http://cdn.ifma.org/sfcdn/membership-documents/green-rating-systems-htg-final.pdf.

243 Taback, H. and R. Ramanan, *Environmental Ethics and Sustainability* (Boca Raton, FL, CRC Press, 2013), p. 120.

244 International Integrated Reporting Community IIRC—International Framework for Integrated Reporting, Accessed April 2017 and available at http://integratedreporting.org/wp-content/uploads/2013/12/13-12-08-THE-INTERNATIONAL-IR-FRAMEWORK-2-1.pdf.

245 Ibid.

246 Ibid.

247 Investopedia, Accounting, Accessed April 2017 and available at http://www.investopedia.com/university/accounting/default.asp.

248 International Monetary Fund (IMF), System of national accounts, Accessed April 2017 and available at https://www.imf.org/en/Publications/Books/Issues/2016/12/31/System-of-National-Accounts-2008-23239.

249 International Monetary Fund (IMF), Monetary and financial statistics manual and compilation guide (MFSMCG), Accessed April 2017 and available at https://www.imf.org/en/~/media/BA1EEFCA3BAD47F291BBFDFA8D99F05D.ashx.

250 United Nations, Accessed April 2017 and available at https://www.ecb.europa.eu/pub/pdf/other/handbookofnationalaccounting2014en.pdf.

251 United Nations, The system of national accounting and environmental extensions, New York, 2014, Accessed April 2017 and available at https://unstats.un.org/unsd/envaccounting/seeaRev/SEEA_CF_Final_en.pdf.

252 Ramanan, R., Environmental performance reporting—State of the art, *Proceedings of the International Interdisciplinary Conference on Sustainable Technologies for Environmental Protection*, ICSTEP2006, Coimbatore, India, 2006.

253 Global Reporting Initiative, Accessed December 2012 and available at https://www.globalreporting.org; Author Dr. Ramanan served as a member of the GRI's G4 Academic Research Group.

254 IPIECA (The global oil and gas industry association for environmental and social issues), API (The American Petroleum Institute), and OGP (The International Association of Oil and Gas Producers), Oil and gas industry guidance on voluntary sustainability reporting, 2nd ed. 2010, Accessed December 2012 and available at http://www.api.org/environment-health-and-safety/~/media/files/ehs/environmental_performance/voluntary_sustainability_reporting_guidance_2010.ashx.

255 Lydenberg, S., J. Rogers, and D. Wood, From transparency to performance: Industry-based sustainability reporting on key issues. The Hauser Center for Non-Profit Organization at Harvard University and Initiative for Responsible Investment, 2010.

256 World Economic Forum, Sustainability data and trends, Accessed December 2012 and available at http://sedac.ciesin.columbia.edu/theme/sustainability.

257 Global Reporting Initiative, Accessed December 2012 and available at https://www.globalreporting.org; Author Dr. Ramanan served as a member of the GRI's G4 Academic Research Group.

258 SASB, Conceptual framework, Accessed February 2016 and available at http://www.sasb.org/wp-content/uploads/2013/10/SASB-Conceptual-Framework-Final-Formatted-10-22-13.pdf.

259 http://www.accounting.com/resources/gaap.

260 Investopedia, Generally acceptable accounting principles, Accessed April 2017 and available at http://www.investopedia.com/terms/g/gaap.asp.

261 US Securities and Exchange Commission, Accessed October 2017 and available at https://www.sec.gov/about/laws/sa33.pdf.

262 US Securities and Exchange Commission, Accessed October 2017 and available at https://www.sec.gov/about/laws/sea34.pdf.

263 FAF, Financial accounting foundation, Accessed April 2017 and available at http://www.accountingfoundation.org/home.

264 FASB, Financial accounting standards board, Accessed April 2017 and available at http://www.fasb.org/home.

265 GASB, Governmental accounting standards board, available at http://www.gasb.org/.

266 Investopedia, Non-GAAP earnings, Accessed April 2017 and available at http://www.investopedia.com/terms/n/non-gaap-earnings.asp.

267 Fields, E., The essentials of finance and accounting for nonfinancial managers.

268 IASB and IFRS, International Accounting Standards Board and International Financial Reporting Standards, Accessed April 2017 and available at http://www.ifrs.org/About-us/IASB/Pages/Home.aspx and http://whatis.techtarget.com/definition/IFRS-International-Financial-Reporting-Standards.

269 Ibid.

270 Investopedia, IFRS, Accessed April 2017.

271 IASB and IFRS, International Accounting Standards Board and International Financial Reporting Standards, Accessed April 2017 and available at http://www.ifrs.org/About-us/IASB/Pages/Home.aspx and http://whatis.techtarget.com/definition/IFRS-International-Financial-Reporting-Standards.

272 SASB, Conceptual framework, Accessed February 2016 and available at http://www.sasb.org/wp-content/uploads/2013/10/SASB-Conceptual-Framework-Final-Formatted-10-22-13.pdf.

273 Global reporting initiative G4 sustainability reporting guidelines, Accessed April 2017 and available at https://www.globalreporting.org/resourcelibrary/GRIG4-Part1-Reporting-Principles-and-Standard-Disclosures.pdf.

274 Ibid.

275 Ibid.

276 Ibid.

277 IPIECA (The global oil and gas industry association for environmental and social issues), API (The American Petroleum Institute), and OGP (The International Association of Oil and Gas Producers), Oil and gas industry guidance on voluntary sustainability reporting, 3rd ed. 2015, Accessed April 2017 and available at http://www.api.org/~/media/Files/EHS/Environmental_Performance/voluntary-sustainability-reporting-guidance-2015.pdf.

278 International Organization for Standardization (ISO), ISO 26000, Accessed April 2017 and available at https://www.iso.org/iso-26000-social-responsibility.html.

279 Ibid.

280 UN PRI, Environmental, social, and corporate governance (ESG) disclosure framework for private equity, March 25, 2013. Accessed April 2017 and available at https://www.unpri.org/download_report/6230.

281 Insead Global Private Equity Initiative, Accessed March 2017 and available at https://centres.insead.edu/global-private-equity-initiative/research-publications/documents/ESG-in-private-equity.pdf.

282 World Economic Forum, Accessed April 2017 and available at http://reports.weforum.org/impact-investment/ and http://www3.weforum.org/docs/WEF_Social_Investment_Manual_Final.pdf.

283 Ebrahim, A. et al., HBS, Accessed April 2017 and available at http://www.hbs.edu/socialenterprise/Documents/MeasuringImpact.pdf.

284 Global Impact Investing Network (GIIN), Accessed April 2017 and available at https://iris.thegiin.org/guide/getting-started-guide.

285 Global Impact Investing Network (GIIN), Accessed April 2017 and available at https://iris.thegiin.org/b-impact-assessment-metrics.

286 Bloomberg, M., Chair, *Task Force on Climate Related Financial Disclosures*, https://www.fsb-tcfd.org.

287 The Economist Intelligence Unit, The cost of inaction: Recognising the value at risk from climate change, 2015, Accessed August 2017 and available at http://www.eiuperspectives.economist.com/sustainability/cost-inaction.

288 Task Force on Climate-related Financial Disclosures (TCFD), Recommendations of the task force on climate-related financial disclosures, Accessed August 2017 and available at https://www.fsb-tcfd.org/wp-content/uploads/2017/06/FINAL-TCFD-Report-062817.pdf.

289 Ibid.

290 TCFD, The use of scenario analysis in disclosure of climate-related risks and opportunities, Accessed August 2017 and available at https://www.fsb-tcfd.org/wp-content/uploads/2016/11/TCFD-Technical-Supplement-A4-14-Dec-2016.pdf.

291 CalPERS, Towards sustainable investment: Taking responsibility, April 2012.

292 Global Sustainable Investment Alliance, Accessed March 2017 and available at http://www.gsi-alliance.org/wp-content/uploads/2015/02/GSIA_Review_download.pdf.

293 Voorhes, M. et al., Executive summary—Fig. B: Growth of SRI $2.7 trillion in 2007 to $3.0 trillion in 2010, in 2010 Report on Socially Responsible Investing Trends in the United States, Social Investment Forum Foundation, Accessed December 2012 and available at http://ussif.org/resources/research/documents/2010TrendsES.pdf.

294 Organisation for Economic Co-operation and Development (OECD), Accessed March 20, 2017and available at http://www.oecd.org/investment/mne/38783873.pdf.

295 Libit, B. and T. Freier, The corporate social responsibility report and effective stakeholder engagement, posted on Harvard Law School Forum based on Chapman publication, December 2013, Accessed April 2017 and available at https://corpgov.law.harvard.edu/2013/12/28/the-corporate-social-responsibility-report-and-effective-stakeholder-engagement/.

296 Ramanan, R. and H. Taback, Environmental ethics and corporate social responsibility. In Dhiman, S. (Ed.), *Spirituality and Sustainability: New Horizons and Exemplary Approaches* (New York: Springer, 2016.)

297 Arjoon, S., Corporate governance: An ethical perspective, *Journal of Business Ethics*, 2005, 61, 343–352.

298 Paine, L.S., *Venturing Beyond Compliance, The Evolving Role of Ethics in Business* (New York: The Conference Board, 1996), pp. 13–16.

[299] Ibid.

[300] President Lyndon B. Johnson's Annual Message to the Congress on the State of the Union January 4, 1965, University of Texas, Accessed June 6, 2007 and available at http://www.lbjlib.utexas.edu/johnson/archives.hom/speeches.hom/650104.asp.

[301] Sims, R.R., *Ethics and Corporate Social Responsibility—Why Giants Fall* (Westport, CT: Greenwood Publishing Group, 2003), p. 99.

[302] Sims, R.R. and J. Brinkmann, Leaders as moral role models: The case of John Gutfreund at Salomon Brothers, *Journal of Business Ethics*, 2002, 35, 327–339; Reidenbach, R. and D.P. Robin, A conceptual model of corporate moral development, *Journal of Business Ethics*, 1991, 10, 273–284.

[303] Carroll, A.B., The pyramid of corporate social responsibility, *Business Horizons*, 1991, 42, 39–48.

[304] Porter, M.E. and M.R. Kramer, Strategy and society, *Harvard Business Review*, 2006, 78–88.

[305] Porter, Creating shared value; Visser, Stages of CSR; Taback, H. and R. Ramanan, *Environmental ethics and sustainability*.

[306] Whyte, W.H., Jr. Groupthink, *Fortune*, 1952, 114–117, 142, 146.

[307] Haidt, J. and S. Kesebir, Morality. In Fiske, S. and D. Gilbert (Eds.), *Handbook of Social Psychology*, 5th ed. (Hoboken, NJ: John Wiley & Sons, 2010), pp. 797–832.

[308] Haidt, J., *The Righteous Mind*, p. 187.

[309] Janis, I.L. (1972). *Victims of Groupthink* (Boston, MA: Houghton-Mifflin), p. 197; Sims, R.R., *Corporate Social Responsibility*, p. 117.

[310] Taylor, H.J., Rotary international code of ethics, Rotary International, Accessed December 2012 and available at http://www.rotary.org/en/aboutus/history/rihistory/pages/ridefault.aspx.

[311] Frede, D., Plato's ethics: An overview. In Zalta, E.N. (Ed.), *Stanford's Encyclopedia of Philosophy*, Accessed December 2012 and available at http://plato.stanford.edu/entries/plato-ethics/#VirStaSou.

[312] Nash, L.L., Ethics without the Sermon, *Harvard Business Review*, 1981, 159, 79–89.

[313] Equator Principles, Environmental and social risk management for project finance, Accessed December 2012 and available at http://www.equator-principles.com/; United Nations, Overview of the UN global compact, Accessed December 2012 and available at http://www.unglobalcompact.org/aboutthegc/thetenprinciples/index.html.

[314] Taback, H., Ethics training: An American solution for doing the right thing. In Wilcox, J. and Theodare, L. (Eds.), *Engineering and Environmental Ethics* (New York: John Wiley Price, 1998), pp. 267–280.

[315] Josephson, M.S., *Making Ethical Decisions* (The Institute of Ethics, 2002), Accessed December 2012 and available at http://josephsoninstitute.org/MED/MED-2sixpillars.html.

[316] Taylor, H.J., Rotary international code of ethics, Rotary international, Accessed December 2012 and available at http://www.rotary.org/en/aboutus/history/rihistory/pages/ridefault.aspx.

[317] Nash, L.L., Ethics without the Sermon, *Harvard Business Review*, 1981, 59, 79–89.

[318] Paine, L., R. Deshpande, J.D. Margolis, and K.E. Bettcher, Up to code: Does your company's conduct meet world-class standards, *Harvard Business Review*, 2005, 83, 122–133.

[319] Sucher, S.J., *Teaching the Moral Leader: A Literature-based Leadership Course: A Guide for Instructors* (London, UK: Routledge, Taylor & Francis Group, 2012), p. 110; Sucher, S.J., *The Moral Leader: Challenges, Tools and Insights* (London: Routledge, 2008).

[320] Paine, L.S., R. Deshpande, J.D. Margolis and K.E. Bettcher, Up to code: Does your company's conduct meet world-class standards, *Harvard Business Review*, 2005, 83, 122–133.

[321] Ramanan, R., How to build sustainability issues in corporate decision making, Net Impact Student Chapter Seminar, IIT Start School of Business, Chicago, IL, March 25, 2011; Ramanan, R., How to build ethics filter in corporate decision making, *104th Air and Waste Management Association Annual Conference and Exposition*, Orlando, FL, June 23, 2011.

[322] Ethics resource Center, Ethics filter, Accessed December 2012 and available at http://www.ethics.org/resource/ethics-filters.

[323] Gandhi, Mahatma Gandhi—A sustainable development pioneer, Govind Singh, Eco Localizer, Accessed October 14, 2008 and available at http://ecoworldly.com/2008/10/14/mahatma-gandhi-who-first-envisioned-the-concept-of-sustainable-development/ in http://www.mkgandhi.org/articles/environment1.htm.

[324] Greenspan, A., Testimony of chairman Alan Greenspan said while presenting the Federal Reserve's Monetary Policy Report Federal Reserve Board, 2002. Archived from the original on June 7, 2011.

[325] Ramanan, R. and H. Taback, Environmental ethics and corporate social responsibility. In Dhiman, S. (Ed.), *Spirituality and Sustainability: New Horizons and Exemplary Approaches* (New York: Springer, 2016).

[326] Demas, M.A. and V.C. Burbano, The drivers of green-washing, *California Management Review*, 2011, 54, CMR494, UC Berkeley.

[327] Taback, H. and R. Ramanan, *Environmental Ethics and Sustainability* (Boca Raton, FL, CRC Press, 2013), p. 120.

[328] Power, D.J., Decision support, Analytics and business intelligence, A business expert press book, 2013.

[329] Rosa, L.P., Munasinghe, M., *Ethics, Equity and International Negotiations on Climate Change* (Cheltenham, UK: Edward Elgar Publishing, 2002), p. 163.

[330] Cavalieri, P. and P. Singer (Eds.), *The Great Ape Project: Equality beyond Humanity* (New York: St. Martins Griffin, 1993), p. 152.

3 Sustainability Analytics and Decision Execution

3.1 SUSTAINABILITY VALUE OPTIMIZATION FOUNDATIONS

3.1.1 CHAPTER OVERVIEW

Sustainability leaders will gain critical insights in this section, including how financial and nonfinancial accounting data drives external investor decisions and internal strategic priorities and investments, how to evaluate corporate sustainability risks and opportunities from a financial perspective, and how to manage/mitigate those risks. This chapter highlights the need and value of analytics and discusses how the nonfinancial metrics on environmental, social, and governance (ESG) and *impact* factors can be used along with traditional financial metrics to assess the risks and opportunities of investment decisions made by institutional investors, venture capitalists, and venture philanthropists, as well as business decisions made by corporate managers. Readers will gain the tools to evaluate, quantify, and assess ESG metrics as a way of differentiating responsible investment choices.

The first subsection introduces the process of conducting a cost-benefit analysis (CBA), including valuation, discounting, impacts of uncertainty, and distributional consequences. The second subsection provides an introduction to revealed preference, stated preference (contingent valuation), and benefits transfer methods of valuing sustainability, especially environmental and social attributes. These two subsections draw extensively from the US Government Office of Management and Budget (OMB)[331] guidance on cost-benefit analysis and benefit valuation for environmental and social regulatory policy analysis and implementation. For instance, a 1992 OMB memo recommends the use of 7% as the discount rate for net present-value analysis.[332] This value affects the weightage given to monetized future benefits.

The third subsection provides a description of the life-cycle analysis techniques for evaluating product environmental and social impacts. The fourth subsection introduces the principles of risk management and its application in prioritizing allocation of community and corporate resources in mitigating risk. Risk-management aspects of emerging areas such as sustainable development and corporate governance are discussed. The conceptual framework of risk-based prioritization of select sustainability initiatives and a basic methodology to analyze and optimize resource utilization to mitigate ESG impacts are presented.

The fifth and final subsection presents a framework for performing an integrated priority assessment of the various options to lead to a rational effective choice.

3.1.2 Cost-Benefit Analysis

CBA is a policy assessment method that quantifies the value of public or corporate policy decisions in monetary terms, including consequences (impacts) on all significant stakeholders. In particular, when markets fail because of externality and resources are limited and must be used efficiently, CBA[333] helps prioritize potential alternative programs, policies, or projects (including the status quo) and helps effective social decision making through efficient allocation of society's resources. Likewise, when the process requires monetization of intangibles under uncertainty, corporate investment decision process is challenged. Uncertainties in science, political sensitivity, complexity, and inconsistency in the methodology of monetizing benefits, emergence of new materials, and discovery of new adverse health effects—and of course the ever-changing geopolitics—are additional compounding factors.[334]

Environmental and sustainability issues are tough sociopolitical choices. Society does not have unlimited resources, so it becomes necessary to prioritize and allocate resources in an optimal manner. Cost-benefit analysis highlights the trade-offs in social and environmental investments and helps answer how much regulation is enough. Society decides if it is willing to pay for the benefits from government policies and if so, how much. Viscusi[335] argues that, if anything, monetizing these human health-risk reductions and environmental protection benefits highlights their economic value and allows policymakers to incorporate them more comprehensively, making more rational policy decisions. Clearly, there are dilemmas involved (e.g., should there be differences in value of human health-risk reduction based on age and potential contribution to society?). While not perfectly or precisely, in the willingness-to-pay valuation methodology citizens get to make those choices. All nature protection scenarios may not draw the same value.

Likewise, allocation of limited society resources between the cleanup of one contaminated site versus another groundwater aquifer can only be made through a benefits-monetization process. These get complicated, not too dissimilar to other public policy choices. For instance, if the society resource needed to avert one premature death through regulating an herbicide level in the water, for instance atrazine/alachlor,[336,337] safe levels in drinking water exceeded US$100 million, and this would exhaust the society's overall resources: several other possible uses of these resources would have to be stalled!

Benefits and costs help to evaluate economic efficiency by comparing favorable and unfavorable effects of policies. Decision makers increasingly make appropriate use of cost-benefit analysis when establishing regulatory priority (e.g., where to focus their resources—which pollutants and health risks and emissions from what sectors should be addressed first), and it is often required for all major regulatory decisions. For nearly two decades, the US OMB, per presidential order, has been assessing the costs and benefits of various regulatory initiatives, and this aids in making a reasoned determination of whether the benefits justify the costs.[338] External reviews and stakeholder engagements improve regulatory analysis. The benefits and costs are quantified, where possible, but presented with a description of uncertainties. Decision makers should not be bound by strict cost-benefit tests. In particular, because equity is a non-economic factor, it is crucial to identify important distributional consequences to ensure that ethical choices are made. In cost-benefit analysis, most market economic approaches do not generally address distributional issues. For instance, net present value using average per capita benefits, by far the most common method used to justify a government policy, does not usually reflect effects on income distribution. This is especially true in such areas as environmental justice—not unduly impacting people of lower socioeconomic strata because they tend to live closer to emissions and discharges from manufacturing facilities that pollute the environment.

A dollar in hand today is worth more than a dollar coming in one or more years later—this is the impact of time on the value of money. Any analysis that requires money or the equivalent benefit or cost flow over multiple years must be brought down to a common year for meaningful comparison. The rate at which future money is discounted to make it comparable to the current year's value of money, to make the basis comparable to money on hand today, is the discount rate. In the policy context, core assumptions regarding the social discount rate, and the monetized value of reducing risks of premature death and of health improvements are required for cost-benefit analysis, which in turn is needed for most environmental and other public policy decisions. A social discount rate to bring benefits and costs to present value for valid comparisons could pose some unprecedented challenges. For instance, some of the benefits in climate change mitigation result in benefits that may be a generation or even a century away. Almost any rate of discount brings the present value to near zero.

3.1.2.1 Policy Options Evaluation Techniques

This subsection draws extensively from the US Government's OMB[339] guidance. Cost-benefit analysis technique is recommended for formal economic analysis of government programs or projects. A less comprehensive technique, cost-effectiveness analysis, may be adequate in situations where a policy decision has been made saying benefits must be provided and competing alternatives found to offer equivalent benefits. The best available control technology for air pollutant reduction is a classic example of applying cost-effectiveness to inform policy decisions for providing an environmental benefit, better air quality. Policies or programs may be justified on efficiency grounds where they address market failure, such as public goods and externalities.

Analyses should include comprehensive estimates of the expected benefits and costs to society. It should be explicit about the underlying assumptions used to arrive at estimates of future benefits and costs. Retrospective studies to determine whether anticipated benefits and costs have been realized are potentially valuable. Both intangible and tangible benefits and costs should be recognized. Distortions may creep in because of externality, monopoly, or taxes and subsidies. Additional guidance for analysis of regulatory policies is provided in Regulatory Program of the US Government, published annually by the US OMB.[340]

3.1.2.1.1 Net Present Value

Net present value (NPV) is a standard criterion to judge a government program on economic principles. NPV is the discounted monetized value of expected benefits less costs. The computation involves monetizing benefits and costs, discounting future benefits and costs using an appropriate discount rate, and subtracting the total discounted costs from the total discounted benefits. Alternatively, a cash flow of benefits minus costs per period can be generated and discounted over the period. Discounting benefits and costs transforms gains and losses occurring in different time periods to a common unit of measurement. Positive NPV usually supports the policy selection.

When monetary values of some benefits or costs cannot be clearly determined, a comprehensive enumeration and quantification (where feasible) of the different benefits and costs, monetized or not, is usually helpful in identifying the full range of program effects, for the instance number of injuries prevented per dollar of cost.

3.1.2.1.2   Cost-Effectiveness Analysis

Cost-effectiveness analysis involves computing life-cycle costs of competing alternatives, all expressed in present value terms for a given amount of benefits. A program is cost-effective if, on the basis of life-cycle cost, it is determined to have the lowest cost. Cost-effectiveness analysis is applicable and useful whenever benefits expressed in monetary terms or physical attributes are equal for all the alternatives; program with the lowest cost to provide the same benefit is preferred. Cost-effectiveness analysis can also be used to compare programs with identical costs but differing benefits. The alternative program with the largest discounted present value of benefits is normally favored.

3.1.2.2   Elements of Benefits and Costs Assessment

It is useful to consider the following guidelines when identifying benefits and costs. Calculation of net present value should be based on incremental benefits and costs. Sunk costs and realized benefits should be ignored. Also, only incremental gains should be recorded as benefits of the policy. Possible interactions between the benefits and costs should be considered. For example, policies affecting agricultural output should not just be subsidized prices. Analyses should focus on benefits and costs accruing to citizens. There are no economic gains from a pure transfer payment. Benefits to those who receive such a transfer are matched by the costs borne by those who pay for it. Therefore, transfers should be excluded from the calculation of net present value.

3.1.2.2.1   Measuring Benefits and Costs

Market prices provide an invaluable starting point for measuring willingness to pay, but prices sometimes do not adequately reflect the true value of a good to society. The economic concept of consumer surplus is the willingness on the part of consumers to pay more than the market price and compares the extra value consumers derive from their consumption with the value measured at market prices. Consumer surplus provides the best measure of the total benefit to society. However, externalities, monopoly power, and taxes or subsidies can distort market prices. When market prices are distorted or unavailable, other methods of valuing benefits may have to be employed. Several ways of measuring value of benefits, in practice today, are discussed in greater detail in the benefits valuation Section 3.1.3.

3.1.2.2.2   Treatment of Inflation

Future inflation is highly uncertain—avoid making an assumption about the general rate of inflation whenever possible. Economic analyses are often most readily accomplished using *real* or *constant-dollar* values—that is, by measuring benefits and costs in units of stable purchasing power.

3.1.2.2.3   Discount Rate Guidance

Computation of net present value requires the discounting of future benefits and costs. This discounting reflects the time value of money. Benefits and costs are worth more if they are experienced sooner. All future benefits and costs, including non-monetized benefits and costs, should be discounted. The higher the discount rate, the lower the present value is of future cash flows. When the benefits and costs are measured in real terms or constant dollars, use a real discount rate that has been adjusted to eliminate the effect of expected inflation. A real discount rate can be approximated by subtracting expected inflation from a nominal interest rate. Market interest rates are nominal interest rates in this sense.

Constant-dollar benefit-cost analyses of proposed investments and regulations should determine net present value and other outcomes by using a real discount rate of 7%. This guidance applies to benefit-cost analyses of public investments and regulatory programs that provide benefits and costs to the general public. This rate approximates the marginal pretax rate of return on an average investment in the private sector in recent years.

3.1.2.2.4   Internal Rate of Return

The internal rate of return of the stream of benefits and costs is by definition the discount rate that sets the net present value of the program or project to zero.

3.1.2.2.5   Shadow Price of Capital

Using the *shadow price of capital* to value benefits and costs is the analytically preferred means of capturing the effects of government projects on resource allocation in the private sector. Shadow pricing is the assignment of a monetary value to an abstract commodity that is not normally quantifiable by a market price.

3.1.2.2.6   Optionality

Optionality is the value of additional optional investment opportunities available only after having made an initial investment. Alternatively, the short-term payoff for this is modest, but the optionality value is enormous.

3.1.2.3   Treatment of Uncertainty

Estimates of benefits and costs, often like many other phenomena, are treated as deterministic; however, typically they are uncertain because of imprecision in both underlying data and modeling assumptions. Characterizing uncertainty involves identifying the key sources of uncertainty, estimating the expected value of the outputs, getting the probability distributions of benefits, costs, and net benefits, application of stochastic methods for insights into the relevant probability distributions, and the sensitivity of results to important sources of uncertainty.

The expected values of the distributions of benefits, costs and net benefits can be obtained by weighting each outcome by its probability of occurrence, and then summing across all potential outcomes. Estimates other than expected values (such as worst-case estimates) may be provided in addition to expected values, but the rationale for such estimates must be clearly stated.

Sensitivity analysis involves varying major assumptions and recomputing net present value and other outcomes to determine how sensitive outcomes are to changes in the assumptions. The dominant benefit and cost elements and the areas of greatest uncertainty of the program must be considered. Variations in the discount rate are not an appropriate method of adjusting net present value for the project specific risks. Use of *certainty-equivalents* involves adjusting uncertain expected values to account for risk. It is the generated return from a risk-free asset that is equally desirable as the return from a risky asset.

3.1.2.4   Distributional Effects

Full compensation of the losers by the gainers is a basic premise of maximizing net present value of benefits. The analysis should indicate if this is the case or not, and if so, should identify relevant gainers and losers from policy decisions. When benefits and costs have significant distributional effects, these effects should be considered along with the analysis of net present value. Distributional effects are generally not relevant to cost-effectiveness analysis where the scope of government activity does not change.

Individuals or households are the ultimate recipients of income; business enterprises are merely intermediaries. Analyses of distribution should identify economic incidence, or how costs and benefits are ultimately borne by households or individuals. Individuals or households could be grouped by income strata, age, industry, or occupation. Where a policy is intended to benefit a specified subgroup of the population, such as the poor, the analysis should consider how effective the policy is in reaching its targeted group. Determining economic incidence can be difficult because benefits and costs are often redistributed in unintended and unexpected ways. For example, a subsidy for the production of a commodity will usually raise the incomes of the commodity's suppliers, but it can also benefit consumers of the commodity through lower prices.

3.1.3   Benefits Valuation

Benefit valuation is often one of the most controversial areas for debate in the public policy arena. Both activists supporting chosen causes and business lobbyists (or advocates as they would prefer to be called) come from divergent perspectives. Economic measure of a benefit of any good or service, is the sum of what all members of a society are willing to pay for it. The economic concept of consumer surplus is the willingness on the part of consumers to pay more than the market price and compares the extra value consumers derive from their consumption with the value measured at market prices. Consumer surplus provides a better measure of the total benefit to society.

Market prices provide an invaluable starting point for measuring willingness to pay, but prices sometimes do not adequately reflect the true value of a good to society. Public goods, unlike private goods, are a source of market failure because of free riders. Free riders have little incentive to voluntarily pay for public goods when they can enjoy the benefits of public goods provided by others. Externalities, monopoly power, and taxes or subsidies can distort market prices. When market prices are distorted or unavailable, other methods of valuing benefits may have to be employed.

The following subsections discuss the various economic, environmental, or social benefit or cost valuation techniques.

3.1.3.1 Economic Benefit Valuation

Economic valuation is anthropocentric; that is, the good or service must be of value to the human being. But not all economic benefits have tangible monetary metrics. For instance, brand reputation, customer satisfaction, or intellectual capital have a significant impact on economic performance but do not have universal monetary measures. According to Ocean Tomo,[341] today intangible assets represent more than 80% of the market value of S&P 500 companies. Valuation of intangible assets becomes a very useful, and probably an inevitable, exercise when investment in sustainability initiatives needs irrefutable evidence for justification. In the context of the fourth bottom line, ethics (now a component of the governance metric in the G4 version of the Global Reporting Initiative (GRI) framework), one may recall the famous words of Warren Buffett, "It takes 20 years to build a reputation and five minutes to ruin it,"[342] and only then do you discover you had a reputation to begin with.

3.1.3.1.1 Intangibles

However hard, these intangibles require economic valuation. Per Chartered Global Management Accountant (CGMA),[343] a joint-venture of American Institute of Certified Public Accountants (AICPA) and Chartered Institute of Management Accountants (CIMA), there are three common approaches to value intangibles. Market approach involves findable market-based transactions of identical or substantially similar intangible assets recently exchanged. Income-based models are best used when the intangible asset is income producing or helps generate cash flow. It converts future benefits (such as cash flows or earnings for a single period or a future stream) to a single discounted amount and then capitalizes on it. But distinguishing the cash flows that are uniquely related to the intangible asset from the cash flows related to the whole company is often very hard. The cost-based analyses are based on the economic principle of substitution; historical cost reflects only the actual cost incurred in development of that intangible asset. *Reproduction cost new* and *replacement cost new* imply the current cost of an identical new property, and of a similar new property, respectively.

CGMA recommends[344] the following economic valuation approach for intangibles (Figure 3.1).

3.1.3.1.2 Commodities

Value gains or losses of small quantities of commodities, such as oil, is measurable by market price. The marginal cost of goods traded and the marginal value to the consumer are both reflected in competitive market prices. For larger changes, supply and demand function for each good is used. The relationship between the quantity

Asset	Primary	Secondary	Tertiary
Patents	Income	Market	Cost
Technology	Income	Market	Cost
Copyrights	Income	Market	Cost
Assembled workforce	Cost	Income	Market
Internally developed software	Cost	Market	Income
Brand names	Income	Market	Cost
Customer relations	Income	Cost	Market

FIGURE 3.1   Economic valuation approach for intangibles.[345]

of the good desired and the price the consumer is willing to pay is the demand function. Similarly, the supply function is the relationship between the quantity and the price of the goods producers provide. The net consumer and producer surplus is the difference between consumer's willingness to pay for a unit of the good or service and society's cost to make it.

3.1.3.1.3   Businesses

Warren Buffett[346] says, "It's far better to buy a wonderful company at a fair price than a fair company at a wonderful price." Common methods for valuation of a business are asset, earning-value, or market-value based. The asset-based approach takes stock of the total assets and liabilities of the business. Going concern, one of the two asset-based approaches, uses the total balance sheet value of its assets and subtracts the value of its liabilities. Liquidation asset-based approach determines the net cash if all assets were sold and all liabilities paid off. The earning-value approach is based on the ability of the business to produce wealth in the future. The capitalizing past-earning approach determines cash flow level using a company's past earnings, normalizing them for unusual expenses or earnings, and then multiplying the normalized cash flow by a capitalization factor. Capitalization factor reflects purchaser's expected rate of return and risk tolerance. On the other hand, the capitalizing future-earning approach uses predicted future earnings instead of past earnings; and the divisor once again, is the capitalization factor. Market-value approach establishes value by comparing it to similar businesses recently sold. This is similar to the way houses are appraised using *comparable*, which is essentially recently sold similar units in the same or close-by neighborhood, adjusting for minor differences.

3.1.3.2   Environmental Benefit Valuation[347]

The value of a public good such as ecosystem is determined by what people are willing to pay or sacrifice for it. Many environmental resources are characterized as public goods and not marketed as a commodity. This makes determining how society values environmental goods such as clean air, very difficult. The use-value component of public goods or services involves a noticeable interaction between the human and the public good such as ecosystem, which may be a consumptive use such as leisure fishing or a nonconsumptive use such as sailing. Non-use value of the public good includes the knowledge of continued existence of pristine areas or near-extinct rare biota and the option to enjoy visiting or seeing them.

The benefits could be direct, indirect, or both. Less damage to vegetation by lowering the sulfur dioxide levels in the atmosphere is a direct benefit; leading people to greater mobility toward cleaner regions is an indirect one. Also, the value of the benefit could either be user driven, such as the ability to swim in a clean lake, or a non-use benefit, such as the simple existence of pristine beauty of a natural waterfall in its grandeur, preserved for the next generation to enjoy. The benefits to society lies in the avoided reduction in value because of physical damage. This could either be estimated scientifically, using an *environmental burden approach* to measure[348], or by using a contingent valuation method based on consumer responses to behavioral surveys, which seek out individuals' willingness to pay.[349]

3.1.3.2.1 Revealed Preference

Many environmental goods are not traded, and hence their market prices are not known. Revealed preference theory, pioneered by American economist Paul Samuelson, is a method of analyzing choices made by individuals, by assuming that the preferences of consumers can be revealed by their purchasing habits. Individual's preferences may be revealed through their buying choices of related products. For instance, damage, resulting from the rise in sea levels, can be measured using the market value of the inundated land plus the cost of building protective sea walls. The change in purchasing habits of individuals—the fall in price—reveals a measure of the impact of sea level rise. Similarly, a shift in the demand or supply of a market good could be measured to obtain the value of an environmental aspect. For instance, the impact of climate change on energy could be assessed through the measurement of the shift in energy resource demand function; or the value of superior environmental performance of a product could be evaluated by the premium consumers are willing to pay for equivalent functionality.

3.1.3.2.1.1 Travel Cost Model Suggested by Harold Hotelling in 1947, the travel cost model exploits the empirical relationship between travel cost and visitation rates to a site as an implicit price. Recognizing that the demand is a function of the price of substitutes, the travel cost model expanded to simultaneously estimate demand for multiple sites. Travel cost models are applied widely to estimate recreational demand. Environmental quality characteristics of multiple sites such as air or water quality could be valued by comparing two sites that differ in only one characteristic. Discrete choice method is useful to model visitor choice as a function of site characteristics.

3.1.3.2.1.2 Hedonic Price Model The basic premise of this model is that the price of a good is the sum of the implicit prices of each of its characteristics. For instance, the price of a home is dependent on its physical traits, neighborhood and environmental quality of the location. Home sales price and characteristics data are used to estimate marginal implicit prices of individual characteristics, such as air quality, or proximity to airport noise or hazardous waste site.

Another variation is where wage data provides a collection of characteristics such as education, training, prestige, benefits, and accidental death or injury. This reveals how much extra wage is sought to bear the additional small risk of accidental death. The compensation given up to avert the additional small risk has been used to value a statistical life. However, the perceived value of the risk of accidental death by workers may not be accurate.

3.1.3.2.1.3   Averting Behavior Model   A willingness to incur private expenditure to avert damages from air pollution and other similar attributes provide at least a partial economic estimate of the value of such damages.

3.1.3.2.2   Stated Preference or Contingent Valuation

Stated preference or contingent valuation[350] (CV) is a survey-based economic technique for the valuation of nonmarket resources, such as environmental preservation or the impact of contamination. Typically, the survey asks how much money people would be willingness to pay (WTP) or willingness to accept (WTA) to maintain the existence of (or be compensated for the loss of) an environmental attribute. The survey creates a hypothetical market, and the objective is to capture behavior similar to that in a market. The survey typically provides a description of the amenity or attribute being valued, asks respondents to provide to place a value (both WTA and WTP) on the amenity, and finally provide data on their socioeconomic attributes.

The method is particularly useful for estimating non-consumptive values (existence/option) that cannot be evaluated by observed consumption. However, respondents lack market experience in environmental goods and hence not calibrated to respond to this aspect of questions, and data becomes skewed by adverse selection of protest responses from activists. Also, response to WTA is commonly multiple times larger than those to WTP, especially true for nonuse values. Improvements include replacement of open-ended questions with closed-ended discrete questions, and focus on WTP questions, avoiding the willingness to accept (WTA) questions.

3.1.3.2.3   Benefits Transfer

The benefits transfer method is used to estimate economic values for ecosystem services by transferring available information from other similar completed studies in another context. For instance, the benefits of enhancing water quality for recreational use could be based on similar studies conducted in the region, which in turn may be based on contingent valuation or travel cost methods, to avoid reinventing the wheel.

3.1.3.3   Social Benefit Valuation

Any social investor is concerned with longevity and purpose and seeks valuation to answers to two fundamental questions: financial health and impact on stakeholders, including society. Implicit in the impact on stakeholders is the potential for scale-up. While current financial accounting techniques could deliver a reliable assessment of financial health, it cannot provide a consistent measure of social (stakeholder) impact because its normative tools are still in a formative stage.

It is worth noting that the subject of economics was for a long time seen simply as a branch of ethics.[351] While a normative definition of morality is a set of absolute values, the descriptive definition refers to the values that a group of people at a given point of time thinks are right. Accountants can more easily act consistent with applied ethics principles, such as do not over claim, be transparent, and verify the result. Social impact measurement has to address normative questions, such as what should the outcomes be? How should the impacts be aggregated across society? Is social value created? Are we valuing the things that matter? Are we including only what is material?

Social Value UK's[352] Guide to Social Return on Investment (SROI) provides a framework for anyone interested in measuring, managing, and accounting for social value or social impact, but the methodology is still being developed. In response to this framework, Fujiwara[353] suggests use of a more emphatic definition for social impact measurement: "A method is a social impact measurement method if and only if it assesses whether the impact of an intervention or action is in the interest of (the individuals that make up) society," and not just on the chosen social indicators.

What is the appropriate social indicator that provides the right balance between personal responsibility and use of social subsidies? This is a normative sociopolitical question. However, normative economics are value-laden. There are two issues involved: ethical bias from dysfunctional groupthink in the selection of the indicator and the proverbial agency problem of *managing by metrics*. Also, measuring well-being through people's desires and preferences faces its own challenges—not everyone is adept at making rational, fully informed decisions; people are not rational automatons, as some economists would have you believe. Finally, the Rank sum rule (where the smallest rank sum is deemed the best), as opposed to equivalent welfare change, in which every individual's well-being has equal weighting in society, is biased against the poor.

There are two definitions of value for nonmarket outcomes:[354]

1. Compensating surplus (CS) is the amount of money, paid or received, which will leave the individual (whole, as in no reduction in well-being) in her initial welfare position following a change in the outcome.
2. Equivalent surplus (ES) is the amount of money, to be paid or received, which will leave the individual (whole, as in no reduction in well-being) in her subsequent welfare position in absence of a change in the outcome.

The relationship between CS, ES, and the more commonly used measurements in contingent valuation of nonmarket priced attributes, WTP and WTA, is presented in the following table, shows gains and losses in well-being in terms of WTP and WTA. Because of risk aversion, typically most surveys find WTA to be much higher than WTP.

	Compensating Surplus	**Equivalent Surplus**
Welfare gain	WTP for the positive change	WTA to forego the positive change
Welfare loss	WTA the negative change	WTP to avoid the negative change

In a recent presentation, Fujiwara suggests that social intervention must be judged only in terms of outcomes and ultimately measured as well-being of human and other species. "Value of a good/service/output equals the amount of money that induces the equivalent change in wellbeing for the individual."[355] For instance, in a contingent valuation survey, data shows that some specific drop in health reduces well-being (measured as life satisfaction) by the same amount as say certain drop in income, then there is an equivalency and monetary measure for good health.

3.1.4 IMPACT ASSESSMENT

3.1.4.1 Products and Services

Commonly used today, the life-cycle assessment (LCA)[356,357] allows one to make informed decisions to minimize environmental or social burden. The steps involved in an LCA are (1) defining objective and scope, (2) analyzing inventory, (3) assessing impact, and (4) interpreting the findings. A LCA measures product or service related social and environmental impacts such as, greenhouse gases, water pollution, ozone depletion and habitat destruction. The findings help select the least socially or least ecologically burdensome of the analyzed products or services. Several software packages are available for conducting life-cycle assessment; but the results are only as good as the data that goes in to the analysis. More information on the basics of life-cycle analysis and on conducting or managing a LCA is available on the US Environmental Protection Agency's (EPA) website.[358]

3.1.4.1.1 Environmental Life-Cycle Analysis

In 1969, the Coca-Cola Company conducted the first known environmental LCA of beverage containers to identify which one had the lowest environmental impact. Environmental life-cycle assessment (E-LCA)[359] is a technique that addresses the environmental aspects and potential impacts of a product or service throughout their life cycle. The procedure considers all internal and external links in the life-cycle chain of a product or service throughout its entire life cycle, from acquiring raw material to manufacturing, use, reuse, and final disposal (e.g., cradle-to-grave). Environmental life-cycle analysis procedures are also part of the International Organization for Standardization (ISO) 14000[360] environmental standards.

Material flow analysis (MFA)[361] or substance flow analysis (SFA), a related tool to LCA, "is a systematic assessment of the flows and stocks of materials within a system defined in space and time."[362] MFA could be seen as a method to establish an inventory for an LCA, and LCA can be an impact assessment of MFA results. MFA methods analyze both the flow of a material through an industry or firm and the effect on ecosystems. The scope of materials flow analysis can be a national scale, a regional scale, a corporate or industrial scale or the life cycle of a product. Other notable applications of MFA in the sustainability area include, balancing industrial input and output to natural ecosystem capacity, dematerializing industrial output, and developing more efficient patterns of energy use. MFA type 1 deals with impacts per unit flow of materials and is applicable for evaluation of product environmental impact. MFA type 2 deals with throughput of sectors and is applicable for providing firm environmental performance data and for derivation of sustainability indicators.

The environmental input-output life-cycle assessment (EIO-LCA)[363] method estimates the materials and energy resources required for, and the environmental emissions resulting from, activities in the economy. The EIO-LCA method was theorized and developed by economist Wassily Leontief in the 1970s based on his earlier input-output work from the 1930s, for which he received the Nobel Prize in Economics. EIO-LCA combines input-output analysis (an economic concept) with the life-cycle analysis (a non-monetary concept) to provide monetized information for evaluation.

3.1.4.1.2   Social Life-Cycle Analysis

A social and socioeconomic life-cycle assessment (S-LCA)[364] is a technique that assesses the social and socioeconomic aspects of products and their potential positive and negative impacts along their life cycle. The procedure considers all internal and external links in the entire life-cycle chain of a product or service, from acquiring raw material to manufacturing, supply chain, use, reuse, and final disposal (cradle to-grave). They may be linked to the behaviors of enterprises, to socioeconomic processes, or to impacts on social capital. S-LCA provides information on social and socioeconomic aspects for decision making, to improve performance of organizations and ultimately the well-being of stakeholders. S-LCA does not provide information on the question of whether a product should be produced or not, nor does it provide a breakthrough solution for sustainable consumption and sustainable living.

Steps involved in an S-LCA are definition of goal and scope, performing a life-cycle inventory, followed by a life-cycle impact assessment, and finally interpreting the S-LCA findings. Questions raised in the setting goals and scope phase of the planned S-LCA are: Why is an S-LCA being conducted? What is the intended use? Who will use the results? What do we want to assess? The objective, for instance, may be learning about and identifying social *hotspots* and about the options for reducing the potential negative impacts and risks. Process chains, which detail the series of operations performed in the making, treatment, use, and disposal of a product, in existing E-LCA models, provide a valuable starting point for new S-LCAs.

ISO 14044[365] states, "The scope shall clearly specify (1) the function, which is the utility, (2) the role that the product plays for its consumers, (3) how to build and model the product system, and (4) how to identify locations and specific stakeholders involved." The definition of the function needs to consider both the technical utility of the product and the product's social utility, which can be described as "a range of social aspects such as time requirement, convenience, prestige, etc."[366]

Functionality may include the main function of the product, technical quality, such as stability and durability, customer support for use and disposal, aesthetics, image, total ownership costs, and environmental and social properties. System boundaries refer to the unit processes that should be included in the system being assessed. The life-cycle inventory phase involves (1) collection of product or service data, (2) collection of data needed for impact assessment, (3) validation of data, development of data relationship diagrams, refining the system boundary, and (4) aggregation of data.

The purpose of life-cycle impact assessment is to provide (1) aggregate inventory data within subcategories and categories and (2) information such as internationally accepted levels of minimum performance. There are two approaches for developing the social and socioeconomic subcategories: top-down approach, which consists of identifying broad social and socioeconomic issues of interest, and a bottom-up approach attempts to provide summaries of inventory information at the organization and process level, which is shared with the stakeholders and their input on what would be relevant summary indicators from their perspective, is captured. The life-cycle interpretation phase comprises (1) significant issues identification, (2) study evaluation for completeness and consistency, (3) stakeholder engagement level, and (4) conclusions and recommendations.

3.1.4.2 Major and Sensitive Projects

Projects proposed by governments, major project funders (investors and lenders), and those impacting a sensitive ecosystem or sociocultural demography are typically subject to scrutiny by public, media, and activists. Global treaties and the potential liability of legal action by the impacted entities are additional drivers. For instance, World Bank faced some major lawsuits prior to the new phase of imposed mandatory environmental and social impact assessments, stakeholder engagements, and defined statements. This pioneering stewardship by World Bank and International Finance Corporation (IFC) has brought into the fold, through UN Principles of Responsible Investment (PRI) and Organization for Economic Cooperation and Development (OECD) Multinational Enterprises (MNE) Guidelines, requirements for due diligence and similar environmental and social impact assessments from all major development and growth efforts. Today, World Bank/IFC supports the capacity development of the banking and financial sector to manage environmental and social risks. Furthermore, nearly a half century ago, the United States mandated the National Environmental Policy Act (NEPA), requiring detailed environmental and social impact evaluation of all significant federal projects.

The following section focuses on World Bank/IFC and US NEPA environmental and social risk and impact assessment, and the principles are applicable much more broadly. Governance, the fourth bottom line has also been addressed within the context of an environmental and social impact assessment (ESIA). For instance, encouraging transparency and disclosure to avoid corruption and exploitation by governments as well as the private investor.

3.1.4.2.1 Environmental and Social Impact Assessment

The sustainability framework of IFC,[367] the financing arm of the World Bank and one of the largest global project lenders for sustainable development, is comprised of:

1. The Policy on Environmental and Social Sustainability defines IFC's commitments to environmental and social sustainability.
2. Performance Standards define clients' responsibilities for managing their environmental and social risks. The desired outcomes are described in the objectives of each Performance Standard. Performance Standards also provide a solid baseline from which to improve sustainability. Managing environmental and social risks and impacts consistent with the Performance Standards is the responsibility of the client and is an important factor in the approval process and the conditions of financing. IFC ensures client's responsible performance through its overall due diligence of financial, environmental, social, and reputational risks.

   The Performance Standards consist of the following:
   a. *Performance Standard 1*: Assessment and Management of Environmental and Social Risks and Impacts. IFC requires investment clients to have a sustainability policy as part of clients' overall environmental and social management system.

b. *Performance Standard 2*: Labor and Working Conditions
c. *Performance Standard 3*: Resource Efficiency and Pollution Prevention
d. *Performance Standard 4*: Community Health, Safety, and Security
e. *Performance Standard 5*: Land Acquisition and Involuntary Resettlement
f. *Performance Standard 6*: Biodiversity Conservation and Sustainable Management of Living Natural Resources
g. *Performance Standard 7*: Free, Prior, and Informed Consent of Indigenous Peoples. IFC will undertake an in-depth review of the process conducted by the client.
h. *Performance Standard 8*: Cultural Heritage

3. The Access to Information Policy articulates IFC's commitment to transparency, good governance on its operations, and outlines the Corporation's institutional disclosure obligations regarding its investment and advisory services.
4. Environmental and Social Categorization uses a process to reflect the magnitude of risks and impacts including those inherent to a particular sector. These categories are:
   a. *Category A*: Business activities with potential significant adverse environmental or social risks and impacts that are diverse, irreversible, or unprecedented.
   b. *Category B*: Business activities with potential limited adverse environmental or social risks and impacts that are few in number, generally site-specific, largely reversible, and readily addressed through mitigation measures.
   c. *Category C*: Business activities with minimal or no adverse environmental or social risks or impacts.
   d. *Category FI*: Business activities involving investments in financial intermediations or through delivery mechanisms involving financial intermediation. The role of financial intermediaries is to channel funds from lenders to borrowers by intermediating. IFC supervises FI compliance with IFC's Environmental and Social Review Procedures.
5. Key components of environmental and social due diligence include:
   a. Reviewing all environmental and social risks and impacts information
   b. Conducting site inspections and relevant stakeholder interviews
   c. Analyzing the compliance with the environmental and social Performance Standards and provisions of the World Bank Group Environmental, Health, and Safety guidelines
   d. Identifying gaps and corresponding additional measures and actions and incorporating them as necessary conditions of IFC's investment.

The techniques to assess environmental, social, and governance risks are similar to other risks for projects. However, the scale, the geopolitical nuances of the location and cultural issues—languages, lack of legal sophistication, cultural practices, lack of baseline environmental and social data, and economic, educational, and technical backwardness—could make them very challenging. For instance, the air dispersion modeling across land, oceans, and borders, thermal plume modeling for water streams with temperature sensitive coasts such as coral reefs, biodiversity evaluation, and monitoring are all rather specialized techniques and rare skills. IFC recognizes the importance of assessment of governance risks to expected benefits where a business activity can have potentially broad implications for the public at large and of the effectiveness of disclosure of information as a means to manage such risks. For instance, select sectors have additional governance and disclosure requirements because of their high-risk significance.

1. IFC, for the Extractive Industry (oil, gas, and mining), encourages governments and corporations to make public disclosure of contracts and their material project payments to the host government (such as royalties, taxes, and profit sharing) for promote transparency of revenue payments to minimize corruption and misuse. Such disclosure shall be made on a project basis or on a corporate basis, depending on what is most appropriate given country taxation laws and corporate arrangements.

2. When IFC invests in infrastructure projects (final retail delivery of essential services, such as the retail distribution of water, electricity, piped gas, and telecommunications) for the general public under monopoly conditions, it encourages public disclosure of information relating to household tariffs and tariff adjustment mechanisms, establishment of service standards to prevent exploitation of uninformed public, and policy to minimize corruption in the privatization process.

IFC also recognizes several key sustainability issues, such as potential impediment of economic and social well-being because of climate change impacts, with low-carbon economic development being only one dimension of a balanced approach, and the significance of business and state responsibility to respect, protect, and fulfill human rights being another. These are incorporated in their due diligence efforts. Finally, for good governance and ethics or integrity, the fourth and very significant bottom-line component of sustainability, IFC has set up compliance advisor/ombudsman (CAO), independent of IFC management and reporting directly to the president of the World Bank Group. The CAO is accountable for addressing concerns and complaints of affected communities in a manner that is fair, objective, and constructive.

3.1.4.2.2 US National Environmental Policy Act

The NEPA[368,369] was signed into law on January 1, 1970. The NEPA requires federal agencies to assess the environmental effects of their proposed actions prior to making decisions. The NEPA process begins when a federal agency develops a proposal to take a major federal action. The environmental review involves three different levels of analysis:

1. The first level provides a categorical exclusion from a detailed environmental analysis if the federal action, based on criteria, is deemed, *individually or cumulatively, to not have a significant effect on the human environment.*

2. The second level calls for using a less intense environmental impact assessment and the equivalent of a simple cost effectiveness analysis by comparing alternative use of resources. The less intense environmental assessment (EA) determines whether a federal action has the potential to cause significant environmental effects. EA comprises of a brief discussion of the need for the proposal, alternative ways to accomplish the project outcomes, and an assessment of the environmental impacts of the proposed action and alternatives. Findings of no significant environmental impact clear the pathway for the project.

3. At the third and highest level, the project encourages federal agencies to prepare an environmental impact statement (EIS) if a proposed major federal action is determined to significantly affect the quality of the human environment. The regulatory requirements for an EIS are more detailed and rigorous than the requirements for an EA. EIS process requires a Notice of Intent—so the public can get involved in defining the range of issues and possible alternatives to be addressed in the EIS. After public comments on the draft EIS, and further analysis if needed, a final EIS is published, which provides responses to substantive comments. The EIS process ends with the issuance of the Record of Decision (ROD).

An EIS, besides administrative details, includes the following:

a. Statement of purpose, need, and the expectation
b. Reasonable alternatives that can accomplish the purpose.
c. Affected environment by the alternatives under consideration.
d. Direct and indirect environmental effects and their significance.

The techniques behind the EIS are very similar to the environmental impact assessment required under International Finance Corporation, the financing arm of World Bank.

3.1.4.2.3   Sustainability Impact Assessment

A sustainability impact assessment could be seen as an integrated assessment of the quadruple bottom lines. It examines economic, environmental, social, and ethical impacts in equal measure. The integrated method identifies synergies and trade-offs across the four domains. Aligned with the requirements of ethical stakeholder engagement similar to what is called for in IFC's EIA and sustainability impact assessment (SIA), the approach must respect open and transparent processes. However, one key additional factor that makes sustainability impact assessments unique, is the long timeframe. Hence, the SIA[370] must examine long-term flows, investments, and effects. As of this date, there is no well-established and broadly accepted framework, format, and process for sustainability impact assessment.

3.1.5   RISK AND OPPORTUNITY MANAGEMENT

The global financial crisis, as well as the recent environmental disasters, such as the US EPA Anima River Toxic Spill, BP Gulf oil spill, and the devastating storms, Sandy, Katrina, and Allison, commonly attributed to climate change, have highlighted the importance of risk management. All organizations, private, multinational, and state-owned corporations, as well as regions and countries, have a need to understand the risks in their operations and how to best mitigate them. It is also critical to recognize and prioritize significant risks and develop suitable informed mitigation strategies. An enterprise-wide comprehensive risk-management (ERM) approach that addresses all stakeholder impacts and all organizational processes would result in mitigating the downside risks and gain from the potential upside opportunity. Although risk and its sibling opportunity are central to business and investment strategies, traditional ERM does not always reveal the opportunities. Risk is pervasive through every step of an organization's value creation chain. Because of its significant impact on performance, risk must be integrated into the culture of the organization, just like ethics has to be to prevent financial, social or environmental tsunamis. A study by Ernst and Young (EY)[371] found that companies with mature risk-management practices (top 20% in risk maturity) outperformed their competitors (bottom 20% in risk maturity) financially, with earnings that were three times higher. "Corporations and leaders … have to endeavor to build new competencies in managing transparency, accountability, stakeholder engagement, ethics culture, and social innovation, which are critical for business success in the Next Economy."[372]

The World Business Council for Sustainable Development's (WBCSD)[373] comparison of material sustainability disclosures of 170-member company sustainability reports and mainstream corporate reports revealed that 71% of sustainability issues that businesses deemed to be material were not disclosed to investors as risk factors. Mainstream and sustainability-risk disclosure alignment was none for 35% of the companies, while 57% had some, and only 8% of companies had full alignment.

The WBCSD[374] believes the barriers to effective management of sustainability risks comprise of both internal organizational forces and innate features of sustainability risks. Sustainability risks are not well understood, often more challenging to quantify, and have a longer timeline. The lack of alignment between sustainability and enterprise risk-management functions causes further breakdown. In large businesses, the role of corporate governance has expanded to include risk disclosure. But the primary objective of corporate legal risk disclosures is to inform investor decision making about the possible material issues that could impact performance. They use different terminology.

Select examples of risk-management papers in the sustainability area by the author Ramanan:

1. Carbon risk management[375]
2. Corporate risk-management strategies[376]
3. Risk sharing for remediation liability management[377]
4. Multidimensional value optimization framework[378]

3.1.5.1 Risk-Management Basics

ISO Guide 73[379] defines risk as "the effect of uncertainty on objectives"; the effect may be negative, positive, or a deviation from the expected. By this definition, risk is linked to objectives. Consequences of a risk materializing may be hazards or opportunities, or there could be increased uncertainty or a deviation from expectations. Risk is a function of likelihood, and consequence and can be mitigated either by reducing the probability (or frequency) of risk or by reducing the potential impact or shifting the burden of consequence. An effective risk-management initiative may affect the probability of likelihood, the severity of consequence, or both.

3.1.5.1.1 Categories of Risks

Risks may be short-term operational risks related to routine activities, medium-term tactical risks associated with projects or mergers and acquisitions, and long-term risks linked to strategy.

1. One way to categorize[380] risks that impact a company is by controllability:
   a. Preventable risks that originate from within the organization are controllable and could, and should, be eliminated or avoided through active administrative or passive technical controls. Privacy of 87 million people were lost because of inadequate technical and governance protection on the Facebook platform. A weak governance could lead to unethical behavior. It is a classic example of preventable risk. Actions such as a well-written code of ethics, development of ethics culture through training, and executive modeling of ethical behavior could significantly fortify governance.
   b. Strategic risks are neither preventable nor are they inherently undesirable; these risks are faced by corporate and society stewards in every endeavor. Here the risk comes from the probability of a positive or negative outcome from a chosen action. A classic situation arises from having to respond to disruptive technology, such as *cloud storage*, Tesla autonomous driving or business model such as Amazon drone delivery, Uber ridesharing. Some classic examples are whether to take the first entry into the market with the current product or choose to make a late entry with an improved product; make your technology transparent for development of multiple applications and gaining from a broader market share or exercise stringent proprietary code controls to gain from high premium and high per unit profit but limited market segment. Other common examples include whether and where to drill, or whether and if so, what entity or adjacency to acquire.
   c. External risks that arise from events outside the company are beyond its influence. Consequences could be catastrophic. Examples include natural disasters, geopolitical upheavals, or financial tsunamis caused by fraudulent failures. These are not preventable, but they should be anticipated and identified, and contingency plans should be put in place to mitigate potential consequences. Adaptation strategies by governments in response to climate change disasters are indeed a model example.

2. Another way to classify risks is to identify the drivers as external or internal and apply the Financial Infrastructure Reputational Marketplace (FIRM) scorecard.

  a. *Externally driven risk components include*: Financial—interest and foreign exchange rates, funds and credit ratings, accounting standards; infrastructure—geopolitics (such as terrorism), supply chain, communication, transport, natural disasters; reputational—corporate social responsibility, regulatory violation and enforcement, product liability and lawsuits, competitor adversarial behavior; and marketplace issues — competition, price wars, employee poaching, disruptive technology or business model, regulations, economy, and customer demand.

  b. *Internal risk drivers comprise of*: Financial—liquidity, cash flow, internal control, and fraud; infrastructure—premises, information technology systems, health and safety, workforce skills; reputational—brand, leadership, work culture; marketplace—intellectual property, contracts, confidentiality, and mergers and acquisition.

3.1.5.1.2   The Seven Rs and the Four Ts

The seven Rs and four Ts of the risk-management process are as follows:

1. Recognition or identification of risks includes capturing relevant information about the risk in a standard format.
2. Risk assessment is the process to evaluate, quantify, and prioritize enterprise risks.
3. Responding to significant risks, commonly known as the four Ts (ISO 31000 uses the phrase risk treatment to include the four Ts), which are:
   a. Tolerate
   b. Treat
   c. Transfer
   d. Terminate
4. Resourcing controls
5. Reaction planning
6. Reporting and monitoring risk performance
7. Reviewing the risk-management framework

3.1.5.2   Risk-Management Frameworks

Aligning Risk with Strategy and Performance (formerly COSO ERM 2004) and International Standards Organization ISO 31000 are the two dominant risk-management frameworks used globally. ISO 73[381] provides guidance on the vocabulary.

3.1.5.2.1   Aligning Risk with Strategy and Performance

Aligning Risk with Strategy and Performance[382] is the new title and current update to the Committee of Sponsoring Organizations of the Treadway Commission

Enterprise Risk-Management standard (COSO ERM 2004), which is linked to the Sarbanes-Oxley requirements. As outlined by COSO,[383] the 2004 framework provides eight components for use when evaluating ERM:

1. *Internal environment*: The internal environment sets the basis for how people in an entity view and deal with risk, including integrity, ethical values, operating environment, risk appetite, and philosophy.
2. *Objective-setting*: ERM ensures that management has a process in place to set objectives. The selected objective must be aligned with the mission of the entity and be consistent with its risk appetite.
3. *Event identification*: Given the objectives, potential events are identified that may affect the achievement of the goal. Internal and external events affecting the achievement are identified, distinguishing between risks and opportunities.
4. *Risk assessment*: Risks are analyzed, considering probability and consequence, as a basis to determine how to manage them. Risks are assessed on both inherent and residual basis.
5. *Risk response*: Management selects risk responses—avoiding, accepting, reducing, shifting, or sharing risk. It develops a set of actions to align risks with the entity's risk appetite and risk tolerances.
6. *Control activities*: Policies and procedures are established and implemented to effectively carry out the risk responses.
7. *Information and communication*: Relevant information is identified, recorded, and disseminated in a timely and clear manner to enable people execute their responsibilities. Effective communication occurs across and up the entity.
8. *Monitoring*: The entire ERM process is monitored, and modified as needed. Monitoring is done through an ongoing management procedure, separate evaluations by a central team or both.

The new 2014 framework, according to COSO[384]:

1. Provides greater insights into strategy and ERMs in setting and executing strategy
2. Enhances alignment between organizational performance and ERM
3. Accommodates expectations for governance and oversight
4. Recognizes the continued globalization of markets and operations and the need to apply a common, albeit tailored, approach across geographies
5. Presents fresh ways to view risk in the context of greater business complexity
6. Expands risk reporting to address expectations for greater stakeholder transparency
7. Accommodates evolving technologies and the growth of data analytics in supporting decision making

3.1.5.2.2   ISO 31000 Standard "Risk Management—Principles and Guidelines"

The International Organization for Standardization (ISO) 31000 standard "Risk Management—Principles and Guidelines," established in 2009, is an international standard that provides a structured approach to implementing enterprise risk management.[385] The reference to enterprises covers both the private and public sectors. AIRMIC (The Association of Insurance and Risk Managers), et al.[386] have developed a valuable guideline that is compatible with both the dominant risk-management frameworks used globally; COSO ERM 2004 and ISO 31000. Some highlights of the ERM process based on this guideline focuses more on ISO 31000, and other practices are presented in greater detail in the following subsections.

3.1.5.3   Risk Assessment

Risk assessment is the identification of risk, usually for evaluation or ranking. Typically, the risk identification information comprises of a title, scope, nature, stakeholders, likelihood and consequences, prior loss experience, if any, and risk owner. Risk assessment techniques span a range starting with structured questionnaires and checklists, workshops and brainstorming, inspections and audits, and qualitative process hazard analysis (PHA) to highly technical methods. The advanced techniques include flow charts and dependency analysis; hazard and operability studies (HAZOP); failure modes effects analysis (FMEA); strengths, weaknesses, opportunities threats (SWOT) analysis; political, economic, social, technological, legal, environmental (PESTLE); and social, technological, economic, environmental, political, legal, and ethical (STEEPLE) analysis.

A simplified method of risk assessment that falls between a qualitative process hazard analysis and a traditional quantitative risk analysis is layer of protection analysis (LOPA).[387] It is a semiquantitative risk screening method that is applied following a qualitative hazard identification tool such as HAZOP. LOPA uses simplifying rules to evaluate initiating event frequency, independent layers of protection, and consequences to provide an order-of-magnitude estimate of risk. It conservatively estimates failure probability, usually to an order of magnitude level of accuracy. The result is by designing an overestimate of the risk. More rigorous quantitative techniques, such as fault tree analysis or quantitative risk analysis, may be required for narrower ranges of risk.

LOPA focuses on one scenario at a time. Each scenario should have only one cause and one consequence. While one or more initiating events may lead to the consequence; each cause-consequence pair is considered a separate scenario. LOPA begins with an event with a potential environmental, health, safety, business, or economic impact. The frequency of the initiating event and the impact or the consequence severity are arrived at using published information, including consequence look up tables, expert opinions, or sophisticated consequence software tools, such as air or thermal plume dispersion models.

Most processes have inherent built-in passive or active safeguards. Each identified safeguard is evaluated for whether it is an independent layer of protection (IPL). "An IPL is a device, system, or action that is capable of preventing a scenario from proceeding to its undesired consequence independent of the initiating event or the action of any other layer of protection associated with the scenario."[388] An IPL is tested for two criteria: Is it effective in preventing the scenario from leading to the consequence, and is it truly independent of the other IPLs, as well as the initiating event? If the safeguard passes both tests, it is an IPL.

LOPA estimates the likelihood of the undesired consequence by multiplying the frequency of the initiating event by the probability of failure on demand (PFDs) for the applicable IPLs that protect against consequence. The LOPA result is a measure for the scenario risk—an estimate of the combination of likelihood and mitigated consequence. This risk estimate can be considered a *mitigated consequence frequency*,[389] where the frequency is mitigated by the independent layers of protection. The risk estimate can be compared to risk tolerance criteria for that particular consequence severity or level to decide if the frequency of the mitigated (that is with the IPLs in place) consequence is low enough. Individual risk criteria and not societal risk measures, which can lead to much more stringent risk reduction measures than needed, should be used. CCPS[390] provides guidance and references on how to develop and use risk tolerance criteria. If not, either process redesign that is more inherently safe or addition of more IPLs is required.

An IPL may prevent the consequence identified in the scenario, but may generate, by design, a different less severe (yet undesirable) consequence. For instance, the proper operation of the rupture disk prevents vessel overpressure or explosion but results in a loss of containment from the vessel to the environment or a secondary containment or treatment system. For instance, while the three relief valves performed their duty and protected the production tower of the BP Texas City isomerization unit, the highly volatile gasoline components overwhelmed the blow down drum and shot through the vent stack to open air. This, in 2005 was the deadliest refinery disaster in the USA. A separate LOPA scenario must be conducted for this scenario and determined if this result also meets the risk tolerance criteria. Combining causes that lead to the same consequence may not be valid, because each or every chain of IPLs may not protect against each initiating event or adequately mitigate consequence. Finally, scenarios that are too complex for LOPA require more detailed risk-assessment tools such as event tree analysis, fault tree analysis, or quantitative risk analysis.

3.1.5.4   Risk Ranking and Mitigation Prioritization

The ranking helps prioritize the risks for the next step, namely risk treatment. "Risk ranking can be quantitative, semiquantitative, or qualitative in terms of likelihood of occurrence and potential consequence or impact."[391] A risk profile could be created from the assessment results, which produce a risk significance rating for each risk, allowing the ranking of the relative importance of the risk. The intensity of the risk treatment effort needs to be proportionate to the level of risk based on size, nature, complexity, and distributional and ethical considerations. "Because of the extensive impact of risk ranking on the allocation of society's resources, ethical behavior in conducting risk management and training to do so takes on a very significant role. For instance, one may debate if one pollutant ranks higher than another; or if any one pollutant pathway deserves greater attention (e.g., mercury or arsenic in water discharges or in stack emissions). It gets increasingly more complex as relevant impacted species, regions of the country, and industry sectors, just to mention a few, are brought into the analysis; and they all compete for allocation of limited resources."[392]

One cost effective way to prioritize and identify the highest risks to be controlled is to use an initial probability/consequence screening matrix approach. The typical steps are to start with assessing where you are now, and then developing a scenario of events including conditional sequence and identifying risk consequences and probabilities. All the risks are located on a simple screening matrix of risk probabilities and consequence severity, typically classified into high, medium, or low buckets. For instance, all high consequence severity events and some high likelihood and medium severity events would fall under the high-risk category.

However, once the risks and opportunities are prioritized, options to mitigate those risks are developed. Options are downstream decisions that can be made to manage the risks. The mitigation options could vary from lowering the event likelihood probability, the consequence severity, or both. Prioritized allocation of resources becomes an integral part of the application of decision analysis.

The next step involves finding what information needs to be gathered and how perfect it needs to be, in order for it to provide at least a marginal value to the decision process. Information gathering is a normal activity in the decision-making process. Sources of information include results of surveys, test results, and expert opinions. Uncertainty is the lack of definite knowledge of something undesirable—and when it could happen—a bad event that may result in a negative consequence. Uncertainties are usually captured as probability distributions. The selection of which distribution is to be applied to describe the variable depends on the characteristics of the random variable; which in turn is assessed by using known information, subjective judgment or expert prior knowledge. New information, during the course of events, may still be useful to revise the current probability distributions through a Bayesian Revision, a statistical procedure. However, information, current or new, has value only if it affects the choice to be made by the decision maker (Figure 3.2).

3.1.5.5 Enterprise Risk Management

The objective of risk management is the assessment of significant risks and the implementation of suitable risk response. It leads to better informed strategic decisions, lower cost of capital, improved image, and enhanced community support. It also helps with the disclosure of the company's *material* risks to investors and to meet regulatory compliance requirements.

Most enterprise risk-management frameworks include a centralized function that performs ERM. Risk management must be integrated into the culture of the organization including its mandate, leadership, and commitment of the board. ERM is an ongoing iterative process. A typical implementation comprises of the following steps:[393]

1. *Plan and design*:
    a. Identify benefits desired and get leadership, preferably the board's mandate.
    b. Scope the proposed ERM and agree on a common set of definitions. ISO 73 provides a good starting point.

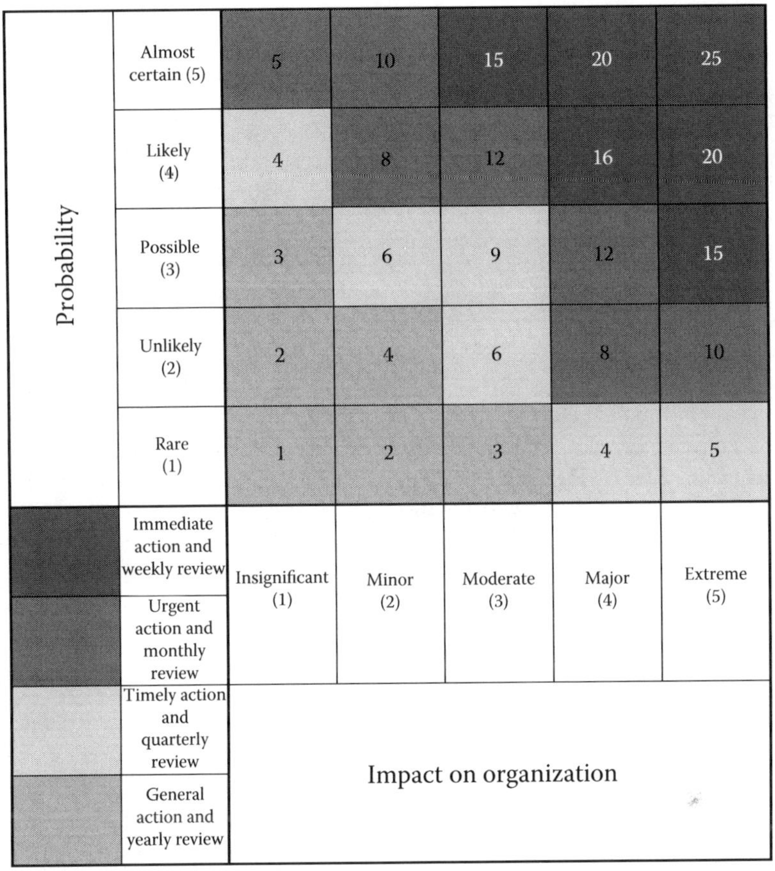

Probability		Insignificant (1)	Minor (2)	Moderate (3)	Major (4)	Extreme (5)
	Almost certain (5)	5	10	15	20	25
	Likely (4)	4	8	12	16	20
	Possible (3)	3	6	9	12	15
	Unlikely (2)	2	4	6	8	10
	Rare (1)	1	2	3	4	5

Immediate action and weekly review
Urgent action and monthly review
Timely action and quarterly review
General action and yearly review

Impact on organization

FIGURE 3.2   Example risk prioritization probability/consequence screening matrix.

    c. Frame strategy, identify process owners, and define the roles and responsibilities.

2. *Implement and benchmark*:
   a. Adopt risk-assessment methods and agree on a risk-classification system.
   b. Determine benchmarks for risk significance and conduct risk assessments.
   c. Establish risk appetite and risk tolerance levels and assess controls in place.

3. *Measure and monitor*:
   a. Confirm cost effectiveness of controls in place and find improvements.
   b. Embed risk management in organization culture and align with current practices.

4. *Learn and report*:
   a. Track and monitor risk performance progress to assess ERM performance.
   b. Report and disclose risk performance as required in word and spirit of the law.

3.2 SUSTAINABILITY DECISION ANALYTICS
AND INFORMATION SYSTEMS

3.2.1 Chapter Overview

Information is now recognized as a critical resource, and models play a key role in deploying this resource to productively obtain insights. This subsection provides an introduction to decision-making models that are relevant to corporate sustainability, public policy analysis, and value management. The applicability and usage of computer-based models have increased dramatically in recent years due to the extraordinary improvements in computer information and communication technologies, including model-solution techniques and user interfaces. However, while decision models are powerful, they are limited tools. The chapter also provides a broad assessment of select sustainability systems and analytical tools currently in the market, and describes how to launch and sustain a sustainability information management system (SIMS) to operationalize the data analytics driven decision process to implement the strategy and outperform the peers.

3.2.2 Sustainability Value Optimization Models

3.2.2.1 Strategic Decision Models

An entity's purpose of being is to create value for stakeholders on a path strewn with uncertainties. Risk is the effect of this uncertainty. Information reduces the level of uncertainty. There is a recognition of the growth of data analytics and its role in decision-making to not only mitigate risk but also enhance opportunities to create value. The 2014 COSO framework has made the enterprise risk management (ERM) and decision-making linkage more explicit, and revised the definition of ERM to state, "The culture, capabilities, and practices integrated with strategy-setting and its execution, that organizations rely on to manage risk in creating, preserving and realizing value."[394]

In particular, per Protivity,[395] revised COSO adds five connected components. The first element is the enhanced significance of risk governance and culture, which also impacts the other four. Risk governance elevates the oversight level of responsibility. Culture brings in ethical values, the fourth bottom-line of sustainability. The second element is the realization that risk in connected to the success of strategy is only one dimension of the mission to create value. But misalignment of strategy with the mission could be a risk in itself. The chosen strategy may succeed as planned, but may not enhance the mission of creating value for the stakeholders. Also, any choice of strategy involves trade-offs, which in turn brings a new risk profile, and presents new risk implications. This is another dimension of risk in any selected strategy. The third component added in COSO framework involves prioritization of identified risks in relation to strategy to ensure effective and efficient allocation of limited resources. The fourth element is to communicate and disclose these risks in a value creating manner. The final component is to address materiality; that is, add more resources to monitor risks that affect performance substantially.

Simulation is the creation of a model that simulates the outcome of interest, incorporates uncertainty, performs evaluations, and tracks the outcomes. Each uncertainty could be simulated or modeled as a probability distribution. Primary objective of a simulation is to define the probability distribution of the outcomes for the chosen decision, while considering the impact of changes in all the random variables, both independent and dependent, simultaneously. Software packages, such as @RISK,[396] for risk analysis using Monte Carlo simulation, and Crystal Ball are simulation models, compatible for use with spread sheets.

Strategic decision models fall broadly under two categories:

1. Models that use probabilities or probability distributions, such as PrecisionTree,[397] for decision analysis using decision trees
2. Models that use priority rankings, such as *multiple-criteria decision-making (MCDM)*[398] *models discussed in the following*

3.2.2.2 Multiple-Criteria Decision-Making Model

MCDM model combines the performance of decision alternatives across numerous, conflicting, qualitative, and/or quantitative criteria and provides an optimal solution. MCDM is capable of considering both technical and nontechnical attributes—useful for policymaking, and evaluation of new technologies and energy sources in the decision-making process. These methods have been successfully applied in a wide range of applications related to energy and sustainability problems.[399]

More commonly used MCDM methods[400] are weighted-sum method (WSM), weighted-product method (WPM), multi-objective optimization (MOO), the analytical hierarchy process (AHP), the technique for the order of preference by similarity to the ideal solution (TOPSIS), elimination et choix traduisant la realité (ELECTRE), and the preference ranking organization method for enrichment evaluation (PROMETHEE). Social, technological, economic, environmental, political, legal, and ethical (STEEPLE), which is an extension that adds ethical factors to the better-known parent political, economic, social, technological, legal, and environmental (PESTLE) analysis, is another model that allows the assessment of the current environment and potential changes.

3.2.2.2.1 *Overview of Select Multiple-Criteria Decision-Making Methods*[401]

3.2.2.2.1.1 Weighted-Sum Method   The WSM is an intuitive process and applicable to single-dimensional problems. It is based on the additive utility hypothesis; that is, the overall value of every alternative is equivalent to the products' total sum. WSM is simple to use if the application has same units' ranges across criteria. The objective is to maximize the sum of the weighted scores. However, when the units' ranges vary, the additive utility hypothesis is violated, and normalization is necessary. WSM steps involve (1) determining the criteria for the decision problem on hand, (2) determining the weight to be assigned to each criterion, (3) assessing the score for each option for each criterion, and then (4) computing the sum of the weighted score for each option. Select the option with the highest weighted score, if maximization of the outcome is the objective.

3.2.2.2.1.2 Weighted-Product Method WPM is similar to the WSM, except that the product of the values, instead of a sum, is used in the method. WPM is applicable for both single and multidimensional cases. Each alternative is compared to the rest through a multiplication of ratios that are related to every criterion. This method compares ratios of alternatives. The optimum solution in a pair-wise comparison is the one that is at least equal to the rest of the alternatives, and more specifically, the best solution is when the ratio of weighted products is >1 for when considering a maximization problem.

3.2.2.2.1.3 Multi-Objective Optimization MOO is apt for problems where there are a number of different and conflicting objectives; it is one of the most commonly encountered types of optimization problems. A MOO problem with constraints will have many solutions in the feasible region. Although it may not assign numerical relative importance to the multiple objectives, it can still classify some possible solutions as dominating and better than others. Solutions that lie along the Pareto line are non-dominated solutions and called Pareto optimal, while those that lie inside the line are dominated because there is always another solution on the line that has at least one objective that is better. The objective in MOO is to find the solutions as close as possible to the Pareto front, where one party's situation cannot be improved without making another party's situation worse. "The final allocation decision cannot be improved upon, given a limited amount of resources, without causing harm to one of the participants. Pareto efficiency does not imply equality or fairness."[402]

3.2.2.2.1.4 Social, Technological, Economic, Environmental, Political, Legal, and Ethical STEEPLE analysis, which is an extension that adds ethical factors to the better-known parent PESTLE analysis, allows the assessment of the current environment and potential changes. Political factors include internal politics such as team cohesiveness, as well as external factors such as employment laws, tax policies, trade restrictions, political stability, and tariffs. Economic factors include microeconomic considerations, such as project viability and project robustness, as well as macroeconomic issues such as economy, labor, and access to credit and financing. The sociological component takes into account all factors that affect the market and the community, including cultural expectations, norms, and population dynamics. Technological factor involves all technical aspects. Legal aspects such as employment, quotas, taxation, resources, imports, and exports are definitely reckoned. Environmental factor takes into consideration a broad range of ecological and environmental aspects. Additionally, ethics is considered in the STEEPLE model.

3.2.2.2.1.5 Technique for the Order of Preference by Similarity to the Ideal Solution TOPSIS is based on the idea that the optimal is as close as possible to an ideal solution, such as the maximum benefits attributes and minimum cost attributes, and simultaneously as far away as possible from a corresponding negative ideal solution, as shown in Figure 3.3. Steps involve standardizing the decision matrix, constructing weighed decision matrix by multiplying

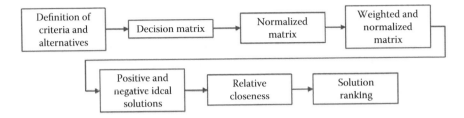

FIGURE 3.3   TOPSIS methodology.[403]

relevant weights to each attribute, and determining positive ideal and negative ideal solutions. Standardizing various attributes transforms them into nondimensional attributes, which allows comparisons across criteria. Two hypothetical alternatives are constructed. The positive ideal solution is the one that has the best values for all attributes. A set of maximum values for each criterion is the positive ideal solution. Likewise, a set of minimum values for each criterion is the negative ideal solution. The negative ideal solution has the worst attribute values. Next the separation from the positive ideal and negative ideal solutions is determined and relative closeness to positive ideal solution is defined and optimal selected. TOPSIS selects the solution closest to the positive ideal solution and farthest from the ideal negative solution.

3.2.2.2.1.6 Analytical Hierarchy Process   AHP[404] solution pertains particularly to problems that involve multiple, often competing or conflicting criteria and both qualitative and quantitative aspects of decision need to be considered. AHP reduces the decision to a series of pair-wise comparisons. The four steps of the AHP method are shown in Figure 3.4.

The first phase involves the hierarchical structuring of the decision problem. The next step involves constructing a pair-wise comparison matrix for each aim. The relative importance between two criteria is measured by a numerical scale. Step two is to compute the matrix of scenario scores, which is to obtain the score of each option for each criterion. The AHP process computes the vectors by the product of the score and the pair-wise comparison weight vector. The next two steps are to rank the scenarios and checking for consistency in the developed set of pair-wise comparisons. Scenarios ranking is done by ordering the weighted scores in decreasing order. The best alternative (for objective maximization) is the one that has the highest value for the sum of weighted comparison pairs.

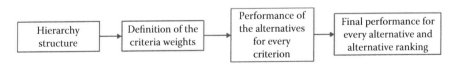

FIGURE 3.4   AHP methodology.[407]

3.2.2.2.1.7 Elimination Et Choix Traduisant la Realité ELECTRE[405] is based on the idea of indifference thresholds, defined by the decision makers. The principle says that when the least favorable options are eliminated, there is greater clarity among the remaining alternatives. Decision problems with fewer criteria and a large number of options find ELECTRE an attractive MCDM method.[406] As options are eliminated the preferences are redefined. One could argue a case for a category between indifference and strict preference, referred to as hesitation zone or weak preference. ELECTRE generates a whole system of binary outranking relations among the alternatives. Sometimes the best may not be identified at all, and instead the system provides a set of leading alternatives.

3.2.2.2.1.8 Preference Ranking Organization Method for Enrichment Evaluation PROMETHEE is an MCDM method that uses the outranking procedure to rank the alternatives. Various versions of the method are practiced based on the type of ranking: partial ranking, complete ranking, interval based ranking, continuous ranking, integer linear programming and net flow, and one that represents the human brain.

Presentation of the decision maker's preference between two actions by a preference function independently is the first step. Step two involves the comparison between a proposed set of alternatives with each other with respect to the preference function. In the third step, a matrix captures the results of the comparisons and value of the criterion for each alternative. Partial ranking issues are sorted out in the fourth step. The fifth and final step finishes the alternative rankings (Figure 3.5).[408]

3.2.2.2.2 Sustainability Applications of Multiple-Criteria Decision-Making Models

MCDM literature reviews show the effectiveness of applying these methods to sustainable and renewable energy applications[409] as well as explore renewable energy scenarios for energy planning.[410] Several studies show the application of MOO,[411] also an MCDM. Following list shows the broad range of MCDM applications[412] in the sustainability area.

1. WSM and WPM to optimize considering technical, social, and economic features.[413]
2. MOO to optimize design of switching converters to be integrated into related renewable technologies with the conflicting objectives of efficiency and reliability,[414] to study photovoltaic systems and electrothermal methods, with the conflicting objectives of efficiency maximization and cost minimization,[415] and in combination with MCDM, for the design process of hybrid renewable energy systems (HRES).[416]

FIGURE 3.5 PROMETHEE methodology.

3. Integer program to optimize ethanol production from several biomass sources, to minimize the cost and the environmentally related issues,[417] for making decisions on trade-offs between two types of crops (that is, food and biofuel), and to balance economic advantages and environmental impacts.[418]

4. STEEPLE or PESTLE for multi-criteria risk prioritization.[419]

5. TOPSIS for sustainable development and renewable energy preferences,[420] for the best islanding detection method for a solar photovoltaic system,[421] and for the selection of the best among three biomass types of boiler.[422]

6. AHP for green supplier selection,[423] for the optimum amongst energy policies in Turkey,[424] and for evaluation of solar water heating systems and viability of renewable energy sources.[425]

7. ELECTRE for optimization of decentralized energy systems.[426]

8. PROMETHEE for a decision support system for renewable energy source (RES) exploitation,[427] to choose the best among four alternative energy exploitation projects,[428] and as a hybrid with AHP to choose the most appropriate desalination system in RES plants.[429]

3.2.2.3 Integrated Sustainability Value Model

Sustainability, driven by the new social contract, is an incipient or emergent societal and economic shift. A principal driver of this societal transformation is the recognition that business is no longer the sole property or interest of a very few. Notably, synchronous interactive connectivity among stakeholders has had a significant role in this change.[430] Globalization, for example, money flow across borders growing to three times the global GDP, and asymptotic growth in digital connectivity, for example, internet and mobile phone users growing to two and five billion respectively, in 2010.[431] have greatly accelerated the sustainability tide. The emergence of Brazil, Russia, India, and China's (BRIC's) economies has intensified competition for natural resources. Global workforce and supply chains have increased potential product safety, human rights, and other geopolitical risks and liabilities.

The current pace of growth of our highly connected global economy threatens the availability of ecosystem services. The triple value model[432] (Figure 3.6) was first developed for the OECD. This framework explicitly defines the linkages and value flows among industrial, environmental, and societal systems and serves as a basis to construct policy-simulation models. Industrial systems use environmental resources to meet societal needs. They extract energy and materials from the environment, and build and deliver products and services to society and plow back wastes to the environment. Economic capital or industrial productive capacity creates and delivers economic and social value, and often environmental value in protecting the environment and restoring natural capital. Societal systems consume the outputs from industrial productive capacity and recycle wastes to the industrial system and discharge generated wastes to the environment. They provide the workforce (human capital), and the consumers (social capital), for the creation and use of the products and services, and also benefit and perhaps suffer in terms of human health and well-being from the economic activities. Societal governance to manage market externalities helps protect the environment and restore natural capital.

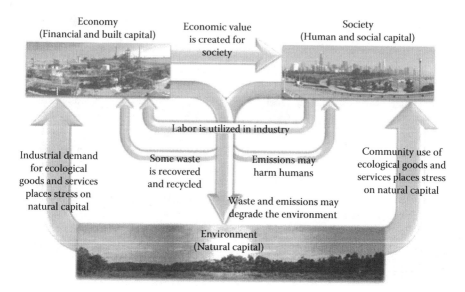

FIGURE 3.6   The triple value model: Environmental, social, and economic linkages.[434]

Environment is the source of all energy and materials, both renewable and non-renewables. Natural capital, the productive capacity of the environment, is finite. If it is degraded at a rate faster than it can resurrect, there are some potential disasters looming. Flow of products and services from the ecosystem delivers value to industrial and societal systems.

The triple value model provides an architecture for modeling, using analytic or simulation techniques to assess trade-offs in intervention. These triple value models enable integrated assessment of the costs and benefits of proposed interventions to fully understand the implications of new policies and practices and to avoid unintended consequences. For instance, the EPA and its regional partners are using this model to develop strategic alternative policies for water-resource management and dynamic-energy economic policy simulation for evaluating energy policy options. Caterpillar Inc. has successfully used this systems approach to pursue a strategy of *dematerialization* using a variety of regulatory and voluntary approaches as a basis for its worldwide remanufacturing business.[433]

The model is based on a stock-and-flow concept to help comprehend the dynamic and complex connections among the economic activities, human health, and the environment. This concept of sustainability assessment is aligned with the UN System of Environmental Economic Accounting for Nations (SEEA) Framework's systems approach to organize the stocks and flows of environmental and economic assets, which allows for the integration of environmental information (often measured in physical terms) with economic information (often measured in monetary terms) because of its capacity to present information in both physical and monetary terms coherently.

The UN SEEA integration in a single measurement system provides an information base on water, minerals, energy, timber, fish, soil, land and ecosystems, pollution and waste, production, consumption and accumulation; for the development of models, such as triple value model; and for detailed analysis of interactions between the economy, human health, and the environment. That, in turn, allows the development of aggregates, indicators, and trends across a range of environmental and economic aspects, such as natural resources, emissions, and discharges to the environment from economic activity, and even the amount of economic activity to protect the environment.

3.2.3 Sustainability Analytics and Information Management System

3.2.3.1 Sustainability Analytics

Deloitte uses the term *sustainability analytics* as a new field of expertise that integrates big-data analytics with sustainability. They define sustainability analytics as "an approach that aims to effectively use technology to collect, disseminate, analyze, and use sustainability-related information across the enterprise."[435] EY[436] states that environmental, health, and safety (EHS) and sustainability analytics programs help companies reduce risks and drive cost savings. The first three components listed by EY, summarization, visualization, and statistical econometric analysis are elements of more traditional analytics. The next three analyses comprising of spatial, human-driven algorithmic, and heuristic machine-learning represent the more advanced analytics applications today.

The first two elements of traditional analytics, summarization and visual representation of vast amounts of data for tracking and comprehending trends, are commonplace in the consumer marketing and consumer finance areas. While mature in retail markets and some other applications, analytics are just taking baby steps in the sustainability arena.

Many market and financial sector leaders have significantly advanced in the application of the third element in traditional analytics, statistical and econometric analysis of data to garner insight for decision-making. This area has also been more fertile for sustainability and EHS because of the benefit potential, and has seen significant advances in the government policy development and industry performance enhancement arena. The need to garner insight for government or corporate policy decisions or academic research has led to the development and use of statistical or econometric analysis using focused manual collection and aggregation of select data elements. The government agencies have routinely performed very complex and intense cost-benefit and risk analysis for most of their significant intervention efforts. The clean Air Act[437] and climate change[438] benefit modeling efforts are very substantive examples. In the corporate sector, Caterpillar Inc., for example, has successfully used the triple value model approach to pursue a strategy of *dematerialization* using a variety of regulatory and voluntary approaches as a basis for its worldwide remanufacturing business.[439]

Significant advances in technology, such as remote sensing, and in software, such as remote data capture, have propelled the application of analytics to a different stratum in select sectors. For instance, *spatial analysis* with real-time feedback provides granular insight into potential opportunity and risk for operational improvement; or *human-driven algorithmic analysis* informs real-time intervention or multi-criteria decision making through predictive and preventive analysis of real-time data.

In the sustainability area, efforts to disseminate best practices have progressed on a selective basis for over two decades. The benefits gained through portfolio risk mitigation were so significant that many corporations implemented these initiatives even in the absence of enterprise resource planning (ERP), environmental, health and safety & security (EHS&S), and enterprise sustainability management (ESM). However, these efforts are very human intense, and automation continues to be challenging in the sustainability area, at least in the near term.

In the corporate sector, for instance, companies like ExxonMobil implemented global programs more than two decades ago to streamline and manage energy efficiency and EHS performance benchmarking across their facilities globally. Value generation,[440] an enterprise-wide initiative in the mid-1990s, was a precursor to best practice networks; in effect, it was a spatial analysis of replicating best practices in the sustainability area by capturing and disseminating them across professionals involved in the relevant function globally. Remediation portfolio risk management involved the development and successful application of a decision aid to inform risk based prioritization of remediation projects for allocation of resources to minimize risk. MobilRITE (Risk Indexing Tool for Environmental, Health, and Safety) was developed as a Mobil internal screening/relative ranking risk-based prioritization decision tool and used to prioritize more than 250 remediation project sites in western Canada based on risk for prioritized cleanup.[441]

Finally, heuristic machine-learning analysis provides unanticipated insight through continuous learning and unbiased thorough analysis of mega quantities of big data. However, today, the level of maturity in the EHS and sustainability in the machine-learning area definitely lags behind.

EHS and sustainability analytics,[442] like any other business analytics, will need to be able to manage large amounts of data, process data faster, and aggregate and clean data effectively to overcome the many technical challenges associated with data collection and analysis. A challenge to automating sustainability data is governance. The wide array of diverse real-time data from global operations with potential legal ramifications continues to be a challenge.

3.2.3.2 EHS&S and Sustainability Management System

A recent global survey by Verdantix[443] shows that ESM software adoption is under 10% at the enterprise level, which implies an even lower sustainability software adoption rate. This is despite a concerted thought-leadership effort to demonstrate value of technology in enhancing sustainability performance, for more than a decade, both at government[444] and corporate[445] forums. Even corporations like ExxonMobil have adopted electronic enterprise-wide EHS&S management information systems, but only in recent years.

3.2.3.2.1 Emergence of Environmental, Health and Safety & Security,
*   and Sustainability Management System*
Historically, most ESM software systems were developed to meet the needs of four purposes: (1) reporting corporate social responsibility (CSR) or sustainability disclosure that emanated from Europe; (2) compliance with US EPA regulations such as Clean Air Act Title V environmental management for refineries, extractives, chemical

manufacture, and combustion; (3) quality requirements of the US Food and Drug Administration for drug companies; and (4) compliance with US Occupational Safety and Health Administration (OSHA) workplace requirements, such as occupational health and industrial hygiene, and incident management.

Sustainability management systems will continue to piggyback on the regulatory mandate-driven EHS&S management system in the United States. In reality, complete convergence of the two within one integrated sustainability platform may be inevitable. Looking further, in the not too distant future, given the move toward a single integrated financial and nonfinancial disclosure and reporting system, one can see a complete integration of the ESM platforms with the enterprise financial platform.

Verdantix[446] categorizes the ESM software markets into three phases: (1) data-warehousing for compliance at site level in the 1990s, (2) data management for analysis and reporting of incidents and compliance at business unit or national level in the years 2005–2014 timeframe, and (3) the addition of performance optimization and operational risk management at the global enterprise-level and development of mobile apps with offline functionality in the next decade.

The Verdantix[447] survey of business case drivers for ESM software showed that reducing operational risk management is the most significant one. The next two factors are mitigation or reduction of regulatory risk and cost savings in reliable data collection and aggregation.

The same Verdantix[448] study also finds that, from a product features perspective, the majority of customers recognize and demand comprehensive integrated ESM platforms, rather than point solutions. Design of user interface and user empowerment to configure software were the next two significant features sought after in the preferred integrated platform. Integrated EHS platforms aggregate data from multiple EHS workflows, such as spills and worker safety incidents. Today, user maturity allows for discerning the advantage of buying an integrated application. For instance, Saab Group, a $3.7 billion defense and aerospace company with a global supply chain and customers in one hundred countries, has invested in Enterprise Resource Planning (ERP) system to integrate data on hazardous material and energy use to utilize analytics to manage its sustainability initiatives.[449]

3.2.3.2.2   Strategies to Select and Implement an Enterprise Sustainability Management

Launching an ESM system is a big commitment and has long-term consequences for the business, because it will be used to inform decision making. This is not the rolling out of a new software; it involves engaging a long-term platform partner.

1. The first step to launch an ESM is to obtain leadership support for getting visibility, empowerment to make decisions, priority participation from peers and staff, and money and human resources to implement and maintain the system.

2. The strategy and planning step starts with defining the business case with a stakeholder committee. Business value may come from averted cost of major incidents, or the value of risk avoidance, prioritizing resources

for risk mitigation, cost saving by streamlining workflows, and improving communications. It could be company specific (for instance, while air emissions reporting may be critical for petroleum industries, it may not be significant for the retail industry). It is best to manage expectations by defining expected value up-front. The project budget is largely defined by the number of users with access, the scope of the implementation and the number of customizations needed.

3. Another key step is the mapping of workflows, the way knowledge is transferred within the organization, which the EHS&S system should reflect. While the EHS&S can improve the process, it is not a substitute. Other aspects that will need attention for a successful launch are no different from any software implementation efforts; and they include, identifying administrators, defining user types, number of users, their frequency of use and levels of training, change management process as well as accountability for validating the data. Segmenting users according to use level, experience and impact of the change is an important consideration.

4. Defining the requirements starts with a distinction between a need and a requirement. Need may be a gap in the business process or an improvement opportunity. Requirement is what has to be done to narrow that gap and gain from that opportunity. The process starts with a strategic discussion about business needs and the biggest EHS&S program challenges. Once needs are identified and an initial list of requirements developed, they have to be prioritized. An *out of the box* system will require configuration to reflect company's unique workflows, desired features, and internal visual branding. A customization, such as an added security feature or a unique organizational structure, requires a change to the source code.

5. Stakeholder input is a key part of the decision-making process and paves the way for change management for system rollout in the company later on. "The selection is where you want to have all of your fights because once it's implemented, you don't want people to not use it; you don't want people to constantly complain about it."[450] System selection begins with not wasting time on vendors that cannot meet requirements. During the demonstration phase, the system vendor's configuration and implementation team should be involved to ensure that they can deliver.

3.2.3.2.3 Select Dominant Environmental, Health and Safety & Security, and Sustainability Management Systems

A recent Verdantix[451] report identifies about half a dozen ESM software systems as quadrant leaders. A brief description of a select set is presented in the following:

1. CRedit360, as a supplier migrating into the ESM market from the sustainability management space, offers a decent baseline capability for environmental reporting. CRedit360 achieves the highest score in ESM and sustainability performance analysis, business intelligence for environmental, and sustainability data. The data explorer tool enables users to generate a wide range of benchmark data at different organizational levels, import external

benchmarks, conduct anomaly analysis, and create intensity metrics. It is one of the few EH&S apps to contain decent forecasting tools and good user interface, and it also offers a robust solution for mandatory greenhouse gas (GHG) emissions reporting regimes. Improvement opportunities include features for environmental impact or occupational health areas out of the box.

2. Enablon provides a comprehensive, integrated software platform across all ESM workflows. Per Verdantix, Enablon scores strongly in many ESM impact areas including air and GHG emissions, and management of audits, permits, and sustainability. It achieves the highest scores in incident, risk, and safety management and helps meet Occupational Health and Safety Assessment Series (OHSAS) 18001, ISO 14001, ISO 31000 requirements. Application architecture shows high scores as well in master and operational data management, data security, and data capture. Improvement areas are occupational health, industrial hygiene, and ergonomics.

3. Gensuite was first developed as an internal ESM application within General Electric but spun off in 2008. It offers a robust and capable technical platform and an integrated suite of cloud-based applications enabling compliance, operational excellence, and risk management. It offers more than 65 applications covering a broad range of solutions spanning the entire EHS and sustainability program spectrum, with parallel solutions for quality, security/crisis management, responsible sourcing, product compliance, and other risk-management programs. Per Verdantix, it got the second-highest score in audit and safety management, chemicals management, and hazardous waste management. Gensuite is also recognized for its intuitive interfaces, real-time trending and data analytics, and integrated multilingual and mobile capabilities.

4. IHS enables unified information strategies for ESM by providing integrated solutions for ESM performance, operational risk, and sustainability such as energy and carbon. IHS engineering workbench unlocks an organization's best practice networks—technical knowledge regardless of document location, format, or language of authorship.

    IHS also provides information and analytics solutions targeting workflows of multiple industries, ranging from economic forecasting, cost modeling, risk assessments, and scenario analysis to techno-economic evaluations. With an intuitive interface and one of the largest sets of data management tools available, IHS econometric modeling software helps create statistical and forecasting equations quickly and efficiently. Designed to make scenario impact analysis accessible across the enterprise, the Global Link Model, which accounts for 95% of global GDP, allows generation of scenarios and prediction of its impacts across the value chain. It helps stress testing of company's performance and building business resilience. It also helps identifying opportunities and making scenario-based demand projections to optimize portfolio.

5. Intelex, per Verdantix, is an industry leader, with the highest overall capabilities. The Intelex Data Service provides a cloud-based data-management software,

built for self-service, social collaboration, and mobility that links real-time data with activities, tasks, and processes. A cloud-based central platform allows consolidation of siloed and discrete systems into one accessible solution. The Intelex safety element helps comply with OSHA and OHSAS 18001. Its risk-management software follows the ISO 31000 risk-management and risk-assessment framework to help reduce risk and drive performance. Real-time dashboards inform decisions for optimal resource allocation; and automated task assignment and follow through drive completion of risk prevention and mitigation tasks. Finally, a preconfigured data connector that uses industry standard transfer protocols allows users to incorporate EHS data into their big data analytics programs.

6. SAP Sustainability Solutions[452] for Corporate Sustainability solutions helps organizations improve sustainability across supply chains and the business value chain by managing controlling sustainability aspects related to products, energy, and personnel, and by streamlining disclosure about social, financial, and environmental performance. The SAP Clear Standard Carbon Impact on-demand solution helps manage sustainability projects across internal operations and the supply chain. It helps asses total carbon emissions inventory across all scopes, prioritize and manage a portfolio of abatement projects, and enables disclosure.

SAP TechniData[453] provides focus on product stewardship and product compliance, and enables users to develop and register products to ensure regulatory compliance. SAP Sustainability Performance Management software automates data collection for credible reporting, as well as cascades strategic goals and insights across the enterprise. It allows management of multiple reporting frameworks, standards, and KPIs—including GRI.

SAP Next-Gen enables SAP customers to seed in disruptive innovation and accelerate through connecting with academic thought leaders and researchers, students, startups, accelerators, venture firms, and other partners in the SAP Next-Gen innovation community. SAP Sustainable Communities solutions help governments improve sustainability performance by prioritizing sustainability initiatives and empowering citizens to participate in decision making processes.

7. SAS[454] provides the capability of integrating sustainability analytics with financial analytics. SAS Sustainability Reporting collects and analyzes the data, identifies the areas of highest impact, and helps make decisions about future impact. It provides valuable insights to find opportunities in sustainability data.

8. Underwriters Laboratories Environmental Health and Safety (UL EHS)-Sustainability solution provides a central location to record training, incidents, observations, and compliance programs that can be cloud or software as a service (SaaS) deployed. Corrective actions can be assigned, underlying causes analyzed, and incidents reported.

3.2.3.3 Big Data Analysis and Analytics Platforms

3.2.3.3.1 Big Data

Big data has been defined by Microsoft as "the term increasingly used to describe the process of applying serious computing power, the latest in machine learning and artificial intelligence, to seriously massive and often highly complex sets of information." Oracle defines big data from a different perspective as "the derivation of value from traditional relational database-driven business decision making, augmented with new sources of unstructured data."[455]

The most widely used definition comes from Gartner: "Big data is high volume, high velocity, and/or high variety information assets that require new forms of processing to enable enhanced decision making, insight discovery and process optimization."[456] The characteristics of volume—exponentially growing amounts of data from megabytes to petabytes velocity—increasing requirements on speed of data processing from batch to real time, and variety—increasing number of types of data, each present challenges as they expand.[457] The big data landscape consists of technologies that store and process data, infrastructure tools that query and analyze data, and applications that turn data into insights. Technologies and infrastructure tools are generic across different applications of big data.

Financial service providers are using big data analytics to improve customer eligibility analysis for equity capital, insurance, mortgage, or credit. Notably, banks like Capital One and credit card companies like Visa are branding themselves as technology companies with a financial edge! Transportation companies are applying big data analytics to improve efficiencies and save costs by real time tracking of fuel consumption and traffic patterns across their fleets. Utility companies are using big data solutions to analyze user behaviors and demand patterns for a better and more efficient power grid.[458]

3.2.3.3.2 Phases in Big Data Analysis

1. Data acquisition and recording phase deals with identifying the data generating source and establishing the process to automate and streamline data flow to gather and store data.
2. Information extraction and cleaning phase extracts data, performs quality analysis and controls, pulls the required information, and expresses it in a format that can be analyzed.
3. Data integration and aggregation calls for homogenizing data for use by analytics applications.
4. The analysis process involves application of statistical and econometric methods. This requires a good comprehension of the data to be evaluated and the acquisition of appropriate tools. Incompleteness is managed by making appropriate assumptions.
5. Interpretation of the results is important for making decisions. This calls for proper documentation of the process and integration of visualization aids to interpret large sets of data.
6. Management of objective performance is achieved through strategies to improve operations and integration of analytics into processes for successful leverage.

3.2.3.3.3 Business Intelligence and Analytics Platforms

Modern business intelligence (BI) and analytics platforms are characterized by easy-to-use tools that support a full range of analytic workflow capabilities and do not require significant involvement from information technology (IT) support. The IT-centric semantic-layer-based development approach of traditional BI and analytics platforms was disrupted by visual-based data discovery. By accelerating data harmonization and visual identification of data patterns, it transformed the market away from IT-centric reporting to business-centric, business-led agile analytics. However, the insight creation tasks have remained largely manual and bias prone. A new innovation wave, smart data discovery, is emerging, and has the potential to be as disruptive. For instance, IBM Watson Analytics and Salesforce leverage machine learning to automate analytics workflow from preparation and exploration of data to explaining discoveries and sharing insights.

Corporations want to expand modern BI usage and make it accessible to everyone in the enterprise and beyond. They want users to analyze a more diverse range and more complex combination of data from sources beyond the data warehouse or data lake, a storage repository that holds a vast amount of raw data in its native format. Simultaneously, interest in deploying BI and analytics platforms in the cloud has jumped to more than 51%. In response, BI and analytics platform vendors are offering cloud deployment and subscription pricing model options.

3.2.3.3.4 Key Critical Capabilities of Business Intelligence and Analytics Platforms

1. Embedded advanced analytics for easy user access to advanced analytics.
2. Ability to create highly interactive dashboards and content with visual exploration and embedded advanced and geospatial analytics to be consumed by others.
3. Interactive visual exploration that enables users to analyze and manipulate the data by interacting directly with a visual representation—display as percentages, bins, and groups.
4. Smart data discovery that automatically finds correlations, exceptions, clusters, links, and predictions in data without requiring them to build models or write algorithms.
5. Mobile exploration and authoring to develop and deliver content to mobile devices in a publishing or interactive mode or both, suiting mobile devices' native capabilities.
6. Embedding analytic content capabilities for users to create and modify analytic content, visualizations and applications, embedding them into a business process.
7. Capability to publish, share, and collaborate on analytic content that allow users to publish, deploy, and operationalize analytic content through various output modes.

3.2.3.3.5 Select Dominant Business Intelligence and Analytics Platforms[459]

1. Microsoft's Power BI suite, offers a broad range of BI and analytics capabili-ties, data preparation, data discovery, and interactive dashboards via

a single design tool. Power BI Desktop can be used as a stand-alone, on-premise option for individual users, or delivered via the Azure cloud. It allows users to manipulate data from multiple data sources—both cloud-based and on-premise, relational as well as Hadoop-based, and including semi-structured content. Microsoft Reporting Services and Analysis Services are traditional enterprise reporting platforms on-premise offerings.

Microsoft Quick Insights is a basic form of smart-data discovery; and its Cortana Intelligence Suite integrates its machine-learning capabilities as part of a complete solution. With recent integration of Power BI with Microsoft Flow and within its business application, Microsoft Dynamics has also moved a step closer to linking insights to actions. Microsoft Power BI is often used in combination with other BI tools because of gaps in functionality; customers may use Power BI as a low-cost option for broadly used, simple dashboards, and then complement it with products from other BI with more robust capabilities.

2. IBM provides BI and analytics offerings such as Cognos, Watson, Predictive and Planning Analytics. IBM's Smarter Planet multi-platform strategy reveals an intelligent, instrumented and interconnected globe. Today, these super computing capabilities can be built into almost anything digital because they live in the cloud. IBM's cognitive business, with Watson at the center, promotes using analytics, natural language processing, and machine learning. Per IBM CEO Ginni Rometty, "redefine the relationship between man and machine," where systems can understand, reason, and learn. IBM is well positioned to leapfrog the current visual exploration market to be a major player in the next-generation machine-learning-enabled BI and analytics.

3. Oracle offers BI and analytic capabilities, such as Oracle Data Visualization in the cloud, on-premise or in hybrid mode, from decentralized as well as centralized deployments. It offers integrated data preparation, data discovery, advanced exploration, and interactive dashboards.

4. Qlik offers governed data discovery and analytics such as Qlik Sense and QlikView, either as a stand-alone application or one that is embedded in other applications; that allow customers to build robust, interactive applications and to visualize patterns in data, not readily available with straight SQL.

5. SAP delivers a broad range of BI and analytic capabilities for both large IT-managed enterprise reporting deployments and business-user-driven data discovery deployments. It offers two distinct platforms, SAP BusinessObjects Enterprise for on-premise and SAP BusinessObjects Cloud for cloud-based deployment. The integrated vision of planning, analytical and predictive capabilities in a unified, single platform and its product vision toward smart data discovery capabilities are promising. SAP's Digital Boardroom is particularly attractive to executives because it includes *what if* analysis and simulations.

6. SAS offers a wide range of BI and analytics capabilities: from interactive discovery, dashboards and reporting for mainstream business users, to specialist tools for data scientists, as well as pre-built solutions for industry verticals. SAS Visual Analytics is available either on-premise or through the cloud in SAS's own data centers or Amazon Web Services (AWS). It provides governed data discovery, dashboards, and advanced analytics, and a graphical user interface for citizen data scientists to refine predictive models while exploring data within. It also provides visual strategy diagrams that create compelling illustrations of the relationship between performance objectives established for various business units.

7. Sisense offers a single platform that allows visual exploration of web-based dashboards but has shortcomings in its product line in terms of cloud and advanced analytics. Notably, Sisense has capabilities that leverage Amazon's Alexa personal digital assistant (PDA) for voice-enabled query and interpretation of results.

8. Tableau offers a highly interactive and intuitive visual-based exploration experience for business users to easily access, prepare and analyze their data without the need for coding. The three primary products are Tableau Desktop, Tableau Server, and cloud offering Tableau Online. Tableau continues to be perceived as the modern BI market leader. It is enhancing its capabilities in machine-learning-enabled data preparation and smart data discovery.

3.2.3.3.6 Future of Big Data and Technology

With expanded and enhanced user access needs, growth of Platform as a Service (PaaS) is inevitable. Forrester Research report[460] summarizes information technology advances that will help meet big data customer need for real-time, predictive, and integrated insights, as follows:

1. Predictive analytics to discover and deploy performance improvement and risk mitigation
2. Not Only SQL databases that include key-value, document, and graph databases
3. Search and knowledge tools to extract new insights from large repositories of data
4. Stream analytics to filter, aggregate, and analyze high throughput live data sources
5. In-memory data fabric to access and process large quantities of data by distributing data across the dynamic random-access memory (DRAM), Flash, or solid state drive (SSD)
6. Distributed files to store data on more than one node, for redundancy and performance
7. Data virtualization to deliver information from sources, in real-time and near-real time
8. Data integration tools for orchestration across solutions such as Amazon and Hadoop

9. Data preparation software to source, cleanse, share, and speedup data use for analytics
10. Data cleansing of large, high-velocity data sets, using distributed stores and databases

3.3   BUILDING THE SUSTAINABILITY BUSINESS CASE AND STRATEGY

3.3.1   CHAPTER OVERVIEW

This chapter starts with a journey on why corporations should be responsible, take on some if not more social responsibility, and be a true private public partner. Building a business case demonstrates normative, anecdotal, as well as some academically robust proof of sustainability gains such as lower average weighted cost of capital and financial outperformance, compared to less responsible peers. The following sections discuss how to make a business case and build a sustainability strategy.

3.3.2   MAKING THE SUSTAINABILITY BUSINESS CASE

Per Nobel laureate economist Milton Friedman, the only purpose of business is to make profits for its shareholders. Also, executives who, in the name of responsibility, go beyond minimum prescribed legal requirements make the company less competitive and fail in their duty to maximize shareholder profit. Friedman goes on to opine that such socially responsible actions in effect turn executives into civil servants, levying "taxes" (in the form of corporate money allocated to social causes) and making "expenditures." This is "the socialist view that political mechanisms, not market mechanisms, are the appropriate way to determine the allocation of scarce resources to alternative uses."[461]

Here are some contrary factors to consider:

1. US corporate law does not prescribe maximization of shareholder wealth.
2. Monetary return is a rather narrow interpretation of shareholder objective. Is the next quarter return critical for the investor, or is it driven by the speculator? "Speculators may do no harm as bubbles on a steady stream of enterprise. But the position is serious when enterprise becomes the bubble on a whirlpool of speculation."[462]
3. Many major corporations are owned by institutional investors, also referred to as *universal owners*. They are typically long-term investors, and the longevity of the company may be far more significant for them. Also, they are more risk averse and unwilling to be subject to sustainability risks, which may be less certain, but could definitely be catastrophic when it occurs.
4. The mission-driven financial sector focused on socially responsible investment has grown from $2.7 trillion in 2007 to $21.4 trillion in 2014.[463] These socially responsible investors, coupled with other philanthropic sponsors, increasingly play a greater role in the resource allocation choices of

companies. Furthermore, not all corporate stewards are focused on just profits. Per Mason, president of Quaker Oats, "Making a profit is no more the purpose of a corporation than getting enough to eat is the purpose of life."[464] I (the author) fondly recall my father's everyday reminder at the dinner table, "We eat to live—not the other way around."

5. Why should society be expected to carve out the high profit potential opportunities to private-sector players, and retain the burden and obligation of cleaning up the environment and serving the underserved?

6. Could CSR be seen as the privatizing some of government's social duties to take advantage of private-sector innovation and scaling? Of course, it comes with profit-making opportunities if managed appropriately. In Friedman's words, a society laws should constrain business from doing bad. Could the doing good be seen as the opposite of not doing bad?

7. A society that provides the private sector access to resources and the license to operate gets to set the rules. For instance, market externality, by definition, addresses public goods that get neglected in the free-market-only approach. Could the corporation be viewed as a societal instrument? Is the free market the only game in town? Is it the sole edict for social progress? Is an optimal hybrid, like the public–private partnership (PPP), or an expanded regulated monopoly a more desirable option?

8. The current executive compensation structure, especially the executives' bonus, is often tied to beating the analyst estimate for the next quarter. This short-term focus reminds one of *cutting the golden goose* parable. Creation of subprime loans and toxic assets that caused the financial tsunami was a classic example of misaligned motif. Both were economically unsustainable.

9. Executives are human, and inherently biased, often backed by dysfunctional groupthink. They have to make investment allocation calls in the absence of perfect information. For instance, a decision to increase the price of a prescription drug may be seen as a choice between higher near-term profits against the risk of loss of opportunity to extract margins before the competing peers jump in, also in the near term. How about the longer-term risk of being charged or even just perceived as price gouging that could lead to loss of trust and loss of much larger number of consumers? For instance, former Turing Pharmaceuticals CEO Martin Shkreli faced price gouging allegations for buying an obscure antiparasitic drug used by AIDS patients and jacking up the price astronomically, Likewise, reducing pollution to at or just below the minimum mandated by law may be a choice based on the lower short-term cost option, however, it ignores the risk of longer-term more stringent laws, loss of consumer preference or loyalty, and risk of lower credit rating and higher cost of capital.

10. Finally, the difficulty in demonstrating the value of sustainability, with the only accepted academic norm—regressions that show statistically significant positive correlation between CSR performance and corporate

financial performance, is not justification enough for inaction or contention that attention to sustainability is at best a distraction from profit making. Finding proof positive of sustainability gains is a challenge; not so much because of its nonexistence but much more so because of its nascent stage, long-term time-frame and less direct correlatability, and that is hardly sufficient reason to discard something critical for the future of humanity.

Porter[465] lays out his criticism of the emerging sequence of responses to the question, "Why should a corporation be socially responsible?"

1. Calling it a moral obligation to be a good citizen and do the right thing is not tenable, as they call for individual corporations to make decisions by balancing competing values (e.g., whether to subsidize today's medicine or to invest in tomorrow's cure).
2. Arguing that it is a license to operate has its share of troubles. The process of obtaining stakeholder permission by meeting their needs has a great inherent danger: a well-organized and articulate interest group could easily get its way at the expense of what may be more socially equitable.
3. While building a corporate image and brand loyalty is invaluable in the marketing space, Karnani[466] notes that corporations resist fundamental transformation, and often indulge in cosmetic campaigns and green-washing. Green-washing, which in one form is lobbying against public interest, takes on a life of its own.
4. Creating consumer purchasing preference or pursuing it to serve as an insurance to temper public wrath in case of company-caused crisis has been frequently observed and even documented in both academic and industry trade literature. However, that is highly uncertain. For instance, it could not fend off the recent jury award of US$110 million, the fourth such award against Johnson & Johnson[467] on talc causing ovarian cancer; an example of unethical behavior by the company in the face of their 1997 internal memo from a company medical consultant that said "anybody who denies" the link between use of hygienic talc and ovarian cancer is "denying the obvious in the face of all evidence to the contrary."[468]
5. Finally, corporate sustainability has been promoted as an act of enlightened self-interest to secure long-term economic performance by avoiding short-term social or environmental cost cutting. But absent goal congruence—where both saving the environment and saving of expenses occur concomitantly—there is a huge gap in this model. Porter says there is no defined framework to validate trade-offs between economic, environmental, and social benefits and optimize value. Furthermore, Ramanan[469] suggests absent ethics as the fourth bottom line, in other words, sans purpose as a moderator, even congruence of the other three bottom lines may be skewed, and not incorporate issues such as distributive justice or intergenerational equity in the decision process.

According to the WBCSD[470] sustainability investments drive financial outperformance. Companies benefit from sustainability by driving down costs through operational performance and efficiency, minimizing business risks, attracting and retaining talent, and increasing employee morale and productivity. Sustainability also generates growth by institutionalizing learning and innovation, creating brand recognition and reputation, building customer loyalty, improving access to capital, improving supply chain management, and providing a long-term strategic focus.

3.3.2.1 Environmental Investments Lower Credit Risk and Cost of Capital[471]

An area of considerable debate is whether environmental expenses are costs with no tangible returns. Does investment in environmental performance pay off in better economic performance? The financial implications of better environmental performance include a lower cost of capital. Sharfman[472] notes Garber's outcome of a robust positive relationship between environmental liabilities (clearly driven by then Superfund fear and notoriety) and costs of equity for large chemical firms. Although higher levels of environmental risk-management permit more debt, they also increase the cost of debt, even after a partial offset by a tax shield. First, lenders see these investments in environmental risk management as inefficient. Second, higher environmental risk-management levels are positively related to leverage, which also increases cost of debt. Yet Sharfman found that "firms that develop a strategy to improve their risk management through improved environmental performance reduce their weighted average cost of capital."[473]

Bauer et al. found environmental performance to have similar financial implications on credit risk.[474] From 1996 to 2006, they analyzed the profiles of around 600 US public corporations testing whether environmental concerns are associated with lower credit ratings and higher cost of debt financing. Firms (borrowers) with poor or inadequate management of environmental risks had a higher default risk for bondholders and hence impaired investments, and if faced with significant liability risk, they filed for strategic bankruptcy to avoid the liabilities and in the process also subordinated the claims of their debt holders. This situation was exacerbated by the far-reaching implications of Superfund or Comprehensive Environmental Response, Compensation, and Liability Act (CERCLA),[475] which is retroactive, and has joint and several strict liabilities for firms and banks. Finally, they observed that the relevance of environmental management issues has also increased for bond investors over the recent decade. Their analysis provides comprehensive evidence that corporate environmental management is a cross-sectional credit risk valuation determinant.

Clark et al. study[476] reveals that "90% of the studies on the cost of capital show that sound sustainability standards lower the cost of capital of companies." Fulton et al.[477] report that 100% of the academic studies (more than 50 academic papers) show clear agreement that high ratings for CSR and ESG factors result in lower cost of capital in terms of debt (loans and bonds) and equity. In effect, the market recognizes the lower risk and rewards.

3.3.2.2 Do Sustainability Environmental, Social, and Governance Investments Enhance Financial Performance?

There is some evidence that shows tangible benefits from sustainability. For instance, companies in Dow Jones Sustainability Index or Goldman Sachs' SUSTAIN outperformed industry averages by about 15% in 16 of 18 categories from May to November 2008.[478] Carroll opines that one incentive to find this elusive business case correlation is "probably a response to Milton Friedman's continuing arguments against the concept (*of CSR*), claiming that business must focus only on long-term profits."[479]

Most academic and professional literature of more than four decades, on value implications of sustainability investments, have not established any significant relationship between sustainability investment and corporate financial performance. Margolis et al.[480] conducted a meta-analysis of more than 250 studies and found a positive but very small empirical link between corporate social performance (CSP) and corporate financial performance (CFP). The weighted average effect size was a zero-order correlation of 0.105, a rather small effect size of association between CSP and CFP. Fulton et al.[481] report that more than 85% of the academic studies (more than 50 academic papers) show compelling evidence that companies with high ratings for ESG factors show market (e.g., stock or bond price, fund returns, Tobin's Q) and accounting based (e.g., return on assets, return on equity, firm value) outperformance. There is a correlation between financial performance and advantageous ESG strategies, over medium to long term (3–10 years). Fulton et al.[482] also report that 88% of the academic studies (more than 50 academic papers) of socially responsible investment portfolio fund returns show neutral or mixed results. These funds use exclusionary screening strategy.

Until recently, the business case for sustainability investments has been quite ambiguous, inconclusive, or contradictory.[483] However, a recent Meier et al.[484] study to find the link between CSP and CFP in Europe demonstrates a strong positive link (+20.36%) between CSR performance and market-based CFP. The study involved 591 European firms and used a new set of data for CSP from Vigeo Eiris, Europe's leading social rating agency. Meier et al's explanations for this finding include reputational effect where CSR is seen as a lever to develop firm image and reputation and generate a positive effect on shareholders, customers, and employees; the stakeholder theory where firm's performance depends on the intensity of its stakeholders' support, given the positive link in Europe between corporate shareholder value and firms' stakeholder management; and the potential relationship between CSR performance and firm's strategic growth options.

Another recent (2015) evaluation by Clark et al. of more than 200 studies[485] finds a remarkable correlation between diligent sustainability business practices and financial (economic) performance and state that "80% of the reviewed studies demonstrate that prudent sustainability practices have a positive influence on investment performance." Companies with strong sustainability scores and investment strategies, which incorporate ESG issues, are less risky and show better operational performance.

Another contemporary, but more extensive study by Friede et al.[486] aggregates evidence from more than 200 company-focused empirical ESG-CFP studies and finds that investing in ESG pays financially, and the positive ESG impact on CFP is stable over time. Also, it finds that each of the three factors, E, S, or G, has a relatively positive effect on CFP.

3.3.2.3 Materiality of Sustainability Investment Matters

It is illusory to believe that the market will reward companies financially for *doing good*. There are very real trade-offs; markets reward only those sustainability initiatives that also enhance financial performance and punish others that depress financial outcome.

Friede et al.[487] note that about 150 portfolio studies exhibit on average a neutral/mixed ESG-CFP performance relationship, but these results may be tainted by systemic and idiosyncratic risks in portfolios. Another important cause for their findings could be not distinguishing material and non-material sustainability investments. In order to harvest the full potential of value-enhancing ESG factors, they call for material ESGs for long-term positive impact on CFP.

Khan et al.[488] report that corporations with superior performance on material sustainability issues significantly financially outperform their sector peers. They also observe that investment in nonmaterial issues, on the other hand, does not lead to financial outperformance. In effect, portfolios based on materiality index outperform those based on total index.

Khan et al. map SASB (Sustainability Accounting Standards Board) industry specific materiality guidance to firm level MSCI KLD (Morgan Stanley Capital International and KLD Research & Analytics Inc.) performance rating, using Bloomberg Industrial Classification System (BICS) for sector and industry allocation. They orthogonalize change in material sustainability score due to size, leverage, profitability, and sector and form a portfolio of firms. They compare the one year ahead abnormal stock return performance of this portfolio of firms to the unexplained residuals in material sustainability score change. This elimination of noise in the total index, which include immaterial issues, provides clarity and much greater predictive power for financial performance.

3.3.2.4 Integration of Environmental, Social, and Governance Factors in Capital Allocation Decisions

Investment policy is a key determinant of an organization's value. Sustainability investment is a newer class of investment. Increasing efficiency of sustainability investment calls for integration of sustainability ESG factors in the capital allocation decisions. This raises the question of which ESG data should be integrated—what is material and matters. All the leading sustainability frameworks, including IIRC, GRI, and SASB, are now focusing on setting standards for nonfinancial factors that help separate material and nonmaterial issues.

These nonfinancial factors supplement financial metrics in understanding how a business is creating value and can throw some light on the long-term sustainability of a business. For instance, investors are increasingly using ESG data from providers including MSCI KLD, ASSET 4, a Thomson Reuter's business, and Bloomberg for a more complete picture. ESG metrics are also used by socially responsible investors to sort undesirable or irresponsible companies from preferred companies with strong metrics. They could influence these select companies through stock purchase incentives and guide efforts of those companies to direct capital towards underserved areas such as affordable housing. "Mainstream investors could use ESG metrics to evaluate company risk elements, incorporate ESG data into traditional financial analysis to get a more complete long-term picture, and identify valuable intangibles that may attract investors, and raise share prices."[489]

When selecting and evaluating asset managers, institutional investors are increasingly assessing how the companies in the portfolio integrate ESG aspects into their processes. The risk and return relationships of different ESG aspects vary. Many investors want a tailored benchmark to reflect their ESG preferences. As part of their stewardship responsibilities, an increasing number of fund managers and pension funds are also engaging in investee company ESG practices.

The ESG ratings can be used as the building blocks for integrating ESG into investments[490] in a variety of ways, including active portfolio management, benchmark construction, and company engagement. It helps assess the range, average and variance of asset manager portfolios with respect to ESG integration. ESG ratings provide a granular and comprehensive data set for research and analysis that allow users to apply them into a proprietary model and develop their own views on how, or how not, to integrate ESG data. It enables investors to define ESG eligibility criteria for an investment universe, identify companies with the greatest ESG exposures, and use alongside conventional risk measures and quantitative models to create unique system indexes to meet custom specifications.

3.3.2.5   Framework to Build and Document a Business Case

3.3.2.5.1   Framework to Build a Business Case

British Standards Institution presents a framework and a tool kit for creating a business case for sustainability. The five-step process from a corporate perspective is as follows:

1. Understand the significant impacts of the corporation's activities on the environment, society, and the economy and identify both real and perceived opportunities and risks they represent.
2. Identify key stakeholder issues, both real and perceived, through consultation to improve ability to maximize opportunities and minimize risks.
3. Make it relevant by linking the opportunities and risks to company's core business and mapping the issues and impacts to company business plan and strategic objectives.
4. Back it up with data and examples to support each opportunity and risk, and include available or estimated financial costs and benefits.
5. Keep it dynamic and updated to ensure it changes with company priorities and stays relevant for the audience.

3.3.2.5.2 Steps to Document the Business Case

While the actual format or structure may vary, one effective way is shown as follows:

1. Develop a summary of the current status of CSR activities relevant to the sector and progress of its adoption by peers and partners.
2. Identify the aspects of CSR that are most relevant to the organization. Follow it with a review of the benefits that may flow from the proposed CSR strategy and activities. These could be presented in one or more of the following four groups:[491]
   a. Reduction in risk and cost deals with the threats from stakeholder demands and their mitigation through enhanced but defined level of CSR performance. These may comprise of lower pollution or higher community involvement.
   b. Gain in competitive advantage is achieved through using improved CSR performance to differentiate offerings from competitors to gain market share or price premium.
   c. Buildup in reputation and legitimacy is accomplished through aligning with stakeholder interests. This gain in reputation and credibility is leveraged for enhanced value creation.
   d. Value creation through win-win synergy focuses on what is known today as material CSR—investing in stakeholder needs that also serve organization's needs, such as, community education contributions also lead to a better talent pool to draw from.
3. Provide a brief discussion on limitations and uncertainties of CSR gains, which may include known common issues, such as consumer inability to impact the market (for instance price of a global commodity like oil), the CSR gains are not perpetual, and CSR gains may not deliver in the short time frame that many small firms and speculators focus on.

3.3.3 BUILDING THE SUSTAINABILITY STRATEGY

Sustainability is a mega-trend, not too different in traits that resulted from electrification, mass production, quality revolution, globalization, and information technology. The quality revolution renovated core tools and methods used in manufacture. Information technology altered nearly every aspect of what and how people do things. Sustainability too is bound to transform the way companies and governments compete to lead the wave; they have to adapt and possibly innovate.

Ramanan's[492] sustainability behavior pyramid, similar to Carroll's[493] ethics pyramid, generically categorizes the state of strategic efforts of corporate institutions. Modest or go-slow approaches yield very little, may be even worse off than those with no effort; costs occur, benefits do not accrue. They tend to languish in the bottom most rung of the pyramid. Top of the pyramid is reserved for those who put in innovative yet realistic levels of efforts across the entire value chain, such as Walmart.

3.3.3.1 Categorization of Sustainability Strategies

Per Zadek,[494] corporations pursue CSR strategies to achieve one or more of the four key objectives: (1) defend threat to reputation, (2) increase surplus of benefits over costs, (3) mainstream with the corporation's broader strategy, and (4) preempt or mitigate risk through learning and innovation. Figure 3.6 shows the author's proposed categorization of sustainability strategies.

The first criterion seeks to determine what sector the strategy is for. Is it global, as in a group of some or most nations, or is it for the government of one nation? Alternatively, is it for the corporate sector, and if so, is it for the financial or nonfinancial segment? Financial corporations are principally engaged in providing financial services, such as banking, project lending, insurance university endowment growth and pension funding services. Clearly there will be overlaps among the state-owned entities that operate like private corporations, as well as financial corporations that invest in portfolio companies. The main purpose of this categorization is to identify common elements in strategy development.

The second set of criteria defines the scope of the sustainability strategy. Is it an overall or broad entity-wide approach, or is it a more specific narrow scope for a focused goal, such as depleting resource?

The third group of criteria starts with the aspect or bottom line being honed in on. For instance, if an organization has a major failure of governance or ethics, the goal may be to resurrect public trust and rebuild an ethics culture. Alternatively, the institution may have recently suffered an environmental disaster, thus attempting to recoup its reputation to enhance credibility and mitigate credit risk.

The next criterion is driven by the objective or the motivation for developing a sustainability strategy. Is the objective to avert a threat or manage a major risk? For instance, a fossil fuel-based power company may be facing significant regulatory pressure to lower its carbon intensity, which is a major threat to its existence. Or an automobile company may be losing market share because of changing customer preference for the electric or hybrid vehicle. For instance, a company like Tesla, which took advantage of the opportunity, is worth more than General Motors today. Other institutions may be facing the risk of depleting natural resource such as timber or fish. Or the risk may come from potential devastation of its assets as a consequence of climate change.

3.3.3.2 Proposed AIM Approach to Build a Sustainability Strategy

Figure 3.7 describes the author's proposed AIM approach (a mnemonic for Adapt, Innovate, and Mitigate) to build a sustainability strategy. All institutions, global, government, corporate and finance, consider building a sustainability strategy to avert an oncoming risk or threat, or recognize a potential opportunity in sustainability, the new mega-trend. My father-in-law often quoted a phrase, he did not know who the author was, that said, give me the courage to change what I can, adapt to what I cannot, and the wisdom to know the difference It is called The Serenity Prayer written by the American theologian Reinhold Niebuhr. The actual quote is "God, grant me the serenity to accept the things I cannot change, Courage to change the things I can, And wisdom to know the difference.

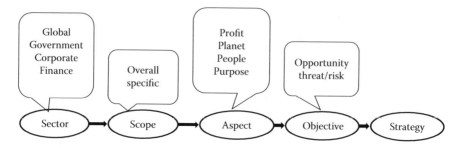

FIGURE 3.7 Strategy categorization flow diagram.

The proposed AIM approach, based on this foundation, is to identify and pursue one of three actions:

1. *Adapt to consequence*: For instance, the climate change-related global warming and heat impact on public health may necessitate nations and governments to develop adaptation strategies. Likewise, electric power companies, a regulated monopoly, is price limited yet required to meet increasingly stringent environmental standards and rising consumer demand. Although not very intuitive, they have adapted to the consequence and joined hands with governments in *un-marketing* their product. These demand-side management (DSM) programs are designed to encourage consumers to modify their energy consumption level and pattern their usage to help defer the need for new sources.

2. *Innovate for opportunity*: Targeting a specific segment of the millennials— environmentally conscious consumers—auto companies have identified the opportunity for electric and hybrid cars. Tesla is a shining example of a startup in an established industry, and it has become the most valuable company in its sector. Similarly, General Electric's (GE) Ecomagination[495] is another innovative initiative to gain from the sustainability mega-trend delivering a revenue of $160 billion from 2005 to 2014. Per IBM, in 2010 alone, the Smarter Planet initiative generated US$3 billion in revenue and double-digit growth from more than 6000 client engagements in areas such as mobile Web, nanotechnology, stream computing, analytics, and cloud.[496] Per Accenture,[497] *circular advantage,* or the opportunities in circular economy, is already a trillion dollar-plus prize. It is a fertile ground for innovation. For instance, a farm-produce company develops its wastes into energy producing ethanol, a renewable fuel, while another sends food waste to an anaerobic digestion facility.

3. *Mitigate risk or threat*: Risk is the effect of uncertainty on objectives and is a function of likelihood and consequence. It can be mitigated either by reducing the probability of likelihood, the severity of consequence, or both. For instance, the World Bank faced some major lawsuits prior to the new phase of imposed mandatory environmental and social impact assessments, stakeholder engagements, and defined statements. This greatly reduces the likelihood of this threat. Similarly, a common risk to reputation comes from

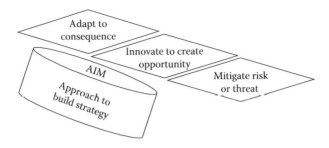

FIGURE 3.8   AIM approach to build strategy.

associating with companies, such as tobacco companies, with a negative
social image. Many institutional investors and most mission driven impact
investors prefer not to invest in funds that carry tobacco stocks in their portfolio. One strategy to mitigate this risk of losing these investors is to eliminate the probability of negative social image by the exclusionary screening
out of tobacco companies, thus mitigating the risk (Figure 3.8).

3.3.3.3   Multistage Strategy Framework[498,499]
Most mega-trends call for a multistage strategy, which in the sustainability context,
comprises of renovate, innovate, and create.

1. *Renovate current operations*: Optimize and do things in better ways to
   perform in more sustainable ways. For example, 3M's pollution prevention pays outperforming competition, and ExxonMobil's Global Energy
   Management Systems industry pace-setting energy conservation.
2. *Innovate new products and services*: Expand into adjacencies that have
   higher sustainability value such as GE's Ecomagination and hybrid automobiles from Tesla and Toyota.
3. *Create new or transform core business* (to differentiate based on sustainability goals): For example, Acumen's move from charitable donations to
   philanthropic capitalism for scaling.

3.3.3.4   Strategy Execution Stages
Typical steps involved in the development and execution of a sustainability strategy are:

1. Select sustainability strategy category and approach using the AIM method
2. Establish methods to assess value, benefits and risks, and weigh options
3. Develop strategy using data and analytics
4. Integrate sustainability goals into core operations or transform core operations to meet sustainability goals[500]
5. Engage stakeholders ethically

3.3.3.5   Example Strategies for Select Sectors or Specific Aspects

3.3.3.5.1   Adaptation Strategies for US Government—
Climate Change Public Health Protection[501]

1. Raising awareness through outreach and education about heat island risks and action plans
2. Providing incentives to spur individual heat island reduction actions such as tax breaks
3. Enacting ordinances and giving grants to communities that promote urban forests
4. Retro-fitting municipal buildings with cool technologies
5. Adding urban heat island mitigation strategies in urban policies or regulations
6. Zoning codes to promote heat island mitigation strategies such as parking lot shade
7. Green building initiatives to place a high priority on human and environmental health
8. Building codes related to energy usage and conservation requirements and standards
9. Weatherization to make low-income family homes energy efficient at no cost to them

3.3.3.5.2   Innovation Strategies for Corporate Private Sector

Sustainable strategies[502,503] to simultaneously boost financial and ESG performance must focus on material ESG issues and produce major innovations in products, processes, and business models. Eccles[504] study of three thousand companies shows that if companies innovate, they can simultaneously improve ESG and financial performance. However, when making a tangible change, the innovation has to be major and involve significant investments and longer payback periods.

Typical strides involved in successful innovation initiative include[505,506]

1. Identify material ESG issues
2. Quantify ESG performance and financial performance relationship
3. Innovate products, processes, and business models
4. Communicate innovations to stakeholders

Representative innovation examples:

1. *Shale gas*: Shale resources and even production techniques to extract it have been known for a while. The hydraulic fracturing process was used in conventional limestone and sandstone reservoirs for more than sixty years before the onset of the shale gas revolution. But even as recently as a few years ago, very little of the resource was considered economical to produce. One major technology often employed in producing natural gas from shale today is horizontal drilling that maximizes the number of natural fractures intersected in the shale that provides additional pathways. Innovative advances—especially in horizontal drilling, hydraulic fracturing, and other well stimulation technologies—have made it possible to recover six hundred trillion cubic feet of shale gas.

2. *Circular economy*: Cisco identified that 80% of the returns, which were being disposed of as scrap at millions of dollars, were in working condition. A value-recovery team identified inhouse, customer service departments and labs that provide technical support, training, and product demonstrations and set it up as a business unit. Cisco's recycling costs fell by 40% and the value recovery team contributed $100 million to the economic bottom line in 2008.[507]

3. *Sustainability technology*: Jeff Immelt, chairman and CEO of GE said, "Ecomagination is one of our most successful cross-company business initiatives. Since its 2005 launch Ecomagination has generated more than $160 billion in revenue. The company's operations have seen a 34% reduction in GHG emissions since 2004 and a 47% reduction in freshwater use since 2006, realizing $300 million in savings."[508] Some of the sustainability technology developments ongoing at GE include:

    a. Innovate and advance solutions to increase power plant efficiencies.

    b. Reduce cost and raise output of wind turbines to lower wind power generation costs.

    c. Evaluate with Statoil if a system can capture CO_2 produced from emissions; reuse the CO_2 to fracture rock formations; and then capture it again for reuse on the next well.

4. Capture previously flared natural gas, propane, and butane for sale and pressurize methane into compressed natural gas (CNG) to power remote fields.

5. Develop with Sasol, a technology that cleans wastewater while producing biogas.

3.3.3.5.3   *Mitigation Strategies for Financial Sector Reputation Sustainability Global Sustainable Investment Alliance (GSIA)*[509]

1. *Negative/exclusionary screening*: Based on ESG criteria, excluding companies from a sector portfolio or even certain sectors entirely from a fund-Blackrock's gun-free retirement fund, in the wake of the school shootings and millennials gun violence concerns is a recent example.

2. *Positive/best-in-class screening*: Including in the investment portfolio, sectors, companies, or projects based on better ESG performance relative to industry peers or compared to other similar sectors.

3. *Norms-based screening*: Selecting investments that meet or exceed minimum standards of business practice based on international norms, usually assessed by a third party.

4. *Integration of ESG factors*: Systematic and explicit inclusion of environmental, social, and governance factors along with traditional financials in investment decisions.

5. *Themed investing*: Investments in assets specifically related to sustainability, such as clean energy, green technology, or sustainable fishery or forestry or agriculture.

6. *Impact or community investing*: Investments, specifically directed to social or environmental purpose or to underserved communities, such as affordable housing.

7. *Corporate activism*: Investing to gain and use shareholder power to influence corporate behavior, including board resolutions and through direct corporate engagement

ENDNOTES

[331] US Office of Management and Budget, Circular No. A-94 revised, Accessed April 2017 and available at https://obamawhitehouse.archives.gov/omb/circulars_a094#8.

[332] Ibid.

[333] Boardman, A.E., D.H. Greenberg, A.R. Vining, and D.L. Weimer, *Cost-Benefit Analysis: Concepts and Practice*, 4th ed.

[334] Taback, H. and R. Ramanan, *Environmental Ethics and Sustainability* (Boca Raton, FL, CRC Press, 2013), 45.

[335] Viscusi, W.K., Monetizing the benefits of risk and environmental regulation, *Fordham Urban Law Journal*, 2005, 33(4), 1003.

[336] US Office of Management and Budget, Report to congress on the costs and benefits of federal regulation, EPA Clean Air Act Section 812 Cost Benefit Analysis, 1997.

[337] US Office of Management and Budget, Regulating risk: The cost-effectiveness of federal efforts to reduce health and safety risks, Accessed April 2017 and available at http://www.rbbelzer.com/uploads/7/1/7/4/7174353/omb_1991__regulating_risk_.pdf.

[338] US Office of Management and Budget, Report to congress on the costs and benefits of federal regulation, EPA Clean Air Act Section 812 Cost Benefit Analysis, 1997.

[339] US Office of Management and Budget, Circular No. A-94 revised, Accessed April 2017 and available at https://obamawhitehouse.archives.gov/omb/circulars_a094#8.

[340] Ibid.

[341] Ocean Tomo, Accessed March 2017 and available at http://www.oceantomo.com/blog/2015/03-05-ocean-tomo-2015-intangible-asset-market-value/.

[342] Time, Warren Buffett's boring, brilliant wisdom, Accessed April 2017 and available at http://business.time.com/2010/03/01/warren-buffetts-boring-brilliant-wisdom/.

[343] CGMA, Valuing intangible assets, Accessed April 2017 and available at https://www.cgma.org/content/dam/cgma/resources/tools/downloadabledocuments/valuing-intangible-assets.pdf.

[344] Ibid.

[345] Ibid.

[346] Time, Warren Buffett's boring, brilliant wisdom, Accessed April 2017 and available at http://business.time.com/2010/03/01/warren-buffetts-boring-brilliant-wisdom/.

[347] Mendelsohn, R. and S. Olmstead, The economic valuation of environmental amenities and disamenities: Methods and applications, Accessed April 2017 and available at https://environment.yale.edu/files/biblio/YaleFES-00000201.pdf.

[348] IChemE UK, Sustainability metrics, Accessed April 2017.

[349] Callan, S.J. and J.M. Thomas, *Environmental Economics and Management* (Mason, OH: Thomson Southwestern, 2010), 159.

[350] Hanemann, W.M., Valuing the environment through contingent valuation, *The Journal of Economic Perspectives*, 1994, 8(4), 19–43, The Exxon Valdez oil spill in Prince William Sound was the first case where contingent valuation surveys were used in a quantitative assessment of damages.

[351] Sen, A., Nobel Prize winning economist, *On Ethics of Economics*, Wiley, 1991.

352 Social Value UK, The SROI guide, Accessed April 2017 and available at http://www.socialvalueuk.org/resources/sroi-guide/.

353 Fujiwara, D., The seven principle problems of SOI, August 2015, Accessed October 2017 and available at https://www.simetrica.co.uk/blog.

354 Hicks and Allen (1934). As quoted in Daniel Fujiwara.

355 Fujiwara, D., Accessed April 2017 and available at https://www.simetrica.co.uk/wwwsimetricacouk-resources.

356 Network for Business Sustainability, Measuring and valuing environmental impacts, Accessed April 2017 and available at http://nbs.net/wp-content/uploads/NBS-Systematic-Review-Impacts1.pdf.

357 International Organization for Standardization, ISO 14040: Environmental management—Life-cycle assessment—Principles and framework, July 1, 2006.

358 Curran, M.A., US EPA "Life cycle assessment: Principle and practice," U.S. Environmental Protection Agency, EPA/600/R-06/060, May 2006.

359 CMU, The economic input-output life cycle assessment (EIO-LCA), Accessed April 2017 and available at www.eiolca.net.

360 ISO, 14000 Environmental Standards ISO 14044:2006 "Environmental management—Life-cycle assessment—Requirements and guidelines," Accessed April 2017 and available at https://www.iso.org/publication/PUB100238.html.

361 MIT OCW, Material flow analysis, Accessed April 2017 and available at https://ocw.mit.edu/courses/engineering-systems-division/esd-123j-systems-perspectives-on-industrial-ecology-spring-2006/lecture-notes/lec14.pdf.

362 Brunner, P.H. and H. Rechberger, *Practical Handbook of Material Flow Analysis*, 2004, Accessed April 2017 and available at https://thecitywasteproject.files.wordpress.com/2013/03/practical_handbook-of-material-flow-analysis.pdf.

363 CMU, The economic input-output life cycle assessment (EIO-LCA), Accessed April 2017 and available at www.eiolca.net.

364 United Nations Environment Programme, *Guidelines for Social Life-cycle Assessment of Products*, 2009.

365 ISO, 14000 Environmental Standards ISO 14044:2006 "Environmental management—Life-cycle assessment—Requirements and guidelines," Accessed April 2017 and available at https://www.iso.org/publication/PUB100238.html.

366 UNEP, *Guidelines for Social Lifecycle assessment for Products* (Griesshammer R. et al., 2006).

367 IFC World Bank, Accessed April 2017 and available at http://www.ifc.org/wps/wcm/connect/7540778049a792dcb87efaa8c6a8312a/SP_English_2012.pdf?MOD=AJPERES.

368 US EPA, Accessed April 2017, available at https://www.epa.gov/nepa/national-environmental-policy-act-review-process and Electronic Code of Federal Regulations, Title 40 → Chapter V → Part 150 http://www.ecfr.gov/.

369 USEPA, What is NEPA, Accessed April 2017 and available at https://www.epa.gov/nepa/what-national-environmental-policy-act.

370 Stevens, C., Sustainability assessment methodologies, Accessed April 2017 and available at http://www.oecd.org/greengrowth/39925248.pdf.

371 EY, Risk management for asset management: EY EMEIA survey 2013, 2013, Available at http://www.ey.com/Publication/vwLUAssets/EY_Risk_Management_for_Asset_Management_Survey_2013/$FILE/EY-Riskmanagement-for-asset-management-survey-2013.pdf.

372 Ramanan, R. and W. Ashton, Green MBA and integrating sustainability in business education, Air and Waste Management Association's Environmental Manager, September 2012, 13–15, Accessed December 2012 and available at http://stuart.iit.edu/about/faculty/pdf/green_mba.pdf.

[373] World Business Council for Sustainable Development (WBCSD) is a global, organization of over 200 leading businesses and a Global Network of almost 70 national business councils working together to accelerate the transition to a sustainable world. Accessed April 2017 and available at http://www.wbcsd.org/.

[374] World Business Council for Sustainable Development, Sustainability and enterprise risk management, Accessed April 2017 and available at http://www.wbcsd.org/contentwbc/download/2548/31131.

[375] Ramanan, R., Corporate carbon risk management—A strategic framework, *Environmental Manager*, October 2010, 20–23.

[376] Ramanan, R., Corporate risk management strategies, *Presentation at the 15th World Clean Air Congress*, International Union of Pollution Prevention and Environmental Protection Association, Vancouver, Canada, September 12–16, 2010.

[377] Ramanan, R., Outsourcing divides owner's risk for remediation projects, *Hydrocarbon Processing*, 1999, 78(5), 101–108.

[378] Ramanan, R., Environmental, safety & health costs and value tracking, *Presented at the Townley Global Management Center for Environment, Health and Safety* (New York: The Conference Board, 1998).

[379] ISO 7 Guide 73, Risk management—Vocabulary—Guidelines for use in standards, Accessed April 2017 and available at https://www.iso.org/standard/44651.html.

[380] Kaplan, R. and A. Mikes, Managing risks: A new framework, *Harvard Business Review*, June 2012, Available at https://hbr.org/2012/06/managing-risks-a-new-framework.

[381] ISO 7 Guide 73, Risk management—Vocabulary—Guidelines for use in standards, Accessed April 2017, available at https://www.iso.org/standard/44651.html.

[382] COSO, Enterprise risk management—Aligning risk with strategy and performance, Accessed April 2017 and available at https://www.coso.org/Documents/COSO-ERM-FAQ.pdf.

[383] North Carolina State University, Accessed April 2017 and available at https://erm.ncsu.edu/library/article/coso-erm-framework.

[384] Knowledge Leader by Protivity, Accessed April 2017 and available at http://info.knowledgeleader.com/bid/163293/what-is-the-coso-enterprise-risk-management-framework.

[385] International Standards Organization (ISO) A Structured Approach to Enterprise Risk Management (ERM) and the Requirements of ISO 31000, The Association of Insurance and Risk Managers, The Public Risk Management Association, and The Institute of risk Management, Accessed April 2017 and available at https://www.theirm.org/media/886062/ISO3100_doc.pdf.

[386] International Organization for Standardization, *ISO Risk Management—Principles and Guidelines: ISO 31000:2009* (Geneva, Switzerland: ISO, 2009).

[387] Center for Chemical Process Safety (CCPS), *Layer of Protection Analysis, Simplified Process Risk Assessment* (New York: American Institute of Chemical Engineers, 2001), available at http://www.wiley.com/WileyCDA/WileyTitle/productCd-0816908117.html.

[388] ibid.

[389] Arthur M. (Art) Dowell, III, P.E. and D.C. Hendershot, Simplified risk analysis—Layer of protection analysis (LOPA), *AIChE 2002 National Meeting Paper 281a*, 1, Accessed April 2017 and available at http://citeseerx.ist.psu.edu/viewdoc/download?doi=10.1.1.198.616&rep=rep1&type=pdf.

[390] Center for Chemical Process Safety (CCPS), *Layer of Protection Analysis, Simplified Process Risk Assessment* (New York: American Institute of Chemical Engineers, 2001), available at http://www.wiley.com/WileyCDA/WileyTitle/productCd-0816908117.html.

391 A Structured Approach to Enterprise Risk Management (ERM) and the Requirements of ISO 31000, The Association of Insurance and Risk Managers, The Public Risk Management Association, and The Institute of risk Management, Accessed April 2017, https://www.theirm.org/media/886062/ISO3100_doc.pdf.

392 Taback, H. and R. Ramanan, *Environmental Ethics and Sustainability* (Boca Raton, FL, CRC Press, 2013) 120.

393 A Structured Approach to Enterprise Risk Management (ERM) and the Requirements of ISO 31000, The Association of Insurance and Risk Managers, The Public Risk Management Association, and The Institute of risk Management, Accessed April 2017, https://www.theirm.org/media/886062/ISO3100_doc.pdf.

394 Protivity, Updated COSO ERM What is new?, Accessed April 2017 and available at the-bulletin-vol-6-issue-2-updated-coso-erm-framework-whats-new-protiviti.pdf.

395 Ibid.

396 Palisade Corp., @RISK simulation, Accessed April 2017 and available at http://www.palisade.com/risk/.

397 Palisade Corp., PrecisionTree, Accessed April 2017 and available at http://www.palisade.com/precisiontree/.

398 Kolios, A., Comparative study of multiple-criteria decision making methods under stochastic inputs, *Energies*, 2016, Accessed April 2017 and available at www.mdpi.com/1996-1073/9/7/566/pdf.

399 Mateo, J.R.S.C., *Multi-Criteria Analysis in the Renewable Energy Industry* (London, UK: Springer-Verlag, 2012); Rogers, M., M. Bruen, and L.-Y. Maystre, *ELECTRE and Decision Support, Methods and Applications in Engineering and Infrastructure Investment* (New York: Springer Science+Business Media, LLC, 2000).

400 Kolios, A., Comparative study of multiple-criteria decision making methods under stochastic inputs, *Energies*, 2016, Accessed April 2017 and available at www.mdpi.com/1996-1073/9/7/566/pdf.

401 Ibid.

402 Investopedia, Pareto frontier, Accessed June 2017 and available at http://www.investopedia.com/terms/p/pareto-efficiency.asp.

403 Kolios, A., Comparative study of multiple-criteria decision making methods under stochastic inputs, *Energies*, 2016, Accessed April 2017 and available at www.mdpi.com/1996-1073/9/7/566/pdf.

404 Saaty, T.L., *The Analytic Hierarchy Process: Planning, Priority Setting, Resources Allocation* (New York: McGraw-Hill, 1980).

405 Roy, B., The outranking approach and the foundations of ELECTRE methods, *Theory Decision*, 1991, 31, 49–73.

406 Triantaphyllou, E., S. Shu, S.N. Sanchez, and T. Ray, Multi-criteria decision making: An operations research approach, *Encyclopedia of Electrical and Electronics Engineering*, 1998, 15, 175–186.

407 Kolios, A., Comparative study of multiple-criteria decision making methods under stochastic inputs, *Energies*, 2016, Accessed April 2017 and available at www.mdpi.com/1996-1073/9/7/566/pdf.

408 Mateo, J.R.S.C., *Multi-Criteria Analysis in the Renewable Energy Industry* (London, UK: Springer-Verlag, 2012).

409 Mardani, A., A. Jusoh, E.K. Zavadskas, F. Cavallaro, and Z. Khalifah, Sustainable and renewable energy: An overview of the application of multiple criteria decision-making techniques and approaches, *Sustainability*, 2015, 7, 13947–13984.

410 Madlener, R., K. Kowalski, and S. Stagl, New ways for the integrated appraisal of national energy scenarios: The case of renewable energy use in Austria, *Energy Policy*,

2007, 35, 6060–6074; Tsoutsos, T., M. Drandaki, N. Frantzeskaki, E. Iosifidis, and I. Kiosses, Sustainable energy planning by using multi-criteria analysis application in the island of crete, *Energy Policy*, 2009, 37, 1587–1600; Behzadian, M., R.B. Kazemzadeh, A. Albadvi, and M. Aghdasi, PROMETHEE: A comprehensive literature review on methodologies and applications, *European Journal of Operational Research*, 2010, 200, 198–215.

411 Adinolfi, G., G. Graditi, P. Siano, and A. Piccolo, Multiobjective optimal design of photovoltaic synchronous boost converters assessing efficiency, reliability, and cost savings, *IEEE Transactions on Industrial Informatics*, 2015, 11, 1038–1048; Ippolito, M.G., M.L. Di Silvestre, E.R. Sanseverino, and G. Zizzo, Multi-objective optimized management of electrical energy storage systems in an islanded network with renewable energy sources under different design scenarios, *Energy*, 2014, 64, 648–662; Mirjafari, M. and R.S. Balog, Multi-objective optimization of the energy capture and boost inductor mass in a module-integrated converter (mic) photovoltaic energy system. *Proceedings of the 2012 Twenty-Seventh Annual IEEE Applied Power Electronics Conference and Exposition (APEC)*, Orlando, FL, February 5–9, 2012.

412 Kolios, A., Comparative study of multiple-criteria decision making methods under stochastic inputs, *Energies*, 2016, Accessed April 2017 and available at www.mdpi.com/1996-1073/9/7/566/pdf.

413 Pilavachi, P., C.P. Roumpeas, S. Minett, and N.H. Afgan, Multi-criteria evaluation for CHP system options, *Energy Conversation and Management*, 2006, 47, 3519–3529.

414 Adinolfi, G., G. Graditi, P. Siano, and A. Piccolo, Multiobjective optimal design of photovoltaic synchronous boost converters assessing efficiency, reliability, and cost savings, *IEEE Transactions on Industrial Informatics*, 2015, 11, 1038–1048.

415 Graditi, G., G. Adinolfi, and G.M. Tina, Photovoltaic optimizer boost converters: Temperature influence and electro-thermal design. *Applied Energy*, 2014, 115, 140–150.

416 Perera, A.T.D., R.A. Attalage, K.K.C.K. Perera, and V.P.C. Dassanayake, A hybrid tool to combine multi-objective optimization and multi-criterion decision making in designing standalone hybrid energy systems, *Applied Energy*, 2013, 107, 412–425.

417 Kantas, A.B., H.I. Cobuloglu, and I.E. Büyüktahtakın, Multi-source capacitated lot-sizing for economically viable and clean biofuel production, *Journal of Cleaner Production*, 2015, 94, 116–129.

418 Cobuloglu, H.I., and I.E. Büyüktahtakın, Food vs. biofuel: An optimization approach to the spatio-temporal analysis of land-use competition and environmental impacts, *Applied Energy*, 2015, 140, 418–434.

419 Kolios, A., G. Read, and A. Ioannou, Application of multi-criteria decision-making to risk prioritisation in tidal energy developments, *International Journal of Sustainable Energy*, 2016, 35, 59–74.

420 Doukas, H., C. Karakosta, and J. Psarras, Computing with words to assess the sustainability of renewable energy options, *Expert Systems with Applications*, 2010, 37, 5491–5497.

421 Datta, A., D. Saha, A. Ray, and P. Das, Anti-islanding selection for grid-connected solar photovoltaic system applications: A MCDM based distance approach, *Solar Energy*, 2014, 110, 519–532.

422 Saelee, S., B. Paweewan, R. Tongpool, T. Witoon, and J. Takada, and K. Manusboonpurmpool, Biomass type selection for boilers using TOPSIS multi-criteria model, *International Journal of Environmental Science and Development*, 2014, 5, 181–186.

[423] Govindan, K., S. Rajendran, J. Sarkis, and P. Murugesan, Multi criteria decision making approaches for green supplier evaluation and selection: A literature review, *Journal of Cleaner Production*, 2013, 98, 66–83.

[424] Kahraman, C. and I. Kaya, A fuzzy multicriteria methodology for selection among energy alternatives, *Expert Systems with Applications*, 2010, 37, 6270–6281.

[425] Mohsen, M.S. and B.A. Akash, Evaluation of domestic solar water heating system in jordan using analytic hierarchy process, *Energy Conversion and Management*, 1997, 38, 1815–1822; Nigim, K., N. Munier, and J. Green, Pre-feasibility MCDM Tools to aid communities in prioritizing local viable renewable energy sources, *Renewable Energy*, 2004, 29, 1775–1791.

[426] Papadopoulos, A. and A. Karagiannidis, Application of the multi-criteria analysis method electre III for the optimisation of decentralised energy systems, *Omega*, 2008, 36, 766–776.

[427] Georgopoulou, E., Y. Sarafidis, and D. Diakoulaki, Design and implementation of a group dss for sustaining renewable energies exploitation, *European Journal of Operational Research*, 1998, 109, 483–500.

[428] Goumas, M. and V. Lygerou, An extension of the PROMETHEE method for decision making in fuzzy environment: Ranking of alternative energy exploitation projects, *European Journal of Operational Research*, 2000, 123, 606–613.

[429] Georgiou, D., E.S. Mohammed, and S. Rozakis, Multi-criteria decision making on the energy supply configuration of autonomous desalination units, *Renewable Energy*, 2015, 75, 459–467.

[430] Taback, H. and R. Ramanan, *Environmental Ethics and Sustainability* (Boca Raton, FL, CRC Press, 2013).

[431] KPMG, Building business value in a changing world, Accessed December 2012 and available at http://www.kpmg.com/Global/en/IssuesAndInsights/ArticlesPublications/Documents/building-business-value.pdf.

[432] Fiksel, J., A systems view of sustainability: The triple value model, Accessed April 2017 and available at https://www.researchgate.net/publication/257742337_A_systems_view_of_sustainability_The_triple_value_model.

[433] Hecht, A.D. and J. Fiskel, Solving the problems we face: The United States Environmental Protection Agency, sustainability, and the challenges of the twenty-first century, *Sustainability: Science, Practice and Policy*, 2015, 11(1), 75–89.

[434] Fiksel, J., A systems view of sustainability: The triple value model, Accessed May 2017 available at http://www.academia.edu/23249479/A_systems_view_of_sustainability_The_triple_value_model.

[435] Deloitte, Analytics for the sustainable business, Accessed April 5, 2014 and available at http://www.deloitte.com/assets/Dcom-UnitedStates/Local%20Assets/Documents/IMOs/Corporate%2Responsibility%20and%20Sustainability/usdsccfbusinessanalytics_011711.pdf.

[436] EY, Using data analytics to improve EHS and sustainability performance, Accessed May 2017 and available at http://www.ey.com/Publication/vwLUAssets/ey-using-data-analytics-to-improve-ehs-and-sustainability-performance/$FILE/ey-using-data-analytics-to-improve-ehs-and-sustainability-performance.pdf.

[437] US EPA, Benefits and Costs of the Clean Air Act 1990–2020, the second prospective study, Accessed May 2017 and available at https://www.epa.gov/clean-air-act-overview/benefits-and-costs-clean-air-act-1990-2020-second-prospective-study.

[438]   US EPA, Climate change in the United States, benefits of global action, Accessed April 2017 and available at https://www.epa.gov/sites/production/files/2015-06/documents/cirareport.pdf.

[439]   Fiksel, J., A systems view of sustainability: The triple value model, Accessed May 2017 and available at file:///C:/Users/ragha/Downloads/A_systems_view_of_sustainability_The_tri.pdf.

[440]   Ramanan, R., Environmental, safety & health costs and value tracking, *Townley Global Management Center for Environment, Health and Safety* (Manhattan, NY: The Conference Board, April 1, 1998).

[441]   Ramanan, R. and A.J. Robb III, Risk-based prioritization: A decision aid, *Proceedings of the Forty-Fifth Annual Southwestern Petroleum Short Course*, Texas Tech University, Lubbock, TX, April 8–9, 1998, 312–15.

[442]   EY, Using data analytics to improve EHS and sustainability performance, Accessed May 2017 and available at http://www.ey.com/Publication/vwLUAssets/ey-using-data-analytics-to-improve-ehs-and-sustainability-performance/$FILE/ey-using-data-analytics-to-improve-ehs-and-sustainability-performance.pdf.

[443]   Verdantix, Green quadrant EHS software 2016 global, Accessed May 2017 and available at http://research.verdantix.com/index.cfm/papers/Products.Details/product_id/873/green-quadrant-eh-s-software-2016-global-/-

[444]   Ramanan, R., *Harnessing Technology to Manage and Report Environmental Performance, Performance Summit* (Arlington, VA: The Performance Institute (a US Govt. Think Tank), 2007).

[445]   Ramanan, R., *Harnessing Technology for Environmental Performance Management* (Phoenix, AZ: ESS EXPO, 2008).

[446]   Verdantix, Green quadrant EHS software 2016 global, Accessed May 2017 and available at http://research.verdantix.com/index.cfm/papers/Products.Details/product_id/873/green-quadrant-eh-s-software-2016-global-/-.

[447]   Ibid.

[448]   Ibid.

[449]   SAAB Group, "…", Accessed May 2017 and available at http://www.computerworlduk.com/tutorial/it-leadership/how-saab-group-uses-analytics-to-create-a-greener-business-3314869/.

[450]   NAEM, Strategies for a successful EHS&S software selection, Accessed May 2017 and available at http://www.naem.org/page/survey_2016_ehsmis.

[451]   Verdantix, Green quadrant EHS software 2016 global, Accessed May 2017 and available at http://research.verdantix.com/index.cfm/papers/Products.Details/product_id/873/green-quadrant-eh-s-software-2016-global-/-.

[452]   SAP, SAP carbon impact, Accessed May 2017 and available at http://www.agdltd.com/Sales/pdflinks/2_SAP_CI_Solution_in_Detail_FINAL.pdf.

[453]   Gartner, SAP to expand EH&S footprint by acquiring TechniData, Accessed May 2017 and available at https://www.gartner.com/doc/1359836/sap-expand-ehs-footprint-acquiring.

[454]   SAS, SAS sustainability reporting, Accessed April 8, 2014 and available at http://www.sas.com/en_us/software/sustainability/reporting.html#section=1.

[455]   TechnologyReview.com, The big data conundrum: How to define it? 2013, Accessed December 8, 2013 and available at http://www.technologyreview.com/view/519851/the-big-data-conundrum-how-to-define-it/.

[456]   Laney, Douglas personal conversation with the author and Beyer, Mark A. and Laney, Douglas, The importance of "big data": A definition, 2012, Accessed October 30, 2017 and available at https://www.gartner.com/doc/2057415/importance-big-data-definition.

[457]   Dalmia, N., Evaluation of the business case for using analytics for corporate sustainability and overcoming the challenges in its execution, MS Theses, MIT, June 2014.

458 Jumiper Networks, Introduction to big data, Accessed May 2017 and available at http://www.one.com.vn/sites/default/files/file-attached/catalog/introduction_to_big_data_-_infrastructure_and_networking_considerations.pdf.

459 Gartner: Sallam, R.L., C. Howson, C.J. Idoine, T.W. Oestreich, J.L. Richardson, and J. Tapadinhas, Magic quadrant for business intelligence and analytics platforms, February 16, 2017, Accessed May 2017 and available at https://www.gartner.com/doc/reprints.

460 Forrester, TechRadar: Big data, Q1 2016, Accessed May 2017 and available at https://www.forbes.com/sites/gilpress/2016/03/14/top-10-hot-big-data-technologies/#2bda93b065d7.

461 Makower, J., Milton Friedman and the social responsibility of business, Greenbiz, Accessed November 24, 2006 and available at https://www.greenbiz.com/news/2016/11/24/Milton-Friedman-and-social-responsibility-business.

462 Visser, W., Ages and stages of CSR. In *The Age of Responsibility*, 1st ed. (Hoboken, NJ: John Wiley & Sons, 2011), 37.

463 Global Sustainable Investment Alliance, Accessed March 2017 and available at http://www.gsi-alliance.org/wp-content/uploads/2015/02/GSIA_Review_download.pdf.

464 Mason, K., President Quaker Oats, *Business Week*.

465 Porter, M.E. and M.R. Kramer, Strategy and society, *Harvard Business Review*, December 2006, 78–88.

466 Karnani, A., Case against CSR, *Wall Street Journal*, August 23, 2010.

467 WSJ, Jury Awards $110 million to Plaintiff in baby talcum powder case, Accessed May 2017 and available at https://www.wsj.com/articles/jury-awards-110-million-to-plaintiff-in-johnson-johnson-baby-powder-case-1493992106.

468 ABC News, Missouri Jury awards $72 million in Johnson and Johnson cancer suit, Accessed February 2016 and available at http://abcnews.go.com/US/wireStory/st-louis-jury-awards-72m-johnson-johnson-cancer-37142765.

469 Ramanan, R. and H. Taback, Environmental ethics and corporate social responsibility. In Dhiman, S. and J. Marques (Eds.), *Spirituality and Sustainability: New Horizons and Exemplary Approaches* (New York: Springer, 2016).

470 Business solutions for a sustainable world, World Business Council for Sustainable Development (WBCSD), Accessed December 2012 and available at http://www.wbcsd.org/.

471 Strandberg Consulting, The business case for sustainability, Accessed December 2012 and available at http://www.corostrandberg.com/ pdfs/Business_Case_for_Sustainability_21.pdf.

472 Sharfman, M. and F. Chitru, How does overall environmental risk management affect the cost of capital, *Strategic Manual Journal*, 2008, 29(6), 569–592. doi:10/1002/smj.678; Feldman, S.J., P.A. Soyka, and P.G. Ameer, Does improving a firm's environmental management system and environmental performance result in a higher stock price? Environmental Group Study (Fairfax, VA: ICF Kaiser International, Inc., 1996). *Journal of Investing*, 1997, 6(4), 87–97; Garber, S. and J.K. Hammitt, Risk premiums for environmental liabilities: Superfund and the cost of capital, *Journal of Environment and Economic Management*, 1998, 36, 267–294.

473 Ibid.

474 Bauer, R. and D. Hann, *Corporate Environmental Management and Credit Risk* (Maastricht University, European Centre for Corporate Engagement), Accessed June 2010 and available at http://www.responsible-investor.com/images/uploads/Bauer__Hann_%282010%29.pdf.

475 US EPA, CERCLA, Accessed October 2017 and available at https://www.epa.gov/superfund/superfund-cercla-overview.

476 Clark, G.L., A. Feiner, and M. Viehs, From the stockholder to the stakeholder—How sustainability can drive financial outperformance, March 2015, Accessed February 2016 and available at http://www.arabesque.com/index.php?tt_down=51e2de00a30f8872897824d3e211b11.

477 Fulton, M., B. Kahn, and C. Sharples, DB climate change advisors, *Sustainable Investing*, Accessed March 2016 and available at https://www.db.com/cr/en/docs/Sustainable_Investing_2012.pdf.

478 Dow Jones sustainability indexes, *SAM Sustainability Investing*, Accessed December 2012 and available at http://www.sustainability-index.com/; GS SUSTINAN, Goldman Sachs, SUSTAIN, December 2012, available at http://www.goldmansachs.com/our-thinking/topics/gs-sustain/index.html.

479 Carroll, A.B. and K.M. Shabana, The business case for corporate social responsibility: A review of concepts, research and practice, *International Journal of Management Reviews*, 2010. doi:10.1111/j.1468-2370.2009.00275.x.

480 Margolis, J.D., H.A. Elfenbein, and J.P. Walsh, Does it pay to be good...and does it matter? A meta-analysis of the relationship between corporate social and financial performance, Accessed May 2017 and available at https://papers.ssrn.com/sol3/papers.cfm?abstract_id=1866371.

481 Fulton, M., B. Kahn, and C. Sharples, DB climate change advisors, *Sustainable Investing*, Accessed March 2016 and available at https://www.db.com/cr/en/docs/Sustainable_Investing_2012.pdf.

482 Ibid.

483 Revelli, C., and E.-L. Viviani, Financial performance of socially responsible investing: What have we learned? *Business Ethics: A European Review*, 2015, 24(2), 158–185.

484 Meier, O., J.-Y. Saulquin, G. Schier, and R. Soparnot, Visiting the corporate social performance-financial performance link in Europe, Accessed May 2017 and available at http://www.vigeo-eiris.com/wp-content/uploads/2016/11/2014_FBS_Schier_Visiting-the-corporate.pdf.

485 Clark, G.L., A. Feiner, and M. Viehs, From the stockholder to the stakeholder—How sustainability can drive financial outperformance, March 2015, Accessed February 2016 and available at http://www.arabesque.com/index.php?tt_down=51e2de00a30f88872897824d3e211b11.

486 Friede, G., T. Busch, and A. Bassen, ESG and financial performance: Aggregated evidence from more than 2000 empirical studies, *Journal of Sustainable Finance & Investment*, 2015, 5(4), 210–233. doi:10.1080/20430795.2015.1118917.

487 Ibid.

488 Khan, M., G. Serafeim, and A. Yoon, Corporate sustainability: First evidence on materiality, *The Accounting Review*, 2016, 91(6), 1697–1724. doi:10.2139/ssrn.2575912.

489 Taback, H. and R. Ramanan, *Environmental Ethics and Sustainability* (Boca Raton, FL, CRC Press, 2013).

490 FTSE, ESG ratings overview, Accessed April 2017 and available at http://www.ftse.com/products/downloads/ESG-ratings-overview.pdf.

491 Kurucz, E., B. Colbert, and D. Wheeler, The business case for corporate social responsibility. In Crane, A., A. McWilliams, D. Matten, J. Moon, and D. Siegel (Eds.), *The Oxford Handbook of Corporate Social Responsibility* (Oxford: Oxford University Press, 2008), 83–112.

492 Ramanan, R., *Climate Hot Spots: Analyzing Emerging US GHG Programs* (San Francisco, CA: IHS Forum, 2007).

493 Carroll, A., The pyramid of corporate social responsibility, *Business Horizons*, July–August 1991, 42, 39–48.

494 Zadek, S., Doing good and doing well: Making the business case for corporate citizenship. Research Report 1282-00-RR (New York: The Conference Board, 2000).

495 GE Ecomagination, Accessed May 2017 and available at http://www.sustainable-brands.com/news_and_views/cleantech/mike_hower/ge_renews_ecomagination_initiative_commits_25_b_clean_tech_rd_20.

496 IBM, Smarter planet, Accessed May 2017 and available at https://www.ibm.com/smarterplanet/us/en/.

497 Accenture, Circular advantage, Accessed April 2017 and available at https://www.accenture.com/t20150523T053139__w__/us-en/_acnmedia/Accenture/Conversion-Assets/DotCom/Documents/Global/PDF/Strategy_6/Accenture-Circular-Advantage-Innovative-Business-Models-Technologies-Value-Growth.pdf.

498 Esty, D. and A. Winston, *Green to Gold* (Hoboken, NJ: Wiley, 2009).

499 Lubin, D.A. and D.C. Esty, The sustainability imperative, *Harvard Business Review*, May 2010.

500 Ibid.

501 US EPA, Climate change adaptation strategies, Accessed May 2017 and available at https://www.epa.gov/arc-x/strategies-climate-change-adaptation.

502 Eccles, R.G. and G. Serafeim, The performance frontier, *Harvard Business Review*, May 2013.

503 Eccles, R.G., I. Ioannou, and G. Serafeim, The impact of corporate sustainability on organizational processes and performance, *Harvard Business School*, July 2013, 1–24.

504 Eccles, R.G. and G. Serafeim, The performance frontier, *Harvard Business Review*, May 2013.

505 Ibid.

506 Eccles, R.G., I. Ioannou, and G. Serafeim, The impact of corporate sustainability on organizational processes and performance, *Harvard Business School*, July 2013, 1–24.

507 Nidumolu, R., C.K. Prahalad, and M.R. Rangaswami, why-sustainability-is-now-the-key-driver-of-innovation, Accessed May 2017 and available at https://hbr.org/2009/09/why-sustainability-is-now-the-key-driver-of-innovation.

508 GE Ecomagination, Accessed May 2017 and available at http://www.sustainable-brands.com/news_and_views/cleantech/mike_hower/ge_renews_ecomagination_initiative_commits_25_b_clean_tech_rd_20.

509 Global Sustainable Investment Alliance, Accessed March 2017 and available at http://www.gsi-alliance.org/wp-content/uploads/2015/02/GSIA_Review_download.pdf.

4 Sustainability Analytics for Global Issues

4.1 SECTION OVERVIEW

This section highlights a select set of significant global issues, relevant treaties, and conventions and discusses the role of sustainability analytics in accomplishing the goals or objectives. It is not meant to be an exhaustive coverage of all sustainability issues and the global treaties or conventions. The objective is to bring awareness—for the corporate and government stewards, in the context of environmental and social impact assessments, and the implications of participation or withdrawals on its stakeholders.

The first two subsections explore climate change, assessment of benefits of global action, the Paris Agreement of 2015, and the sustainable development of clean energy. They provide a foundation in the strengths and weaknesses of shale gas and wind and solar energy and emphasize the policy and market aspects of green energy.

The third section starts with environmental justice and the Aarhus Convention and highlights the import of ethical dimension in climate change decision frameworks to ensure distributive justice and raises a red flag that, "Carbon-share[510] could become the most contentious distributive justice issue globally." Also, metrics that are relevant and material to climate change distributive justice are discussed. The fourth section on industrial accidents highlights some of the major mishaps in the nuclear and industrial facilities and the responses thereof. The fifth section deals with biodiversity, the lifeline of our ecosystem, and the metrics for its measurement.

The sixth section discusses the impact of integrity failure and how to resurrect public trust and establish an ethical culture. Next section presents the Basel convention for hazardous waste and the emerging practice of circular economy and recycling and green chemistry and design. The eighth section discusses the centrality of water and its global dimensions and the potential war for water! This section starts with an overview of the centrality of water in human life and its global dimension, followed by the global human water footprint and the threats to human water security. It goes on to cover the impact of climate change on water, the nexus between water and energy—especially significant in the context of climate change, and presents some use-specific ways to value water. The ninth and final section covers the MARPOL Convention and oil spill responses.

Figure 4.1 lists the key global treaties and conventions in place.

Aspect	Global Treaties and Conventions
Atmosphere	Framework Convention on Climate Change (UNFCCC), New York, 1992, including the Kyoto Protocol, 1997, and the Paris Agreement, 2015
	Stockholm Convention on Persistent Organic Pollutants, Stockholm, 2001
	Vienna Convention for the Protection of the Ozone Layer, Vienna, 1985, Montreal Protocol on Substances that Deplete the Ozone Layer, Montreal, 1987
Freshwater	Convention on the Protection and Use of Transboundary Watercourses and International Lakes (ECE Water Convention), Helsinki, 1992
Marine	International Convention for the Prevention of Pollution of the Sea by Oil, London, 1954, 1962, and 1969
	Convention on the Prevention of Marine Pollution by Dumping of Wastes and Other Matter (London Convention), London, 1972
Wetlands	Ramsar Convention on Wetlands of International Importance, especially as Waterfowl Habitat, Ramsar, 1971
Land	Convention to Combat Desertification (CCD), Paris, 1994
Forests	International Tropical Timber Agreement (ITTA), Geneva, 1994
Biodiversity	Convention on Biological Diversity (CBD), Nairobi, 1992
	Convention on the Conservation of Migratory Species of Wild Animals (CMS), Bonn, 1979
	Convention on the International Trade in Endangered Species of Wild Fauna and Flora (CITES), Washington, DC, 1973
Social	Aarhus Convention on Access to Information, Public Participation in Decision-making and Access to Justice in Environmental Matters, Aarhus, 1998
	World Heritage Convention Concerning the Protection of the World Cultural and Natural Heritage, Paris, 1972
	Espoo Convention on Environmental Impact Assessment in a Transboundary Context, Espoo, 1991
Hazardous substances	Convention on the Control of Transboundary Movements of Hazardous Wastes and their Disposal, Basel, 1989
Industrial accidents	Convention on the Transboundary Effects of Industrial Accidents, Helsinki, 1992
Industrial noise pollution	Working Environment (Air Pollution, Noise, and Vibration) Convention, 1977
Nuclear safety	Comprehensive Test Ban Treaty, 1996

FIGURE 4.1   Select global treaties-the Canada Research Chair in International Political Economy.[511]

4.2   CLIMATE CHANGE

In 2004, when author Ramanan had lunch with Nobel laureate Mario Molina, who discovered the root cause of stratospheric ozone depletion, discussions turned to who parallels his discovery in the climate change arena. What surfaced quickly was the name of Nobel laureate Svante Arrhenius, and his 1896 paper describing how carbon dioxide could affect the temperature of the earth. Recent NOAA[512] data shows a strong linkage between the surface temperature of the earth and the carbon dioxide level of the atmosphere.

In 2013, Nobel laureate Rajendra Pachauri, chair of the Intergovernmental Panel on Climate Change (IPCC), at a dinner with the author Ramanan, said that scientific consensus among the thousand-plus scientists was a tough task, not as much because of differences in scientific views but more due to the political pressure of the interest groups they served. Last year, Pope Francis, leader of the Catholic faith with a following of more than one billion people, drew the world's attention to one of the mega-issues of sustainability when he said, "Climate change is a dire threat that humans have a moral responsibility to address."[513]

4.2.1   GLOBAL WARMING, CLIMATE CHANGE ISSUE, AND IMPACT

Global warming and the resultant climate change is the issue and the way the change in environment affects is the consequential impact. A National Oceanic and Atmospheric Administration (NOAA)[514] simulation demonstrating the link between the surface temperature of the earth and the carbon dioxide level of the atmosphere is one of the most complex applications of analytics in the sustainability arena. Further integrated assessment modeling and analysis, which has the consensus of most scientists in the world, show that human activity is, at least partially, responsible for this increase in carbon dioxide level and the resultant warming of the planet. While many gases, termed greenhouse gases (GHGs), contribute to global warming, the gases of most concern, based on their abundance and potential to impact global warming (more than 99%), are carbon dioxide (CO_2), methane (CH_4), nitrous oxide (N_2O), and fluorinated gases. Anthropogenic carbon dioxide is by far the most dominant GHG known to cause global warming. GHG emissions are expressed in terms of carbon dioxide equivalents, and the term *carbon* has been used to represent that here.

Carbon is a natural resource, but unlike other dwindling substances, its use has to be constrained to contain the generation of carbon dioxide. But the near omnipresence of carbon in human life today makes any mitigation or reduction very challenging. Complexity comes from who should bear the burden, reduce consumption, tolerate increase in cost of production, or pay more for the same exact functionality.

Uncertainties in science, political sensitivity, complexity, and inconsistency in the methodology of monetizing benefits, the emergence of new materials and discovery of new adverse health effects are additional complicating factors. Furthermore, "With different countries likely to undertake different levels of climate-change mitigation, the concern arises that carbon intensive goods or production processes could shift to countries that do not regulate greenhouse gas emissions."[515] When coupled with currency exchange rates and other geopolitical uncertainties, the problems compound and confound exponentially.

Climate change, *the two-degree classic*, also truly tests how intergenerational equity and distributive justice is incorporated in making ethical choices. Traditional cost-benefit analysis fails because discount rates beyond a very short period become meaningless. For instance, some of the benefits in climate change mitigation result in benefits that may be a generation or even two centuries away. Almost any rate of discount brings the present value to near zero. Distributive justice, an ethical mandate requires that all human beings get equal share of public goods—earth's atmosphere. Absent purpose as a moderator, powerful stakeholders could skew the objective through the inherent bias of self-interest. The impact of global warming and climate change and the benefits of averting this catastrophe are mirror images covered in the following.

4.2.2   Paris Climate Change Agreement, 2015

The first world Climate Change Conference was held in Geneva in 1979; and the first assessment report was produced by the IPCC in 1990. In response, a global treaty, the UN Framework Convention on Climate Change (UNFCC) was adopted in 1992 in New York and went into effect in 1994. UNFCC signatories met in Kyoto in 1997 and agreed to the Kyoto Protocol reduction targets; however, the developing nations then did not commit to any GHG emission reduction. IPCC findings in their fourth assessment report in 2007 received general scientific consensus, and that same year they received the Nobel Peace prize. But meetings since IPCC consensus, 2007 (Bali), 2009 (Copenhagen), and 2010 (Cancun) failed to result in any binding agreements. Finally, the Paris Agreement 2015 on Climate Change led to binding targets.[516] The Paris Climate Change Agreement has to be ratified by individual countries. This year (2017), the United States decided to withdraw from the Paris Agreement. Highlights of the agreement are:

1. *Global temperature goal*: The goals are to keep global temperature rise well below 2°C above preindustrial temperatures while pursuing efforts to limit it to 1.5°C; increase the ability to adapt; and make finance flows aligned toward low emissions and climate-resilient development. Starting in 2023, conduct five-yearly reviews to assess collective progress.
2. *Mitigation goal and nationally determined contributions* (NDC): The goal is to achieve a balance between anthropogenic greenhouse gas (GHG) emissions by sources and GHG removals by sinks in the second half of this century. All countries are encouraged to formulate and communicate Low Emission Development Strategies and NDCs in 2020, and the plans to strengthen them, based on their national abilities. It encourages conservation and enhancement of biomass, forest, oceanic, and other greenhouse gas sinks and reservoirs as part of mitigation. Parties communicate their NDCs when they join the Agreement.
3. *Adaptation goal*: The Agreement establishes a notional and aspirational *global goal on adaptation* to enhance adaptive capacity, strengthen resilience, and reduce vulnerability to climate change. Adaptation is recognized as a key component of the long-term global response to climate change

and as an urgent need of developing country Parties. All countries communicate adaptation priorities, support needs, plans, and actions, and the collective adaptation efforts are to be reviewed at a five-year frequency.

4. *Loss and damage basis*: To address loss and damage from climate change impacts, the Agreement incorporates the Warsaw International Mechanism for Loss and Damage, and calls for its strengthening. The loss and damage text contains the cryptic words "does not involve or provide a basis for any liability or compensation," reflecting the concern by some that it could be construed as an admission of liability for climate change-related damage and could potentially result in claims for compensation.

5. *Compliance mechanism*: A compliance mechanism is established, to facilitate implementation and promote compliance in a transparent and non-punitive manner. Developing countries will receive support to implement transparency measures.

6. *Capacity building support*: It stipulates that developed countries will provide financial support to developing countries to assist them with capacity building, which includes ability to implement adaptation efforts, take mitigation actions; develop, transfer, disseminate, and deploy mitigation technology; access climate finance; educate, train, and raise public awareness; and enable transparent, timely, and accurate communication of information.

7. *Broader scope*: The path to the common goal must reflect equity and differentiated capability based national responsibility. It is not solely an environmental problem—it cuts across and affects all areas of society. Must respect and promote: human rights; the right to health; the rights of indigenous peoples, local communities, migrants, children, persons with disabilities, and people in vulnerable situations; the right to development; gender equality; the empowerment of women; and intergenerational equity.

8. *Exchange mechanisms*: Market-based and nonmarket based mechanisms are established to allow parties to voluntarily cooperate in mitigation and adaptation to implement their NDCs. Internationally transferred mitigation outcomes (ITMOs) between two or more parties, similar to joint implementation under the Kyoto Protocol used for project-based trading between developed (Annex I) countries; and a centralized global sustainable development mechanism (SDM), likely to mirror the Kyoto Protocol's Clean Development Mechanism, are in place. Nonmarket-based approaches promote mitigation and adaptation ambition and enhance public and private sector participation in implementing NDCs.

9. *Finance*: The issue of differentiation in financing was sorted by stating that developed countries *shall* provide climate finance for developing countries, while developing countries are *encouraged* to provide support voluntarily. Developed countries *should* also take the lead in mobilizing climate finance *from a wide variety of sources*. Public or private finance issue was resolved with adding *the significant role of public funds*. Developed countries will report their public financial contributions to developing countries every two years. Developing countries may also voluntarily provide support. Finance will be part of the global five-year review.

4.2.3 CLIMATE CHANGE ANALYTICS—ASSESSMENT AND VALUATION

One of the first comprehensive estimates of multi-sectoral integrated assessment of the impacts of climate change in the European economy using a consistent framework was the European Commission's Projection of Economic impacts of climate change in Sectors of the European Union based on boTtom-up Analysis (PESETA).[517] project. The EU addressed five sectors that are highly sensitive to climate change. These sectors include agriculture, river floods, coastal systems, tourism, and human health. The approach enables a comparison between the impact categories and helps rank the relative severity of the damage inflicted. The two different time frames considered are the 2020s and the 2080s. PESETA underestimates the impacts of climate change in Europe to a large extent, because it does not include other key impacts on sectors, such as forestry, ecosystems, biodiversity, and catastrophic events.

Proceedings of the National Academy of Sciences (PNAS)[518] summarizes the physical and economic consequences of climate change in Europe. The overall agricultural impact on EU for the 2020 scenario would be a yield gain of around 15%. River flooding would affect 250,000–400,000 additional people per year in Europe by the 2080s, and the additional property damage per year from river floods in the 2080s is estimated to range between 7.7 billion € and 15 billion €, more than doubling the annual average damages over the 1961–1990 period. Without adaptation, the number of people affected annually by flooding in the 2080s increases significantly in all scenarios, in the range of 775,000–5.5 million people, compared to 36,000 people in 1995. However, when adaptation is taken into account (dikes and beach nourishment), the number of people exposed to floods is significantly reduced. Tourism in Europe, measured in bed-nights, is projected to increase by 15%–25%. The consequences of climate change in the aforementioned four market impact categories can be valued in monetary terms.

For the climate scenario of 2080s, the annual damage of climate change to the EU economy in terms of GDP loss is estimated, in today's currency value, to be between 20 billion € for the 2.5°C scenario and 65 billion € for the 5.4°C scenario. It should be noted that while GDP impact may increase, given the computation methodology of adding repairing of damages to buildings due to river floods, it does not increase consumer welfare.

Per PESETA, by the 2080s, the increase in heat-related mortality ranges between 60,000 and 165,000 persons, while the range of estimates for the decrease in cold-related mortality is between 60,000 and 250,000 persons. Adaptation efforts, if taken into account, reduces these estimates by about 80%.

A comprehensive coverage of the potential impact of global warming and climate change in the United States and valuation of benefits from global action to mitigate global warming and avert climate change is available in a recent (2015) US Environmental Protection Agency's (EPA) "Climate Change in the United States: Benefits of Global Action."[519] This report presents the expected physical and monetary benefits from averted impacts on the United States, relative to status quo, because of global GHG mitigation efforts. The benefits are presented by select major sectors.

This study estimates the physical and economic risks of unmitigated climate change and the potential benefits to the United States of reducing global GHG emissions. Global action to mitigate GHG emissions are assumed to reduce and avoid impacts in the United States. Impacts of climate change and benefits of reducing GHGs on select sectors are captured, and the findings highlighted as follows:

1. Human health, the US economy, and the environment will be profoundly impacted by unmitigated climate change. Impacts will not be uniform across the United States, will show a complex pattern across regions.
2. Long-term risks and impacts cannot be avoided unless there is near-term action to reduce GHG emissions significantly. Global GHG mitigation reduces the risk of some extreme weather events and their subsequent impacts on human health and well-being substantially.
3. Adaptation, especially in the infrastructure sector and more so in the coastal property section, can substantially reduce the estimated damages of climate change, highlighting the need for concurrent mitigation and adaptation actions.

Comprehensive and quantitative estimates of climate change impacts are needed to evaluate the benefits of GHG mitigation as well as to evaluate the cost-effectiveness of adaptation responses, and to support climate and energy policies decisions. Appendix E presents the benefits table from the US EPA[520] report. Key benefits from that report are as follows:

1. Improved air quality will avert 13,000 premature deaths valued at US$160 billion in the year 2050, and 57,000 fewer premature deaths valued at US$930 billion in the year 2100. These estimates do not include the benefits from improved air quality because of co-control of air pollutants.
2. Avoidance of extreme temperatures, both high and low, will prevent 1,700 premature deaths valued at US$21 billion in the year 2050, and 12,000 fewer premature deaths valued at US$200 billion in the year 2100.
3. Averting a water supply and demand gap will eliminate water shortage damages ranging in value from US$3.9 to $54 billion in the year 2050 and from US$11 to $180 billion. These were estimated using two climate models that show very different future precipitation patterns.
4. Not having to fortify bridges and adapt roads is estimated to save US$1 to $4 billion in the year 2050 and US$5 to $9 billion in the year 2100. These were estimated using two climate models that show very different future precipitation patterns.
5. Avoided damages in agriculture, forestry, and coral reeves is estimated at US$12 to $15 billion in the year 2050 and US$8 to $14 billion in the year 2100.
6. Depending on the climate model used, carbon storage ranges from a decrease in carbon stocks of 0.5 billion metric tons to an increase in carbon stocks of 1.4 billion metric tons by the year 2100. The economic value of these changes in carbon storage ranges from $9 billion in disbenefits to $120 billion in benefits.

4.2.4   Climate Change Response

Climate change response, given its complexity and potentially catastrophic consequence, requires a multipronged attack. The following approaches, in no particular order, have to be applied in various combinations. One approach is to sequester carbon dioxide coming out of stacks or expand carbon sinks by reforestation. Another approach could be regulatory intervention and voluntary treaties to constrain GHG generation, such as stringent GHG emission limits on GHG emission sources. Allowing GHG emission to be traded as a commodity is a valuable hybrid of supplementing government's market externality intervention with an international GHG emission trading framework and trading exchanges. A third approach recognizes that energy use and power generation for consumption are the biggest contributors to the GHG emission load. Conservation of energy use, a novel demand-side management approach, is indeed a cost-effective way to save energy, reduce GHG emissions, and save money. Innovation to discover clean energy technology, especially renewables, may be inevitable for a sustainable solution (Conservation and clean energy are addressed in the energy section.) Finally, it is best to be prepared to adapt, just in case the consequences or remnants after mitigation and other approaches are still above a threshold.

4.2.4.1   Carbon Sequestration

Carbon dioxide sequestration is complicated by the diverse nature of sources. Roughly 60% of emissions originate from large stationary facilities, such as power plants, cement production, and petroleum refineries. These emissions are commonly a mixture of gases, and often require separation prior to CO_2 storage. There are three major types of carbon storage: geological storage, ocean storage, and mineral carbonation.[521] (1) Geological storage involves the injection of CO_2 into porous rock. CO_2 storage reservoirs may be sedimentary basins, depleted oil reservoirs, or uneconomic coal beds. The cap rock has to be impermeable because CO_2 density is generally less than that of water. (2) Ocean storage involves the injection of captured CO_2 into the ocean, and to isolate it from the atmosphere, usually at depths greater than 1000 meters. CO_2 would eventually dissolve into the ocean and become part of the global carbon cycle. This storage method is not fully developed. (3) Mineral carbonation creates stable carbonate minerals such as magnesite ($MgCO_3$) and calcite ($CaCO_3$) by reacting CO_2 with silicate minerals containing magnesium and calcium. Such minerals are stable over geologic timescales, so sequestration by this method minimizes risk of leakages. There may be some potential for increasing the amount of CO_2 that is stored as biomass in forests and soils. CO_2 can also be injected into oil field reservoirs to enhance oil recovery (EOR).

4.2.4.2   Reforestation and Carbon Sink

Forests play a significant role in the global carbon cycle and have gained attention because of their role as carbon sinks to meet national carbon emission reduction commitments in global climate change treaties. National and global programs have to balance-development support, which may cause deforestation, and reforestation help, through tax provisions. Policies must reckon that leakage, wood supply shifts to other locations, or replacement by more energy intensive products may exacerbate global climate change impact.

Plants use sunlight to photosynthetically convert carbon dioxide, water, and nutrients into sugars and carbohydrates, which accumulate in leaves and roots and release carbon dioxide through respiration. Plants, at death, release their stored carbon to the atmosphere quickly or to the soil slowly, increasing soil carbon levels. Processes and diverse rates of soil carbon change are not yet well understood.

Both afforestation and deforestation impact carbon storage. When some forest vegetation is cut to enhance growth of desired trees, the enhanced growth stores more carbon, but cutting vegetation releases carbon dioxide. Despite stimulation from rising atmospheric carbon dioxide, tree growth may be limited by unavailability of other nutrients. In this context, timber harvesting is an especially controversial forestry practice. The quantitative relationships between net carbon storage and factors that impact, such as soil effects, treatment of residual forest biomass, proportion of carbon removed from the site, and duration and disposal of the products, are yet to be fully established.[522]

4.2.4.3   Greenhouse Gas Emission Trading

In 1997, industrial countries and some emerging economies agreed to legally binding emission targets at the Kyoto Conference and negotiated a legal framework as a protocol to the Convention—the Kyoto Protocol (UN Framework Convention on Climate Change, 1997). The Protocol entered into force in 2006 after being ratified by at least 55 Parties representing at least 55% of the total carbon dioxide emissions of the 1990 Rio Convention Annex I countries.

Countries with commitments under the Kyoto Protocol to limit or reduce GHG emissions must meet their targets primarily through national measures. As an additional means of meeting these targets, the Kyoto Protocol introduced three market-based mechanisms, thereby creating what is now known as the *carbon market*. However, countries have to demonstrate that their use of the mechanisms is *supplemental to domestic action* to achieve their targets. Businesses, nongovernmental organizations (NGOs), and other legal entities may participate in the three mechanisms under the authority and responsibility of governments. Because emission trading separates the issue of who pays for control from who implements the emission control, it offers the possibility of reaching the environmental goals at a lower cost than would be possible if each country were limited to reduction options within its borders.[523] Tradeable permits also facilitate the mobilization of private capital for controlling global warming; private capital is likely to be a critical component. Tradable commodity market will facilitate the development and implementation of innovative approaches to climate change control. Countries differ with regard to their marginal abatement costs because of the differences in dependence on GHG emitting production activities, relative resource efficiency and access to energy sources. The key purpose of these market mechanisms is to stimulate sustainable development through technology transfer and investment, help countries meet their targets by reducing emissions or removing carbon from the atmosphere in other countries in a cost-effective way, and encourage the private sector and developing countries, to contribute to emission reduction efforts.

The Kyoto mechanisms are discussed in the following subsections.[524]

4.2.4.3.1   Clean Development Mechanism

The Clean Development Mechanism (CDM), defined in Article 12 of the Protocol, allows a country with an emission-reduction or emission-limitation commitment under the Kyoto Protocol (Annex B Party) to implement an emission-reduction project in developing countries. Such projects can earn saleable certified emission reduction (CER) credits, each equivalent to one ton of CO_2, which can be counted toward meeting Kyoto targets. It is the first global, environmental investment and credit scheme of its kind, providing a standardized emission offset instrument, CERs. A CDM project activity might involve, for example, a rural electrification project using solar panels or the installation of more energy-efficient boilers. The mechanism stimulates sustainable development and emission reductions, while giving industrialized countries some flexibility in how they meet their emission reduction or limitation targets.

4.2.4.3.2   Joint Implementation

Joint implementation (JI), defined in Article 6 of the Kyoto Protocol, allows one Annex B Party to earn emission reduction units (ERUs) from an emission-reduction or emission removal project in another Annex B Party, each equivalent to one ton of CO_2, which can be counted toward meeting its Kyoto target. The investing parties get a flexible and cost-efficient means of fulfilling a part of their Kyoto commitments, while the host Party benefits from foreign investment and technology transfer.

4.2.4.3.3   International Emission Trading

International Emissions Trading, Article 17 of the Kyoto Protocol, allows countries that have emission units to spare—emissions permitted to them but not *used*—to sell this excess capacity to countries that are over their targets. Thus, GHG emissions, a new commodity, was created in the form of emission reductions or removals. Since carbon dioxide is the principal greenhouse gas, people speak simply of trading in carbon. Carbon is now tracked and traded like any other commodity. This is known as the *carbon market*.

Transfers and acquisitions of these units are tracked and recorded through the registry systems under the Kyoto Protocol. An international transaction log ensures secure transfer of emission reduction units between countries. The other units, which may be transferred under the scheme, each equal to one ton of CO_2, may be in the form of (1) a removal unit (RMU) on the basis of land use, land-use change, and forestry (LULUCF) activities such as reforestation, (2) an emission-reduction unit (ERU) generated by a joint implementation project, and (3) a certified emission reduction (CER) generated from a clean development mechanism project activity.

4.2.4.4   Climate Adaptation

Climate change requires two types of responses: (1) mitigation or reduction of GHG emissions, and (2) adaptation action to deal with the unavoidable impacts. "Adaptation means anticipating the adverse effects of climate change and taking appropriate action to prevent or minimize the damage they can cause, or taking advantage of opportunities that may arise."[525] A strategic approach is needed to ensure that timely and effective adaptation measures are taken, ensuring coherency across different sectors and levels of governance.

4.2.4.4.1   Climate Change Adaptation Framework

The EU climate change adaptation framework is based on four pillars of action: (1) building a solid knowledge base on the impact and consequences of climate change for the EU; (2) integrating adaptation into EU key policy areas; (3) employing a combination of policy instruments (market-based instruments, guidelines, public-private partnerships) to ensure effective delivery of adaptation; and (4) stepping up international cooperation on adaptation. Expanding on incorporation of adaptation strategies into policies, three key questions[526] have been identified that must be addressed in each policy area:

1. What are the actual and potential impacts of climate change in the sector?
2. What are the costs of action/inaction?
3. How do proposed measures influence and interact with policies in other sectors?

EU's Framework for Action on Adaptation (FAA)[527] describes the various elements. The EU highlights the following as the significant impact sectors: agriculture, forests, fisheries and aquaculture, coasts and ecosystem (including marine ecosystems and biodiversity), energy, infrastructure, tourism, water resources, human health, and animal and plant health.

The US Center for Disease Control (CDC)'s Building Resilience Against Climate Effects (BRACE) framework is a five-step process that allows health officials to develop strategies and programs to help communities prepare for the health effects of climate change. The five steps of BRACE are

1. Ascertain the scope of climate impacts, associated potential health consequences, and identify populations and regions vulnerable to these health impacts.
2. Quantify incremental burden of health consequences of climate change.
3. Recognize the most suitable interventions for the health impacts of greatest concern.
4. Develop a climate health adaptation plan, disseminate, and oversee its implementation.
5. Evaluate impact of process, determine value, and improve quality of activities.

4.2.4.4.2   Climate Change Adaptation Strategies

1. Public health and extreme heat[528]
   Climate change may cause a broad range of excessive heat related public health risks:
   a. Raising awareness through outreach and education about heat island risks and action plans.

    b. Providing incentives to spur individual heat island reduction actions including below-market interest-rate loans, tax breaks, product rebates, grants, and giveaways.

    c. Enacting ordinances and giving grants to communities that promote urban forests.

    d. Retrofitting municipal buildings with cool technologies.

    e. Adding urban heat island mitigation strategies in policies or regulations for development and conservation that occurs within its planning jurisdiction and prescribing minimum requirements, especially for new construction.

    f. Promoting inclusion of heat island mitigation strategies such as parking lot shading requirements as part of the zoning codes process.

    g. Supporting green-building initiatives to place a high priority on human and environmental health and resource conservation over the life cycle of a building. Many local, state, and federal governments have adopted green-building programs, or standards, which capture heat island reduction strategies.

    h. Building codes related to energy usage, and conservation requirements and standards, to include cool roofing to save energy, particularly during peak loads.

    i. Sponsoring weatherization to make the homes of qualifying residents, generally low-income families, more energy efficient at no cost to the residents.

2. Public health and air quality

Climate change may cause a broad range of air quality-based public health risks:

    a. As summertime temperatures rise, the rate of ground-level ozone formation, or smog, increases and will cause premature deaths, hospital visits, lost school days, and acute respiratory symptoms. By lowering temperatures, urban heat island mitigation strategies such as urban forestry and cool roofs can help reduce ground-level ozone. Expected performance of ozone standards under changing climate conditions need reassessment.

    b. Wildfires emit fine particles and ozone precursors that in turn increase the risk of premature death and adverse chronic and acute cardiovascular and respiratory health outcomes. Climate change is projected to increase the number and severity of wildfires. Current wildfire management capabilities and monitoring efforts require re-evaluation.

    c. Rising temperatures, altered precipitation patterns, and increasing concentrations of atmospheric carbon dioxide, and airborne allergens will increase asthma episodes.

    d. Vulnerability of extensive forest or rangeland to wildfires or ozone will need reassessment.

    e. Identification of groups vulnerable to these impacts due to preexisting health concerns, sensitive life stages, and so on, or extensive time spent outdoors due to profession or trade will be become critical.

3. Public health and water quality

Climate change may cause a broad range of water quality-based public health risks such as

a. Changing water temperatures may mean that waterborne Vibrio bacteria and harmful algal toxins will be a threat in the water at different times of the year, or in new places.

b. Waterborne pathogens (bacteria, viruses, and parasites such as Cryptosporidium and Giardia) and toxins produced by harmful algal and cyanobacterial blooms in the water may all be exacerbated by increased runoff, warmer temperatures, and point source discharges.

c. Extreme weather events and storm surges will increase the failure risk of drinking water, wastewater, and storm-water infrastructure. The risk of exposure to water-related pathogens, chemicals, and algal toxins will increase in receiving waters and in the drinking water.

d. Identification of groups vulnerable to these impacts due to preexisting health concerns, sensitive life stages, and so on, or extensive time spent in waters for swimming or fishing will become important.

4. Preparing for prolonged power outages

The increased frequency and duration of electrical power outages should be considered due to the increased frequency and severity of storms anticipated with climate change. Methods of ventilating buildings and maintaining acceptable thermal conditions, as well continued quality performance of infrastructure such as drinking water treatment facilities, using resilient or passive design strategies should be included in building and infrastructure design or modification strategies.

4.3 ENERGY

4.3.1 CLEAN ENERGY

The last several years have marked a turning point for clean energy such as solar, wind, and shale gas as well as carbon dioxide sequestration. But a host of complications make sustaining and growing the transition to a green economy challenging. These complications involve technology, finance, politics—and simply the low cost and convenience of our existing fossil fuel-based infrastructure.

4.3.1.1 Wind Energy

Kinetic energy of the flowing wind acts on rotor blades and converts it into torque or rotational energy. It is used either within a generator to produce electricity or directly driving equipment such as pumps or mills. By nature, wind power is intermittent and typically requires either a storage or a backup system. Wind energy systems are classified into (1) grid connected electricity generating; (2) stand-alone electricity generating that are either complemented by other electricity producing energy systems, or supported by storage systems such as batteries or water tank; and (3) mechanical systems.

4.3.1.2 Solar Energy

Energy from the sun is captured and converted either into electric or thermal energy. It is used to either convert sun light directly to produce electricity or use the sun's thermal energy directly to heat, dry, evaporate, or cool. Solar energy systems are classified into (1) solar electric or photovoltaic (PV) systems and (2) solar thermal systems. Photovoltaic devices convert sunlight directly into electric energy; the energy that can be produced is dependent on the solar intensity. The rate is lower in winter and on cloudy days. PV has a seasonal as well as diurnal variation in its capacity to produce electricity. PV devices use the chemical electrical interaction between sun's light and a semiconductor to produce DC electricity.

Most solar cells are made of silicon. The most common technologies to make the solar cells are (1) monocrystalline silicone cells made of silicon wafers cut from one homogenous crystal, (2) polycrystalline silicon cells, which are poured and usually cheaper but less efficient than the monocrystalline silicon cells, (3) thin film cells produced by depositing an extremely thin layer of PV materials on less expensive backing, and (4) multiple-junction cells that use multiple layers of different materials to use a wider spectrum of radiation. Solar thermal systems use sun's heat energy directly. Complex concentrating solar collectors produce temperatures high enough to produce steam, which, in turn, drive steam turbines to produce electricity. Applications, besides electric power generation, include cooking, cooling, distillation, drying, and heating.

4.3.1.3 Geothermal Energy

It is based on the heat in the form of hot water or steam emitted from within the earth. The geothermal heat is sourced either from the radioactive decay of isotopes or the original heat produced by gravitational collapse at the time of the formation of the earth. High temperature resources are used for electricity generation, while the low temperature sources are used for direct applications such as drying or district heating. Energy is extracted from geothermal aquifers with naturally occurring deep porous rocks via a production bore hole and disposed of via an injection hole. Natural variations in geothermal resources occur only over very long periods. Production of highly corrosive brine and sometimes noxious gases such as hydrogen sulfide are potential challenges.

4.3.1.4 Hydro Energy

It is the extraction of energy from flowing water. When water falls from a high to low altitude and passes through a water turbine or water wheel, it converts the kinetic energy of water into mechanical and subsequently electrical energy. Less efficiently and far less frequently, hydro energy is extracted using zero headwater current turbines and is placed directly in a river current. While limited by natural geographical availability, hydropower may be very reliable and cost-effective, such as Niagara Falls. Small hydro units are best suited for base, peak, and standby power generation and select standalone direct drive applications.

4.3.1.5 Bio Energy

This refers to energy derived from organic materials of plant and animal origin. Generally, it is used for renewable energy sources, such as wood, agricultural, and human wastes. There are high variations in their physical and chemical characteristics such as biogas, liquid biofuel, and solid biomass.

Ethanol fuel[529] is produced by breaking down the starch present in corn into simple sugars (glucose), feeding these sugars to yeast (fermentation), and then recovering the main product (ethanol) and byproducts (e.g., animal feed). Wet milling and dry grinding are the two major industrial methods for producing ethanol fuel in the United States. The EPA requires that the gas supply must include fifteen billion gallons of ethanol each year.[530] Most gas sold in the United States contains about 10% corn ethanol. Recent studies show that corn ethanol uses significant amounts of fossil fuels, and hence does not displace that much gasoline.

Biofuels from photosynthetic[531] algae show some promise. The biofuels used around the world today are largely derived from crops like sugar cane and corn to make ethanol, or from vegetable oils like soy to make biodiesel. Applying sustainability analytics, for instance, using lifecycle GHG emissions to compare, ExxonMobil has determined that "algae offer some of the greatest promise for next-generation biofuels,"[532] and opted to invest heavily in photosynthetic algae. Algae interacts with sunlight, carbon dioxide and water to form more algae that contains increasing amounts of hydrocarbons. These algae are processed through bio-oil production and bio-oil conversion to get gasoline and diesel.

Key benefits of using algae for biofuels production include (1) grows on land and water unsuitable for food production—does not compete with food; (2) yields, potentially, greater volumes of biofuels per acre than other biofuel sources; for instance, it can yield five times more biofuel per acre than biofuels from sugar cane or corn; (3) takes energy from sunlight and carbon from carbon dioxide to grow and potentially produce lipids that resemble today's transportation fuels; (4) consumes carbon dioxide when growing and its life-cycle GHG emissions are much lower than corn ethanol; and (5) can grow in brackish water, including seawater, and that eliminates strain on freshwater resources.

4.3.1.6 Shale Gas

Shale gas is natural gas, primarily methane, found in shale formations.[533] Methane that developed from the organic matter and escaped into sandy rock layers formed easy to extract conventional natural gas. But some of it remained locked in the tight, low permeability shale layers, becoming shale gas. Shale gas formations are *unconventional* or low permeability reservoirs.

The reservoir must be mechanically *stimulated* to create additional permeability and to free the gas for collection. This typically involves hydraulic fracturing,[534] where fractures are created by injecting pressurized fluids down a wellbore into the target rock formation to stimulate or fracture shale formations and release the natural gas. Hydraulic fracturing fluid commonly consists of water, proppant (typically sand and sometimes engineered materials), and chemical additives that open and enlarge fractures within the rock formation. Wells are drilled hundreds to thousands of feet vertically below the surface, as well as in horizontal or directional extensions. Sand pumped in with the fluids (often water) helps to keep the fractures open. The reuse of flowback fluids for subsequent hydraulic fracture significantly reduces the volume of wastewater generated, which is typically injected underground for disposal.

Some of the challenges related to shale gas production and hydraulic fracturing include increased consumption of fresh water and induced seismicity from shale flowback water disposal.

4.3.2 ENERGY CONSERVATION

Energy efficiency, driven by multiple drivers, with some overlap and some synergy, has grown by orders of magnitude in the past half century. Regulatory intervention in the form of building codes and appliance standards was the first stimulator. The next driver, known as demand-side management, was the push by electric utility companies, which are typically regulated monopolies in the United States, to encourage consumers to modify their level and pattern of electricity usage to help defer the need for new sources of power. The third driver is the increasing energy-conscious consumer adoption of end-use energy consuming devices and appliances that respond to consumers' decisions.

As innovations bring achievable potential closer to technical potential, enhanced connectivity will lead to larger end-use data sets, increased controls, and new opportunities for utilities to manage and control end-use devices leading to greater grid flexibility. The Electric Power Research Institute (EPRI)[535] estimates that absent appliance standards and building codes, electricity consumption could be 3.5% greater in 2035 and an additional 3% higher without current demand side energy efficiency activity and market driven efficiency. Per EPRI,[536] energy efficiency programs have the potential to reduce US electricity consumption in 2035 by 488–630 billion kW·h (kilowatt hour).

1. The electrical energy savings from the application of appliances energy standards is indeed significant. For instance, refrigerator standards have lowered average energy use of a refrigerator from about 1,800 kW·h per year in 1972 to less than 500 kW·h per year in 2016.[537]
2. Adopted by individual states, building energy codes include energy efficiency measures to make buildings less energy intensive. Buildings consume about a third of the energy and about two thirds of the electricity in the United States. The IECC (International Energy Conservation Code)[538] and ASHRAE (American Society of Heating, Refrigerating, and Air-Conditioning Engineers) Standard 90.1[539] serve as the technical baseline standards for the design and construction of new buildings.
3. Demand-side management (DSM) programs are designed to encourage consumers to modify their level and pattern of electricity usage. Historically, the primary objective of most DSM programs was to help defer the need for new sources of power, including generating facilities, power purchases, and transmission and distribution capacity additions. DSM refers to utility administered programs that attempt to modify total energy and peak load patterns. It does not include changes arising from the normal operation of the marketplace or from government-mandated energy-efficiency standards. However, because of changing business ambiance, electric utilities are also using DSM as a way to enhance customer service. Consortium for

Energy Efficiency (CEE)[540] estimates that the utility spending in the United States and Canada on DSM programs rose to $9.9 billion in 2014, demonstrating the value of DSM.

4. Increasing energy-conscious consumer adoption[541] of end-use energy consuming devices and appliances that respond to their decisions is the current wave. These end-use energy consuming devices and appliances consume less energy because of market-driven efficiency, and that too in the absence of any government mandated codes and standards or without any incentives from utilities (for instance, transition to small battery powered mobile devices from grid connected desktop computers and monitors). Home energy management systems are typically incorporated with one or more of these wireless communications technologies, such as Bluetooth, so that various devices and appliances can be managed by the end user via a central interface or device. These mobile devices with smart technology provide remote connectivity that helps consumers take the next step in consumer-driven efficiency.

4.3.3   MULTI-CRITERIA DECISION MODEL FOR RENEWABLE ENERGY OPTIMIZATION

Because of the inherent nature of clean energy applications, often on the cusp of policy and technology uncertainty, it calls for application of multi-criteria decision-making models (MCDM). An extensive review of MCDM applications in a wide spectrum of sustainable development, such as technology selection, risk prioritization, or multi-conflict criteria has been captured by Kolios et al.[542] For instance, biomass energy applications of MCDM include biomass type selection,[543] biofuel production viability,[544] and biofuel *food-land-use* competition.[545] Some typical solar energy applications of MCDM include domestic water heating,[546] photovoltaic boost converter design,[547] and anti-islanding selection.[548] There are numerous applications of MCDM for renewable energy multi-source optimization.[549]

4.4   ENVIRONMENTAL JUSTICE

4.4.1   ENVIRONMENTAL JUSTICE—ISSUE AND IMPACT

"Climate change is a dire threat that humans have a moral responsibility to address"—Pope Francis.[550] Everyone has equal rights to use and a duty to protect the Earth's atmosphere. But the near omnipresence of carbon makes allocation of GHG emission rights very complicated and packed with very *contentious global distributive justice*[551] issues. Distributive justice requires that all human beings get an equal share of public goods—Earth's atmosphere—and the onus is placed on those who want more shares to show morally supportable grounds. Another major challenge is the equitable allocation of damage and mitigation responsibility based on historical accountability and ability to pay.

Distributive justice is an ethical mandate. Ethical rationales could be *right and wrong* and the action that delivers the best outcome as agreed by common consent.[552] Based on their mission, interest groups have promoted different aspects as the fourth bottom line: governance (by incorporating ethics and integrity metrics within);[553] future orientation (by highlighting the significance of legacy and importance of the needs of future generations);[554] Indigenous people (by differentiating their way of life and its harmony with the environment);[555] employee treatment (by justifying special consideration of people working within the organization);[556] and spirituality (by claiming it to be a monozygotic twin of sustainability—both having the human being at the center).[557]

The triple bottom line of sustainability—people, planet, and profit—must be expanded with a fourth component, purpose, as a balancing force in every climate change decision framework to preempt the disastrous pitfalls of *economics without ethics*. Ethics, or purpose, must be the fourth bottom line to achieve and ensure climate change distributive justice. In Taback and Ramanan's book,[558] the authors advocate for ethics or purpose as the most effective fourth bottom line, in all sustainability decision frameworks and disclosures, including climate change.

The author's proposal is to develop climate change-specific metrics for distributive justice. *Carbon-share* is defined as the amount of carbon associated with meeting any human need for products or services. Carbon-share is a challenging sociopolitical choice where science and society offer direction and regulators intercede. Ethical allocation of carbon-share needs distributive justice metrics.

4.4.2   THE AARHUS CONVENTION, 2001

The Convention on Access to Information, Public Participation in Decision-Making, and Access to Justice in Environmental Matters[559] (Aarhus Convention[560]) entered into force in 2001.[561] It bestows rights to the public and imposes on Parties, and their public authorities, binding obligations regarding access to environmental information, public participation in environmental decision-making, and access to justice in environmental cases. The public comprises all natural and legal persons, regardless of citizenship, nationality, or domicile and such public cannot be penalized, persecuted, or harassed in any way for their involvement. Generally acknowledged as the world's foremost international instrument that links environmental and human rights, it contributes to governmental accountability, transparency, and efficiency.

Most countries require an environmental impact assessment prior to issuing a license to operate. Aarhus Convention sets out key elements of public participation in the environmental decision-making process; these have become de facto minimum standard for effective participatory environmental decision-making regarding specific projects to general rules and regulations. The public participation requirements include (1) access to relevant information, (2) early and ongoing involvement of the public (stakeholder engagement) in decision-making, (3) a broad scope of participation, (4) a transparent and user-friendly process, (5) an obligation on authorities to take due account of public input, and (6) a supportive infrastructure and effective means of enforcement/appeal.

Although the Aarhus Convention does not apply to legislative bodies, it does apply to the development of legally binding rules and regulations that could have a potentially significant impact on the environment. The key driver is the convention bestowed access to environmental information, public participation in environmental decision making, and access to justice in environmental cases.

The Protocol on Pollutant Release and Transfer Registers to the Aarhus Convention (Protocol on PRTRs),[562] entered into force in 2009, obliges the Parties to make publicly available and free of charge information on emissions (releases) and transfers of pollutants, including GHGs, heavy metals, and toxic chemical compounds. The objective is to increase corporate accountability and facilitate public access to information on pollutants, which helps public participation in environmental decision-making and prevention and reduction of pollution.

In summary, the Aarhus Convention provides access to environmental justice in three contexts: with regard to information requests, public participation in decision making, and enforcement of national law. The procedures are required to be *fair, equitable, timely, and not prohibitively expensive*. Protocol on PRTRs has similar access to justice.

4.4.3 ENVIRONMENTAL JUSTICE POLICY IN THE UNITED STATES

4.4.3.1 Environmental Justice Considerations in the National Environmental Policy Act Process

Federal agencies must consider environmental justice in their activities under the National Environmental Policy Act (NEPA).[563] In light of Executive Order 12898, the White House Council on Environmental Quality (CEQ) issued Environmental Justice; Guidance[564] Under the National Environmental Policy Act. This guidance includes six principles for environmental justice analyses to determine any disproportionately high and adverse human health or environmental effects to low-income, minority, and tribal populations. The principles are to

1. Consider the composition of the affected area to determine whether low-income, minority, or tribal populations are present and whether there may be disproportionately high and adverse human health or environmental effects on these populations.
2. Consider relevant public health and industry data concerning the potential for multiple exposures or cumulative exposure to human health or environmental hazards in the affected population, as well as historical patterns of exposure to environmental hazards.
3. Recognize the interrelated cultural, social, occupational, historical, or economic factors that may amplify the natural and physical environmental effects of the proposed action.
4. Develop effective public participation strategies.
5. Assure meaningful community representation in the process, beginning at the earliest possible time.
6. Seek tribal representation in the process.

4.4.3.2   Equitable Development and Environmental Justice

Equitable development[565] is an approach for meeting the needs of underserved communities through policies and programs that reduce disparities while fostering places that are healthy and vibrant. An effective place-based action creates strong and livable communities. Equitable development differs from place-based approaches in that it is driven by priorities and values, as well as clear expectations that the outcomes from development need to be responsive to underserved populations and vulnerable groups, in addition to using innovative design strategies and sustainable policies. It acknowledges and understands that both are necessary for sustaining environmental justice. Lower-income citizens and minorities can successfully, through active participation, guide the changes that occur within their communities rather than react to them.

4.4.3.3   Environmental Justice for Tribes and Indigenous Peoples

In July 2014, the EPA completed its Policy on Environmental Justice for Working with Federally Recognized Tribes and Indigenous Peoples,[566] others living in Indian country as well as Indigenous peoples throughout the United States. The seventeen principles of the Policy help the EPA protect the environment and public health and address environmental justice concerns in Indian country. The EPA is working to incorporate the seventeen environmental justice principles in the following four areas:

1. Direct implementation of federal environmental programs in Native American country.
2. Work with federally recognized tribes/tribal governments on environmental justice.
3. Work with Indigenous peoples (state recognized tribes, tribal members, and so on) on environmental justice.
4. Collaborate with federal agencies and others on environmental justice issues of tribes, Indigenous peoples, and others living in Native American country.

4.4.3.4   Environmental Justice 2020—Action Agenda

The EPA's Environmental Justice 2020 (EJ 2020)[567] focuses on the environmental and public health issues and challenges confronting the nation's minority, low-income, tribal, and Indigenous populations. EJ 2020 consists of eight priority areas and four significant national environmental justice challenges; each of these areas have laid out the agency's objectives, the plans for achieving them, and how to measure success.

4.4.4   Environmental Justice Analytics—Assessment and Response

4.4.4.1   Carbon-Share Allocation and Distributive Justice Metrics

The author's proposal is to develop carbon-share as a climate change specific metric for distributive justice. Carbon-share is the amount of carbon associated with meeting any human need for products or services. Carbon-share, given its market externality, is a socio-political choice where science and society provide direction and regulators intervene. While initial assignment does not matter for economic efficiency,[568] it impacts equity and leads to contentious distributive justice concerns. For instance, climate change, *the two-degree classic* truly tests how intergenerational equity is incorporated in making ethical choices.

Carbon-share, based on equal per-capita GHG emissions rights, is consistent with the ethical idea to *treat others as you wish to be treated* and with human rights principle of *duty to prevent climate change*. The per capita lens on GHG emissions, at various moments in history, could address contentious ethical issues like liability for future excess emissions and responsibility for past emissions. One approach is to assign responsibility for reducing GHG emissions based on consumption, not production; it makes is easier to assign liability.

4.4.4.2 Emission Rights and Other Social Justice Metrics

1. One Human One Emission Right[569] recognizes that every human has to emit some GHGs and the same emission rights exist inter-generationally, globally, and permanently. The One Human One Emission Right approach is consistent with the author's concept of carbon-share and distributive justice concerns, but does not address market efficiency or metrics.

2. In the *personal carbon allowances*[570] (PCAs) proposal, states issue PCAs to citizens and requires them to surrender PCAs to become GHG neutral. The use of PCAs is a command and control approach.

3. Sno-Caps[571] proposes an independent *people's* carbon cap-and-trade system, under which individuals sign up and start with their own carbon share with no monetary value; but concerns of no buyers and system abuse exist. In the author's proposed *carbon-share* system, some of the Sno-Cap concerns are preempted by recognizing carbon-share as a socio-political choice where science and society provide direction and regulators intervene.

4. Global Reporting Initiative's G4 (GRI)[572] incorporates ethics within governance as a fourth set of disclosures, expanding the triple bottom line measures. However, in its current form, this does not address issues of equity and future orientation—two key components of distributive justice in climate change decision analysis.

5. One approach gaining traction, contraction and convergence (C&C),[573] requires agreement on atmospheric GHG concentration, calculation of global GHG emissions budget, and a convergence date to reach equal per-capita GHG emissions rights (e.g., a target of 1.5 tons per person by 2030). Narrow focus on climate change justice issues makes it simple yet flexible enough to include historical emissions, economic wealth of nations, and ability of nations to pay in the decision framework. Many people are far below their *equal* per capita GHG emissions rights today. C&C does not address development of human need-based per capita GHG emissions. Also, C&C does not leave adequate GHG emissions margins to allow developing nations to rise to sustenance levels.

4.4.4.3 Baseline Per-Capita "Carbon-Share" and Sustenance Margins

1. Similar to price parity basket of food and essentials, develop a base human-needs basket.
2. Establish base-level energy per capita for select nations for select energy raw materials.
3. Similar to Maximum Achievable Control Technology (MACT),[574] calculate lowest GHG emissions from select major energy sources.
4. Project population growth and improvement in select GHG emission reduction technologies.
5. Calculate daily activity based GHG emissions, extrapolate to get per capita *carbon-share*.
6. Establish GHG emissions margins to allow developing nations to rise to sustenance levels.
7. The data can be synthesized from basic components and extrapolated with some assumptions. Alternatively, in today's world of big data, patterns can be assessed and models developed.

4.4.4.4 Limitations of the "Carbon-Share" Concept

This is based on readily available published materials. While the subject scope is global, the analysis is limited to the major economies and larger populations. Finally, the metrics and materiality require validation and acceptance. Further research and global collaboration is mandatory to address the issues of global distributive justice—in particular legal framework, advocacy and disclosure/transparency.

4.4.4.5 Distributive Justice Needs Ethical Filters in Decision Frameworks

Absent ethical filters in the decision framework, one could easily distort one's judgment—caring only for people who mimic us, protecting only parts of the ecosystem that overtly serve us, and, of course, generating profits only for a subsection of the stakeholders. Ethics or purpose as the fourth bottom line takes into consideration all relevant stakeholders. Distributive justice needs ethical filters in decision frameworks and ethical allocation of carbonshare needs distributive justice metrics. Carbon-share based on equal per-capita GHG emissions rights meets most ethics principles.

4.5 INDUSTRIAL ACCIDENTS

Industrial accidents could result in worker and plant damage and in some situations, severely impact public health. The two classic examples are the nuclear fallout at Chernobyl and the Union Carbide disaster in Bhopal, India.

4.5.1 CHERNOBYL NUCLEAR FALLOUT AND IMPACT ASSESSMENT

Nearly three decades ago, a sudden surge of power during a reactor systems test destroyed Unit 4 of the nuclear power plant at Chernobyl, Ukraine, in the former Soviet Union.[575] The reactor fallout and the fire that followed released massive amounts of radioactive material into the environment.

1. *Emergency response*: Helicopters poured sand to extinguish the fire and boron on the reactor debris to prevent additional nuclear reactions. A few weeks later, to limit further release of radioactive material, they covered the damaged unit completely in a temporary concrete structure, called the *sarcophagus*. Also, a square mile of pine forest near the plant was cut down and buried to reduce radioactive contamination at and near the site. About 115,000 people were evacuated from the most heavily contaminated areas in 1986, the year of the accident, and another 220,000 people in subsequent years.

2. *Health effects from the accident*: Severe radiation effects killed two workers within hours, and 28 more of the site's 600 workers in the first four months after the event. Another 106 workers received high enough doses to cause acute radiation sickness.

   Only a small fraction of the 600,000 cleanup workers were exposed to elevated levels of radiation. Wide areas of Belarus, the Russian Federation, and Ukraine inhabited by five million residents were contaminated; however, they received only a very small radiation doses comparable to natural background levels (0.1 rem per year).

   Many children and adolescents in the area in 1986 drank milk contaminated with radioactive iodine, which delivered substantial doses to their thyroid glands; and to date, about 6,000 thyroid cancer cases have been detected among these children. While 15 died, the rest have been treated successfully. There is no evidence of adverse pregnancy outcomes, delivery complications, stillbirths, or overall health of children among the families living in the most contaminated areas. These health effects are far lower than initial speculations of tens of thousands of radiation-related deaths.

3. *Lessons learned and US nuclear regulatory commissions' (NRC) response*: The Soviet nuclear power authorities presented their initial accident report to an International Atomic Energy Agency (IAEA) meeting in Vienna, Austria, in August 1986. The NRC continues to conclude that plant design, broader safe-shutdown capabilities, and strong structures to hold in radioactive materials ensure US reactors can keep the public safe.

   The NRC's post-Chernobyl assessment emphasized the importance of (1) proper design of reactor systems; (2) correct implementation during construction and maintenance; (3) proper procedures and controls for normal operations and emergencies; (4) well-trained, competent, and motivated plant management and operating staff; and (5) appropriate backup safety systems to deal with potential accidents. The NRC concluded that the lessons learned from Chernobyl fell short of requiring immediate changes in the NRC's regulations.

4. *Sarcophagus transformation*: The concrete sarcophagus to cover the destroyed Chernobyl reactor was considered a temporary fix to filter radiation out of the gases from the destroyed reactor before the gas was released to the environment. Ten years later, in 1997, the G-7, and the European Commission agreed to jointly help Ukraine transform the existing sarcophagus into a stable and environmentally safe system that will protect workers, the nearby population, and the environment for decades from the very large amounts of radioactive material still in the sarcophagus; and is designed to last at least 100 years.

4.5.2   CONVENTIONS ON SAFETY OF NUCLEAR PLANTS, SPENT FUEL, AND RADIOACTIVE WASTE

The convention on Nuclear Safety,[576] adopted in 1994 and entered into force in 1996, commits participating states operating land-based civil nuclear power plants to maintain a high level of safety by setting international benchmarks to which states would subscribe. Based on the Parties' common interest to achieve higher levels of safety, which will be developed and promoted through regular meetings. Parties are obligated to submit reports on the implementation of their obligations for *peer review* at meetings at IAEA headquarters.

Joint Convention on the Safety of Spent Fuel Management and on the Safety of Radioactive Waste Management, adopted in 1997 and entered into force in 2001, is the first legal instrument that covers spent fuel and radioactive waste management safety on a global scale by setting international benchmarks and a *peer review* process. It applies to spent fuel resulting from the operation of civilian nuclear reactors and to radioactive waste resulting from civilian applications. It also applies to spent fuel and radioactive waste from military or defense programs if transferred permanently to and managed within exclusively civilian programs; it also covers planned and controlled releases into the environment of liquid or gaseous radioactive materials from regulated nuclear facilities.

4.5.3   MAJOR HAZARD INDUSTRIAL ACCIDENTS ASSESSMENT AND BHOPAL GAS TRAGEDY

4.5.3.1   Major Hazard Industrial Accidents Assessment

There have been several accidents at industrial facilities over the past two centuries. By far, the worst disaster, assessed on human impact, was the Bhopal gas tragedy that killed more than 2,000 people and injured tens of thousands. For comparison, the other major hazard incidents listed on the UK Government's Health, Safety, and Environment website,[577] Nypro Chemical (1974, Flixborough United Kingdom) had 28 deaths and 36 injuries; BP Refinery (2005, Texas City, near Houston, United States) had 15 fatalities and many more injuries; and Icmesa Chemical (1976, Seveso, near Milan, Italy) had no deaths but several people fell ill.

At the Nypro chemical plant (Flixborough), a crack in a reactor was leaking Cyclohexane, when a bypass system ruptured and resulted in the formation of Cyclohexane vapor cloud and led to a large fire and explosion. This incident led directly to the introduction of the Health and Safety at Work Act 1974. At Seveso, Italy, about 15 miles from Milan, a bursting disc on a chemical reactor ruptured. A dense white cloud of considerable altitude, containing a small deposit of TCCD (Dioxin), a highly toxic material, was seen to issue from a vent on the roof, and drifted offsite. This incident led to the Seveso Directive, following the need for consistency throughout Europe, and led to the formation of the Control of Major Accident Hazards Regulations in 1984.

At the BP Texas City refinery, about 40 miles from Houston, the raffinate splitter, which distills chemicals, in the isomerization unit was being restarted after a shutdown. The splitter was overfilled and overheated. Liquid subsequently filled the overhead line, and the relief valves opened. This caused excessive liquid and vapor to flow to the blow-down drum and the vent at top of the stack, and an explosion occurred. An independent Safety Review Panel, established to make a thorough, independent, and credible assessment of the company's five US refineries and of the company's corporate safety culture, published its report in January 2007. This report pushed the entire industry to improve its process safety management, to develop site specific lagging and leading indicators of process safety performance.

4.5.3.2 Bhopal Gas Tragedy

The severity of the Union Carbide Bhopal India pesticide plant accident, which released deadly methyl isocyanate gas into the nearby housing of the industry town, makes it the worst recorded industrial accident within the chemical industry. The exact numbers of dead and injured are uncertain, as people have continued to die of the effects over a period of years. Some estimates put the total death toll at more than 15,000, with as many as 500,000 injured. Approximately 2,000 people died within a short period and tens of thousands were injured, overwhelming emergency services. Further compounding was the fact that the hospitals were unaware as to which gas was involved or what its effects were.

1. *Leading to the incident*: An operator had noticed the pressure inside the storage tank to be higher than normal, but not outside the working pressure of the tank. At the same time, a methyl isocyanate (MIC) leak was reported near the vent gas scrubber (VGS). The pressure inside the storage tank was rising rapidly, so the operator went outside to the tank. Rumbling sounds were heard from the tank and a screeching noise came from the safety valve. Radiated heat could also be felt from the tank. Attempts were made to switch on the VGS, but this was not in operational mode. A relief valve on a storage tank containing highly toxic MIC lifted. A cloud of MIC gas was released, which drifted onto nearby housing.
2. *Failings in technical measures*: Union Carbide had the most advanced safety systems in the world at the time. An investigation commissioned by Union Carbide claimed that the accident resulted from sabotage by a disgruntled employee. Subsequent inquiries, though, revealed that a failure of safety systems, poor maintenance, reduced staffing, and a lack of contingency planning contributed to the scale of losses. The real problem was the safety systems in Bhopal had not been maintained, staff had been reduced, and money was not being spent on the facility.
3. *Recipe for a perfect storm*: There was a recipe for a perfect storm of failure of safety systems and emergency response:[578]
   a. The fact that the flare was a critical element within the plant's protection system was not recognized, and it was out of commission for three months prior to the accident.
   b. While hazards associated with runaway reactions in a chemical reactor are generally understood, such an occurrence within a storage tank had received little attention.

    c. The ingress of water, a dangerously incompatible material, caused an exothermic reaction with the process fluid. The exact point of ingress is uncertain though poor modification/maintenance practices may have contributed.

    d. Maintenance procedure training and competence levels of staff were questionable or inadequate.

    e. The plant modification change procedure was inadequate. The decommissioning of the refrigeration system was one plant modification that contributed to the accident. Without this system, the temperature within the tank was higher than the design temperature of 0°C.

    f. The emergency response from the company to the incident and from the local authority suggests that the emergency plan was ineffective. During the emergency, operators were uncertain about when to use the siren system. No information was available regarding the hazardous nature of MIC and what medical actions should be taken.

4. *A catalyst for change*: Unfortunate as it was, the Bhopal tragedy did serve as a catalyst to widespread changes in safety management, legislation, and the insurance market. It also had a significant impact on health and safety planning and regulations, resulting in better preparation and prevention, emergency response and community right-to-know regulations.

    a. This led to the Responsible Care program. Introduced in 1988, it is a code of conduct that governs accident preparedness, community awareness, process and occupational safety, and environmental protection, companies are required to conduct independent audits of their Responsible Care management system every three years.

    b. Bhopal and the 1989 explosion at a Phillips Petroleum Co. plant in Pasadena, Texas led the US Occupational Safety and Health Administration (OSHA) to set up its process safety management program in 1990.

    c. Bhopal alerted insurers to the potential enormity of industrial accidents, giving rise to the total pollution exclusion from the more-limited pollution exclusions, leading to specialty insurance covers for environmental impairment in the US and London insurance markets.[579]

4.5.4 THE EMERGENCY PLANNING AND COMMUNITY RIGHT-TO-KNOW ACT, 1986

The Emergency Planning and Community Right-to-Know Act (EPCRA)[580] was passed in 1986 in response to concerns regarding the environmental and safety hazards posed by the storage and handling of toxic chemicals. These concerns were triggered by the 1984 Bhopal (India) gas tragedy, caused by an accidental release of methyl isocyanate, which killed or severely injured more than 2,000 people and a similar incident that occurred at the Institute, West Virginia, six months later. These two events highlighted the need for local preparedness for chemical emergencies and the importance of availability of information on hazardous chemicals.

The EPCRA requirements cover emergency planning and *Community Right-to-Know* reporting on hazardous and toxic chemicals. The Community Right-to-Know provisions help increase the public's knowledge and access to information on chemicals at individual facilities, their uses, and releases into the environment. States and local communities, working with facilities, can use the information to improve preparedness and response to similar accidents and protect public health and the environment.

Key Provisions of the Emergency Planning and Community Right-to-Know Act are addressed in the sections described briefly:

1. Per Sections 301 to 303 on emergency planning, local governments are required to prepare chemical emergency response plans and to review these plans at least annually. State governments are required to oversee and coordinate local planning efforts. Facilities that hold Extremely Hazardous Substances (EHS) on-site in quantities that exceed a threshold have to participate in emergency response plan preparation.
2. Section 304 on emergency notification mandates facilities to immediately report accidental releases in excess of prescribed Reportable Quantities of EHSs and *hazardous substances* defined under the Comprehensive Environmental Response, Compensation, and Liability Act (CERCLA).
3. Per Sections 311 and 312. Community Right-to-Know Requirements facilities handling or storing any hazardous chemicals, defined under the OSHA have to submit Material Safety Data Sheets (MSDS) to the relevant authorities. MSDS describe the properties and health effects of these chemicals. Facilities must also submit an inventory form for these chemicals to relevant authorities.
4. Section 313. Toxics Release Inventory (TRI) requires facilities to submit a toxic chemical release inventory form (Form R) annually. Form R must be submitted for each of the more than six hundred listed TRI chemicals that are manufactured or otherwise used above the applicable threshold quantities.
5. Per Section 322. Trade secrets: facilities are allowed to withhold the specific chemical identity from the reports filed under sections 303, 311, 312, and 313 of EPCRA if the facilities submit a claim with substantiation to EPA.

4.5.5 RISK-MANAGEMENT PROGRAM, 1996

Risk-Management Program (RMP) Rule[581] is the EPA regulations and guidance for chemical accident prevention at facilities that use certain hazardous substances. The objective of seeking RMP information from facilities is to help local fire, police, and emergency response personnel prepare for and respond to chemical emergencies, and the purpose of making it available to public is to foster communication and awareness to improve accident prevention and emergency response. The EPA rule exempts flammable substances used as fuel and fuels sold at retail facilities. Most oil and gas production facilities, as well as most retail gas stations and propane retailers, are not subject to the rule because the flammable substances at these facilities are excluded from threshold determinations.

1. The rule includes a list of regulated substances under section 112(r) of the US Clean Air Act, including their synonyms and threshold quantities (in pounds) to help assess if a process is subject to the RMP rule. These regulated substances are also subject to the requirements of the general duty clause. Facilities use prescribed list of regulated Toxic or Flammable Substances for Accidental Release Prevention to develop an RMP to submit to EPA.

2. Facilities holding more than a threshold quantity of a regulated substance in a process are required to comply with EPA's Risk-Management Program regulations. The regulations require owners or operators of covered facilities to implement a risk-management program and to submit an RMP to EPA. The plans are revised and resubmitted to the EPA every five years.

3. The EPA provides a general guidance.[582] Each facility's program is required to address

    a. Hazard assessment that details the potential effects of an accidental release, an accident history of the last five years, and an evaluation of worst-case and alternative accidental releases.

    b. Prevention program that includes safety precautions, maintenance, monitoring, and employee training measures.

    c. An emergency response program that spells out emergency health care, employee training measures, and procedures for informing the public and response agencies (e.g., the fire department) should an accident occur.

4.6 BIODIVERSITY

4.6.1 Biodiversity—Issue and Impact

Biological diversity, or biodiversity, is a term used to describe the variety of life on Earth. "Biodiversity is the degree of variation of life forms within a given species, ecosystem, biome, or an entire planet. Biodiversity is a measure of the health of ecosystems."[583] Per International Union for Conservation of Nature (IUCN), "it (bio-diversity) refers to the wide variety of ecosystems and living organisms: animals, plants, their habitats, and their genes. It is the foundation of life on Earth and is crucial for the functioning of ecosystems that provide humans with the products and services without which we could not live. Oxygen, food, fresh water, fertile soil, medicines, shelter, protection from storms and floods, stable climate, and recreation all have their source in nature and in healthy ecosystems."[584]

4.6.2 Convention on Biological Diversity, Nairobi, 1992

The Convention on Biological Diversity (CBD) is a global treaty that entered into force in December 1993. It has three main objectives: "conservation of biological diversity, sustainable use of the components of biological diversity, and the fair and equitable sharing of the benefits arising from of the utilization of genetic resources."[585] Other very relevant global treaties to protect biodiversity are the Convention on Wetlands of International Importance, also called the Ramsar Convention, the Convention on International Trade in Endangered Species of Wild Fauna and Flora (CITES) and the Convention on the Conservation of Migratory Species of Wild Animals (also known as CMS or Bonn Convention).[586] Ramsar is an intergovernmental treaty that provides the framework for national action and international cooperation for the conservation and wise use of wetlands and their resources. Wetlands hold a significant number of species. The Ramsar Convention is the only global environmental treaty that deals with this particular ecosystem. The treaty was adopted in the Iranian city of Ramsar in 1971, and the Convention's member countries cover all geographic regions of the planet. CITES is an international agreement between governments to ensure that international trade in specimens of wild animals and plants does not threaten their survival. Concerned with the conservation of wildlife and habitats on a global scale, CMS aims to conserve terrestrial, aquatic, and avian migratory species throughout their natural habitats.[587] In the United States, the Environmental Impact Report (EIR)[588] is expected to identify those impacts and try to bring together a compromise that will satisfy the majority of the stakeholders.

4.6.3 BIODIVERSITY ANALYTICS—ASSESSMENT AND VALUATION

Biological diversity, comprising plants, animals, and microbes interlinks land, water, and the atmosphere both chemically and physically to enable humans and millions of other species to coexist in an interdependent manner. It is everywhere, both on land and in water, and it includes all organisms, from microscopic bacteria and viruses to more complex plant and animal species. According to the Millennium Ecosystem Assessment, the total numbers of species on Earth range from 5 to 30 million; and only 1.7 to 2 million species have been formally identified.[589]

4.6.3.1 Biodiversity Assessment

Although many data sources and some tools have been developed, biodiversity remains difficult to track and measure precisely. But we do not need precise figures and answers to devise an effective comprehension of where biodiversity is located, how it is changing over space and time, what the drivers responsible for this change are, what the consequences for ecosystem services and human well-being are, and the available options for response. A common currency measure of biodiversity is *species richness*, which is a count of the number of different species with no consideration for its abundance or level. However, for a more holistic view, it needs to be supplemented by other metrics. For instance, *species diversity*, another measure, takes into account both species diversity and species profusion.[590]

4.6.3.2 Biodiversity Valuation

Biodiversity, which comprises a large number of ecosystems and natural processes, is a classic example of market externality, and the services provided by many of these ecosystems are not traded in the commodity market. Because these services have no price tag attached to them, financial markets ignore them.

The Economics of Ecosystems and Biodiversity (TEEB) study claims that while biodiversity is economically valuable, it is not built into private and public policy decisions. Exploding global population and the associated increase in land use and urbanization are the major reasons for loss of biodiversity. Better understanding of local and global benefits as well as the total cost of revival of biodiversity is required to arrest further loss. Loss of biodiversity and consequent degradation of associated services could be stemmed if these values are incorporated in policy decisions. TEEB study is working toward this. "The study aims to develop mechanisms to assess the value of nature, drawing attention to the global economic benefits of biodiversity and highlighting the growing costs of its loss."[591]

4.6.4 Biodiversity Response—Conservation

Biodiversity conservation work is carried out by the IUCN through its various programs including water and forests.[592] The IUCN maintains a *red list* of endangered species.[593] Conserving biodiversity involves addressing three categories: (1) species and their subpopulations; (2) genetic diversity; and (3) ecosystems. Hundreds of projects are underway around the world aimed at saving species and ecosystems and providing the knowledge needed for successful conservation action.

Biodiversity could become a major ethical issue. While ethicists like Peter Singer[594] value wildlife and wild animals with equal status, people in general are anthropocentric and give humans a strong preference over other species. The sustainability stewards now have to resolve ethical biodiversity issues to everyone's satisfaction, applying the fourth bottom line, namely ethics. This can be time consuming and costly but the right thing to do; following the precautionary principle calls for preserving the existing ecosystem to the greatest extent possible, while achieving a balance for the good of all. Clearly this is not always an easy task.

4.7 INTEGRITY FAILURE

Gandhi said, "Earth provides enough to satisfy every man's needs, but not every man's greed." Extreme greed, whether for money or nature's resources, has disastrous consequences. Capitalism in general and the American dream in particular interprets greed to be a healthy trait. Defining US corporations as legal persons (Citizens United v FEC), mandates the relentless pursuit of shareholder benefits thus exemplifying corporate greed; however, there is no such mandate in US corporate law. Greed has become pervasive in business from executives, corporations, banks, and financial markets. "Greed is good—[it] has marked the upward surge of mankind," is the mantra of the early business age.[595] The only target stakeholder for a business in the greed stage is the shareholder. This mantra, along with an obsession with the primacy of shareholder interests, has driven most early entrepreneurial efforts to privatize gains and socialize costs.

The 2008 tsunami of toxic assets highlights banking greed at its peak, soon after deregulation. These dried up market mortgage-backed securities and Enron-like vaporware, financial derivatives are termed Wall Street neutrons. They are speculative bets in which most trades are speculations outside real economy. As Visser puts it, "Speculators may do no harm as bubbles on a steady stream of enterprise. But the position is serious when enterprise becomes the bubble on a whirlpool of speculation."[596] The speed and scale of these greed-driven actions make them potentially catastrophic. Enron and other such fiascos led to the birth of Sarbanes-Oxley!

4.7.1 Integrity Failure—Issue and Impact

4.7.1.1 Integrity Failure and Evaporation of Enron

Unethical behavior from any sphere provides a lesson. In the case of Enron, the company evaporated; the Enron fiasco was a classic example of how executive greed could destroy a major corporation within a short span of time. Enron executives, just prior to the failure of the company, took in average bonuses of $50 million while, at the same time, firing employees with a severance of $50,000. The leaders were using short-term shareholder wealth creation as a veil and creating an illusion that the primary standard was doing what was best for the company, while the executives were running a risk-free company for themselves.

The culture became one in which ethical issues were treated as optional, not obligatory! Collusion—partnership in crime—became rampant. Employees accepted bonuses for supporting unethical deals; speculators gobbled up high-risk, high-reward Enron stocks; banks, to earn huge profits, invested in high-risk partnerships that Enron used to hide debts; and then Enron auditors, Arthur Anderson, played along, making more than $50 million on the side as consultants. Then Enron attorneys, Vinson and Elkins, wrote creative special purpose partnerships. The result was that true investors looking for long-term value, as well as long-term employees, saw their investments and lifetime savings evaporate. The WorldCom disappearance was another similar fraudulent accounting disaster.

4.7.1.2 Integrity Failure and Banks Fined for Antitrust Collusion

The Sarbanes-Oxley Act of 2002 (SOX) was the US Congress's response to the huge accounting scandals of the time, including Enron and WorldCom. However, what surfaced in September 2008 was unprecedented. Several "too big to fail" financial institutions like AIG, Fannie Mae, Freddie Mac, Lehman Brothers, Merrill Lynch, and Washington Mutual fell like dominoes—some of the largest failures in US history. This was many times faster, much more obscure and global in reach compared to any, including the Great Depression of 1929.[597]

Yet again, starting in 2007, over the course of five years, five major banks—Barclays, Citibank, JPMorgan Chase, the Royal Bank of Scotland, and UBS,[598] as a cartel, colluded to manipulate US dollar and Euro spot market's exchange rates, through a private electronic chat room using coded language to conceal their collusion, benefiting their trading positions but harming countless consumers and investors around the world. They acted as partners—rather than competitors—in an effort to push the exchange rate favorable to their banks but detrimental to many others.

Global regulators levied nearly $6 billion in fines on five major banks—Barclays, Citibank, JPMorgan Chase, the Royal Bank of Scotland, and UBS[599] for violations of the Sherman Antitrust Act, London Interbank Offered Rate (LIBOR) manipulation, and unsound practices in the foreign exchange (FX) market. Penalties included prohibition of continued employment of involved employees and improvement of their senior management oversight, internal controls, risk management, and internal audit policies and procedures for their FX trading activities and controls over their sales practices.

4.7.1.3 Omnipresent Bribery

Per the World Bank, more than USD $1 trillion is paid in bribes each year.[600] Impacts are disastrous from creating failed states by eroding political stability, increasing the cost of business, and preventing mission-oriented charities from reaching the intended underserved recipients. In addition, at a macro level, it creates a significant risk and at the micro level highly negative impact on employee morale. Despite national laws and global treaties, such as the UN Convention against Corruption, bribery persists.

4.7.2 Integrity Failure Analytics—Assessment

In response to being fined heavily by global regulators, CEOs of the five major banks—Barclays, Citibank, JPMorgan Chase, the Royal Bank of Scotland, and UBS, in their apology, conveyed how such behavior was in contrast to their values and that an ethical culture was obligatory to mitigate the failure of integrity—a key element of the quadruple bottom line. Some relevant extracts of their apologies include

1. Behavior was an embarrassment to our firm, and stands in stark contrast to Citi's values. Fostering a culture of ethical behavior has been and continues to be a top priority.
2. The conduct of a small group of employees, or of even a single employee, can reflect badly on all of us, and have significant ramifications for JPMorgan.
3. The conduct of a small number of employees was unacceptable and we have taken appropriate disciplinary actions at UBS. Recidivism will be punished.
4. The misconduct at the core of these investigations is wholly incompatible with Barclays' purpose and values and we deeply regret.

4.7.3 Integrity Failure Response

4.7.3.1 Resurrection of Salomon Brothers

Warren Buffet's turnaround of Salomon Brothers provides a classic example of the value of ethical compliance beyond just legal compliance, and how to restore organizational reputation after catastrophic integrity failure. The steps to restore reputation, especially after a catastrophic failure of integrity, are succinctly captured by Fomburn and noted by Sims:[601]

1	Take public responsibility	7	Hire independent investigators
2	Convey concern to all stakeholders	8	Reorganize for greater control
3	Full cooperation with authority	9	Establish strict procedures
4	Remove negligent incumbents	10	Eliminate infraction practices
5	Appoint credible leaders that represent all interests	11	Revise practices and pay systems
6	Dismiss tied suppliers/agents	12	Monitor compliance

Warren Buffett, when he took over as CEO of Salomon Brothers in 1991, instilled a new corporate culture in order to resurrect the company:

1. First, he acknowledged the straying from the path to the entire workforce: "In some way, we had lost our way... A bravado was attached to the taking of risk and the making of money. As a result, we were inattentive to shareholders and external constituencies."[602]
2. Next, he made it crystal clear that not only should one not be afraid of blowing the whistle, it is in fact a mandate that one do so: "you are each expected to report, instantaneously and directly to me any legal violation or moral failure on behalf of any employee of Salomon... You are to make reporting directly to me your first priority."[603]
3. Finally, he also went public with a compliance note to major print media like *The Wall Street Journal*, *The New York Times*, and *The Financial Times*:[604] "[we will] be guided by the test that goes beyond the rules... simply want no part of any activities that pass legal tests but that we as citizens would find offensive."[605]

4.7.3.2 Sarbanes-Oxley, 2002 and Dodd Frank Wall Street Reform, 2010

The Sarbanes-Oxley Act of 2002 (SOX)[606] was to protect investors from fraudulent accounting activities by corporations, mandating strict reforms to improve financial disclosures from corporations and prevent accounting fraud. It was created in response to accounting malpractice by Enron Corporation, Tyco International PLC, and WorldCom, which shook investor confidence in financial statements. Section 302 is a mandate that requires senior management to certify the accuracy of the reported financial statement. Section 404 is a requirement that management and auditors establish internal controls and reporting methods on the adequacy of those controls. In addition, it also outlines requirements for information technology (IT) departments regarding electronic records. Finally, SOX promulgated a whistleblower program.

Dodd-Frank Wall Street Reform and Consumer Protection Act[607] established a number of new government agencies to oversee the banking system. The Financial Stability Authority monitors the financial stability of major firms whose failure could have a major negative impact on the economy (companies deemed "too big to fail"). Similarly, the new Federal Insurance Office is supposed to identify and monitor insurance companies considered "too big to fail." The Consumer Financial Protection Bureau (CFPB) is expected to prevent predatory mortgage lending.

The Volcker Rule, of Dodd Frank, restricts the ways banks can invest, limiting speculative trading and eliminating proprietary trading; in effect, it separates the investment and commercial functions of a bank. Banks are not allowed to be involved with hedge funds or private equity firms, as these kinds of businesses are considered too risky. In an effort to minimize possible conflict of interests, financial firms are not allowed to trade proprietarily without sufficient "skin in the game." The Volcker Rule is a movement in the direction of the Glass-Steagall Act of 1933, a law that first recognized the inherent dangers of extending commercial and investment banking services at the same time. The act also contains a provision for regulating derivatives to control the risk they posed to the greater economy.

Finally, Dodd-Frank strengthened and expanded the SOX whistleblower program and established a mandatory bounty program under which whistleblowers can receive 10% to 30% of the proceeds from a litigation settlement.

4.7.3.3   The United Nations Convention against Corruption, 2004

The UN Convention against Corruption[608] is a legally binding universal anticorruption instrument. The Convention's far-reaching approach and the mandatory character of many of its provisions make it a unique tool for developing a comprehensive response to a global problem. The vast majority of UN Member States are parties to the Convention.

The Convention entered into force in December 2005 and covers five main areas: preventive measures, criminalization and law enforcement, international cooperation, asset recovery, and technical assistance and information exchange. The Convention covers many different forms of corruption, such as bribery, trading in influence, abuse of functions, and various acts of corruption in the private sector. However, it does not directly address the sensitive and controversial issue of political campaign contribution including interference in the elections of other sovereign nations.

4.7.3.4   ISO 37001 Anti-Bribery Management Systems

International Organization for Standardization (ISO) 37001, Anti-bribery Management Systems[609] is designed to help build an anti-bribery culture within an organization and implement appropriate controls, and to detect bribery and preempt its occurrence in the first place. It covers bribery in the public, private and not-for-profit sectors, including that through or by a third party. It spells out how to establish, implement, maintain, and improve an anti-bribery management system and measures progress in adopting an anti-bribery policy, appointing someone to oversee policy compliance, vetting and training employees, conducting project and partner bribery risk assessments, implementing financial and commercial controls, and instituting reporting and investigation procedures. It can also provide evidence in the event of a criminal investigation that the organization or the person responsible has taken reasonable steps to prevent bribery.

4.7.4   Ethics Monitoring and Enforcement

Sustainability analytics, for the fourth bottom line, calls for establishment of a set of codes of ethics, policies governing access and oversight on employee practices, and most importantly, developing a culture of ethics. Ethics monitoring and enforcement is a decisive factor for the continued success of institutionalizing and building an ethics culture in any organization. In addition to the distribution of written codes, policies, and guidelines to all concerned parties, reinforcement through broad communication by executives, and assigning advisors, hotlines, and ombudsmen; it is essential to develop training and implement programs geared to help apply policy to everyday work situations. This is best accomplished through frequent participatory workshops.

Ethical values, like most others, cannot be taught; rather, they must be lived. Employees will do what they see. Values must be simple, easy to articulate, realistic, and applicable; they must apply to internal and external operations. These are first and best communicated in the selection of employees. Ideally, values should be in agreement with the diverse stakeholders, show obsession/passion for fairness and equity, and seek not just collective but individual responsibility.

4.8   HAZARDOUS WASTE AND PRODUCT TOXICITY

4.8.1   HAZARDOUS WASTE AND PRODUCT TOXICITY—ISSUE AND IMPACT

Chemicals have always been, but more so in recent years, an integral part of our everyday life. But when these chemicals are used inappropriately or disposed of improperly after use, they can have harmful effects on humans, plants, and animals. Even when used properly, many chemicals can still harm human health and the environment. When these hazardous substances are discarded, they become hazardous wastes. Hazardous wastes typically have one or more of the following characteristics: corrosive, ignitable, reactive, or toxic. Hazardous waste is also very often a byproduct of a manufacturing process—material left after products are made. Unless it is disposed of properly, it can create health risks for people and damage the environment. Health effects could range from minor irritation to serious illnesses, such as cancer, organ failure, or even death. Some chemicals may affect reproductive systems, resulting in genetic mutations.

When a toxic substance or hazardous waste is released in the air, water, and/or on land it can spread, contaminate more of the environment, expose more people, and pose even greater threats to public health and the environment. For example, when rain falls on soil at a waste site, it can carry hazardous waste deeper into the ground and underlying aquifer. Mercury emitted into the air from a power plant can contaminate the air we breathe. A toxic or hazardous substance can cause injury or death to a person, plant, or animal if a large amount is released at one time, if a small amount is released many times at the same place, or if the substance is very toxic (e.g., arsenic). The effects of coming into contact with a substance depend on how the substance is used and disposed of, who (the very young or old or the sick are more vulnerable) is exposed to it, and the level of the concentration, or dose, of exposure, and how long or how often someone is exposed. Also, exposure may be acute as in a short-time at high-level exposure, or chronic or long-term because of lower level but repeated exposure.

Exposure to hazardous substance could happen through multiple pathways; inhalation, ingestion, or dermal contact. Inhalation occurs from breathing polluted air, hazardous gases, or hazardous liquid vapors; ingestion comes from eating contaminated fish, fruits, and vegetables, or meat or drinking contaminated water; and dermal exposure is caused by absorption by the skin through direct contact.

4.8.2 Hazardous Waste, Product Toxicity Regulations, and the Basel Convention

There are many regulations in place worldwide for managing toxic substances and hazardous waste and preventing their release into the environment. In the United States, these span an alphabet soup of regulations that range from those aimed at prevention of new toxic substances entering the environment, such as the Toxic Substances Control Act (TSCA),[610] to those aimed at tracking the release and minimizing the release of toxic substances such as the TRI.[611] Other regulations target the handling, treatment, and safe disposal of toxic substances and hazardous wastes, including the remediation of contaminated sites. These include the Resource Conservation and Recovery Act (RCRA), Hazardous and Solid Waste Amendments (HSWA),[612] and the CERCLA, commonly known as Superfund,[613] among several others. CERCLA in particular deserves special attention for its construct in *retroactive and several* liability that transformed the environmental insurance industry and has drawn attention from the financial sector for reporting material environmental risks.

The Registration, Evaluation, Authorisation and Restriction of Chemicals (REACH) is the European Community Regulation of chemicals and their safe use and deals with the registration, evaluation, authorization, and restriction of chemical substances. The law entered into force on June 1, 2007.[614] Also, the Protocol on PRTRs to the Aarhus Convention,[615] entered into force in 2009, obliges the parties to make publicly available and free of charge information on emissions (releases) and transfers of pollutants, including GHGs, heavy metals, and toxic chemical compounds.

Deliberate dumping of hazardous waste to circumvent the rules is unfortunately not uncommon. Public resistance to the disposal of hazardous wastes in the better informed developed world had resulted in what became known as the NIMBY (Not in My Back Yard) syndrome. Dwindling supply of potential sites led to an escalation of disposal costs. Seeking cheaper disposal options for hazardous wastes, operators made a move to export toxic wastes to places with minimal regulations and enforcement mechanisms. The Basel Convention on the Control of Transboundary Movements of Hazardous Wastes and their disposal was adopted in response to a public outcry following the discovery of imported deposits of toxic wastes.[616] Adoption of the Basel Convention had the purpose of combating the *toxic trade.* The overarching objective of the Basel Convention, which came into force in 1991, is to protect human health and the environment against the adverse effects of hazardous wastes. It was a classic example of reaction to an unethical act.

4.8.3 Hazardous Waste and Product Toxicity Assessment and Response

4.8.3.1 Hazardous Waste and Product Toxicity Assessment

4.8.3.1.1 Human Health Risk Assessment of Chemicals

Risk assessments establish links among emission sources, human exposures, and adverse health effects and form the basis for environmental public health policy decisions. "Human health risk assessment is a process intended to estimate the risk to a given target organism, system or sub-population, including the identification of attendant uncertainties, following exposure to a particular agent, taking into account the inherent characteristics of the agent of concern as well as the characteristics of the specific target system."[617]

Human health risk assessment of chemicals refers to methods and techniques that apply to the evaluation of hazards, exposure, and harm posed by chemicals. The five steps[618] are

1. *Problem formulation*: Establishes the scope and objective of the assessment
2. *Hazard identification*: Identifies the type and nature of adverse health effects
3. *Hazard characterization*: Qualitative or quantitative description of inherent properties of an agent having the potential to cause adverse health effects
4. *Exposure assessment*: Evaluation of duration, pathway, concentration or amount of a particular agent that reaches a target population
5. *Risk characterization*: Advice for policy and other decision making

A detailed description of risk assessment, including technical issues, is available at van Leeuwen.[619] Human health risk assessments of chemicals can help evaluate past, current, and future exposures to any chemical found in air, soil, water, food, consumer products, or other materials. They can be quantitative or qualitative in nature. Risk assessments are often limited by a lack of complete information. Regardless, chemical risk assessments rely on scientific understanding of pollutant behavior, exposure, dose, and toxicity. In general terms, risk depends on (1) amount of a chemical present in an environmental medium (e.g., soil, water, air), food or a product; (2) pathway amount of contact (exposure) a person has with the pollutant in the medium; and (3) toxicity of the chemical.

Often, risk assessments require that estimates or judgments be made in the absence of some of the data. Consequently, risk assessment results have associated uncertainties.

Despite these uncertainties, human health risk assessment of chemicals can help answer[620] basic questions about potential dangers from exposure to chemicals to help policy decisions, such as

1. What chemical exposures pose the greatest risks? Can the risks be ranked to allow risk-based prioritization of resource allocation?
2. What are the risks to the public of drinking water from a specific source? Should drinking water be provided from a different, safer source?

3. Is this chemical spill dangerous? What is the appropriate emergency response?
4. Is it *safe* to build homes on this old hazardous waste site? Should we clean up this contaminated soil, or is the risk of moving much higher than in-situ treatment?
5. What, if any, limits on chemical exposure should be established in occupational settings, in consumer products, in environmental media, and in food?
6. Should there be limits set for chemical emissions from industrial, agricultural, or other human activities?

4.8.3.1.2 Toxicity Assessment for Chemical Mixtures

Chemicals legislation is based predominantly on toxicity assessments carried out on individual substances. Since humans and their environments are exposed to a wide variety of substances, there is increasing concern in the general public about the potential adverse effects (*cocktail-effects*) of the interactions between those substances when present simultaneously in a mixture. The EPA[621] and the UK Committee on Toxicity of Chemicals in Food, Consumer Products, and the Environment[622] provide guidance for conducting cumulative risk assessments. Risk assessments in the EU deal mainly with individual substances with the exception of *complex substances* falling under the REACH regulation, pesticide and biocidal formulations, and cosmetic products. The REACH guidance allows the use of approaches for predicting the overall risk based on information on the individual components.

The hazard of chemical mixtures can be assessed by a whole-mixture or a surrogate mixture of reasonably similar composition approach. These approaches have the advantage of accounting for any interactions. If the components of a mixture are known, a component-based approach is usually performed. But chemical composition of all the mixture components in a product is often unknown and the levels may vary with time and environmental conditions.

Human health exposure to mixtures may occur from specific pathways or chemical products. Aggregate exposure from all sources to multiple chemicals may happen as well. In occupational settings, exposure assessment may be relatively simple through direct measurements and known models. However, for the general public, in which exposure may occur via multiple pathways, this becomes complex.

Exposure assessments of mixtures generally use emissions data, measurement of the components or a lead component in environmental media, and biomarker information. Fate and transport of the mixture components in the environment, routes of exposure and pharmacokinetics of components once in the body may all be considered in the exposure assessment. For a *worst case* estimate, it may be necessary to assume maximum exposure to each component of the mixture based on the assessment of daily exposure from all sources. Frequently used methods[623] for dose/concentration addition are the hazard index (HI), the reference point index (RfPI), also known as point of departure index (PODI), the relative potency factor (RPF), or the toxic equivalency factor (TEF). Environmental exposure is the result of complex patterns depending upon widespread emissions and point releases, means of contact-inhalation, ingestion or dermal, and the fate and the distribution of chemicals in the different pathways, such as water, sediment, air, soil, and biota (food).

In a recent study, the European Commission[624] has reached the following conclusions:

Humans and environmental species are exposed to an infinite combination of chemicals, so an initial filter, to allow a focus on only mixtures of potential concern, is necessary. Lack of exposure information is a major gap in the assessment of chemical mixtures. Chemicals could act jointly in a way that the overall level of toxicity is affected. These may be synergism, where the combined effect is greater than the sum total of each; potentiation, where adding a non-harmful chemical makes a harmful chemical more toxic; and antagonism, where one chemical interferes with the other. The effects are described by dose/concentration addition method, which is preferred over the independent action. Interactions (such as synergies) usually occur at medium- or high-dose levels. At low-exposure levels, they are toxicologically insignificant. No robust evidence is available that exposure to a mixture of independently acting chemicals, at or below their zero effect levels, is of health or environmental concern.

4.8.3.2 Circular Economy and Recycling

Circular economy is conceptually[625] a transformation of a two-century-old mindset, from a linear take-make-throw economy to a customer-centric circular Cradle-to-Cradle model that

1. Applies closed-loop recycling to eradicate waste systematically throughout the life cycles and uses of products and their components
2. Innovates business models that build on the interaction between products and services as well as product design that utilizes the economic value retained in products after use in the production of new offerings.

Economies have been living on borrowed time for more than 250 years, enjoying abundant and inexpensive natural resources and ignoring the impact on the environment. Companies were able to extract raw materials, manufacture products, sell to growing customer base who discarded them after use. Many non-renewables cannot keep up with demand, the regenerative capacity of renewables is strained to its limits, prices and uncertainty of scarce natural resources are going up and environmental foot prints are impacting intangibles such as reputation and interest rates on borrowings. It has now become mandatory to decouple growth from scarce resource use. Per Accenture,[626] *circular advantage* is already a trillion-dollar-plus prize.

Accenture[627] has identified five circular business models that companies can leverage:

1. Circular supplies model replaces resources with renewable (avoiding scarcity), recyclable (lowering demand), or biodegradable *materials* (reducing disposal). For instance, a farm-produce company develops its wastes into energy producing ethanol, a renewable fuel.
2. Resource recovery model recovers and reuses resource outputs, such as sending food waste into an anaerobic digestion facility.

3. Product life extension model extends the lifecycle of their products and assets through remanufacturing, repairing, upgrading, or remarketing. Several consumer-durable makers take back used products, reprocess them, and put it back to productive use, such as the auto industry selling certified pre-owned cars. Caterpillar Inc. has internalized the rebuilding of returned used farm equipment for several years.

4. Sharing platforms model is based on better utilization of idle resources, such as renting of personal cars and lodging out vacant homes.

5. Product as a service business model offers the option to lease a product or pay-for-use. While these have been in use for long, such as leasing in the auto and some capital equipment industry, mainstreaming them in the consumer business is gaining ground, such as lights or carpets, where the supplier retains the ownership. This eliminates disposal and ensures recycling.

In circular economy models, companies become highly involved in the use and disposal of products, find ways to generate revenue from selling the functionality and dematerializing, and/or optimizing performance along the entire value chain. Planned design obsolescence is discarded, and massive realignment of customer and business incentives occurs. Rather significantly, there may be a great opportunity to service the bottom of the pyramid market.

4.8.3.3 Sustainable Design and Green Chemistry

It is worth refreshing our memory of Albert Einstein's guidance to scientists and engineers, "Concern for the man himself and his fate must always form the chief interest of all technical endeavors... Never forget this in the midst of your diagrams and equations"[628] was one of the first calls to be socially responsible. It continues to serve as the philosophical basis for sustainable design and green chemistry.

4.8.3.3.1 Roots of Green Chemistry in Pollution Prevention

The EPA's Pollution Prevention Act[629] (1990) is at the root of not polluting in the first place and arguably forms the foundation of today's green chemistry and green design. One may recall the 3Ps slogan of 3M Company: "Pollution prevention pays." Chemicals are less hazardous to human health and the environment if they are (1) less toxic to organisms, (2) less damaging to ecosystems, (3) not persistent or bioaccumulative in organisms or the environment, and (4) inherently safer to handle and use because they are not flammable or explosive.

Source reduction is defined as any practice that avoids, prevents, or eliminates the release of hazardous substances and, in effect, reduces the hazards to public health and the environment associated with the release of such substances, pollutants, or contaminants. Source reduction could be accomplished by modifications to equipment or technology, process or procedures, redesign of products, and the substitution of raw materials as well as improvements through in housekeeping, maintenance, training, or inventory control. Prevention of pollution covers reduction in the amount of any hazardous substance, pollutant, or contaminant entering any waste stream or otherwise released into the environment (including fugitive emissions) prior to recycling, treatment, or disposal. Isolation of contaminant sources has been commonly practiced in the progressive companies, both to reduce treatment costs and to avert showing large pollutant discharges compared to peers.

4.8.3.3.2   The Avoid-Prevent-Eliminate Principles of Green Chemistry

The Pollution Prevention Act also establishes a pollution prevention hierarchy that says pollution should be (1) prevented or reduced at the source, (2) recycled, (3) treated, and (4) disposed of or otherwise released into the environment only as a last resort. Green chemistry aims to design and produce chemical products and processes that are at the top of the pollution-prevention hierarchy—source reduction and prevention of chemical hazards by (1) designing them to be less hazardous to human health and the environment; (2) making them from renewable feedstocks, reagents, and solvents that are less hazardous to human health and the environment; (3) choosing syntheses and other processes with reduced or even no chemical waste; (4) using less energy or less water; (5) designing chemical products for reuse or recycling; and (6) if feasible, by reusing or recycling chemicals, treating chemicals to render them less hazardous before disposal as a last option.

Figure 4.2, provides the green chemistry aspects in the context of the avoid-prevent-eliminate (APreE) principles. Avoid refers to the avoidance of hazards by using processes that inherently preempt hazards and derivatives, as well as inefficiency in atomic yields and non-catalytic conversions. Prevent refers to the source reduction component that prevents waste of material or energy, pollution, and accidents. The eliminate principle eliminates product, solvent and auxiliary toxicity, product permanence or non-degradation, and depletion of non-renewables to ensure sustainable production. Circular economy is largely a further enhancement of this evolution.

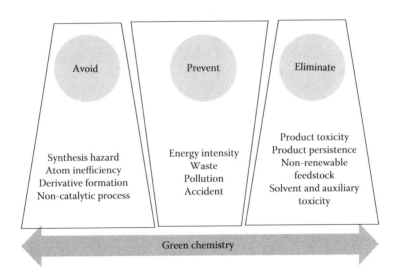

FIGURE 4.2   Avoid-prevent-eliminate (APreE) principles of green chemistry. (From US EPA, Green chemistry, https://www.epa.gov/greenchemistry/basics-green-chemistry#bookmarks, April 2017.)[630]

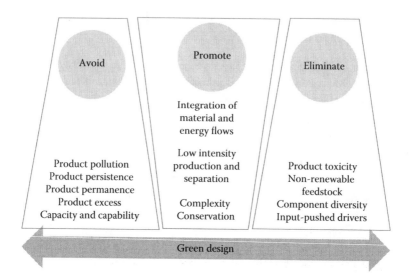

FIGURE 4.3 Avoid-promote-eliminate (AProE) principles of green design. (From US EPA, Green chemistry, https://www.epa.gov/greenchemistry/basics-green-chemistry#bookmarks, April 2017[631]; ACS, Green engineering, https://www.acs.org/content/acs/en/greenchemistry/ what-is-green-chemistry/principles/12-principles-of-green-engineering.html, April 2017.[632])

4.8.3.3.3 The Avoid-Promote-Eliminate Principles of Green Design

Similar to green chemistry, green design or green engineering has its roots in pollution prevention. Figure 4.3 provides the green design aspects in the context of the avoid-promote-eliminate (AProE) principles. Avoid refers to the avoidance of (1) product pollution; (2) persistence in the environment with no degradation; (3) permanence or immortality instead of targeted useful life; and (4) bountiful capacity and capability way beyond what is useful. The second set is promote (instead of prevent, as used in the case of green chemistry). It refers to promoting low intensity production and separation processes, integration of material and energy flows, and conserving the value of the embedded entropy and complexity while choosing recycle, reuse, or beneficial disposal. Eliminate principle eliminates product hazard, depletion of non-renewables to ensure sustainable production, component diversity to allow disassembly and value retention, and input-pushed drivers to enhance output pulled products, processes, and systems. Once again, circular economy is largely a further enhancement of this evolution.

4.9 WATER—A GLOBAL CRISIS

Countries have gone to war, many times, to gain control of extractive natural resources. That day is not out of the realm when countries may go to war for water!

4.9.1   THE CENTRALITY OF WATER AND ITS GLOBAL DIMENSIONS

Water is widely recognized as the most essential natural resource for Earth's eco-systems and human society. Yet the relationship between water and society is com-plex. Water is a multifaceted resource that is important to all economic sectors and across a range of spatial scales from local to global. Water is also frequently a hazard; rising sea levels, floods, droughts, and contaminated water are formidable threats to human well-being. To deal with this seemingly dual nature of water, people have long modified the water cycle through engineering schemes like dams, reservoirs, irrigation systems, and inter-basin transfer systems as well as through land use and land-cover change.

Given the multitude of ongoing human activities impacting the water cycle, a robust plan is needed to manage and govern water. However, most water governance plans address the issue at a local, national and river basin scale; matching supplies to demands. Most water plans focus on meeting user demands, without consider-ing rationale for the demand or ways to conserve water. First, the integrated water management strategy must balance human resource use and ecosystem protection. Second, in light of significant global trading of water-intensive commodities, global dimensions of water also need more intense and immediate attention.

4.9.2   GLOBAL WATER FOOTPRINT OF HUMANITY

China, India, and the United States, the three most populous countries in the world, have the largest total in-country water footprint accounting for nearly 40% of the global water footprint. The average water footprint is 2,482 cubic meters per year per capita in the United States, and just under 1,100 cubic meters per year per capita for China and India. The water footprint for humanity may be classified broadly as con-sumptive use of rainwater (green water) 74%; ground and surface water (blue water) 11%; and polluted water (gray water) 15%. India has the largest blue water (24% of the global blue), and China has the largest gray water (26% of the global gray).[633] These footprints were calculated using the Global Water Footprint Standard developed by the Water Footprint Network.[634]

Agriculture accounts for 92% of the water footprint; cereal 27%; meat 22%; and dairy products 7%. Virtual water flows, measured based on trade in agricultural and industrial commodities, is about 25% of the total water footprint. Given the large flow of virtual waters through global trade, many countries depend on foreign water resources and have significant impact on foreign water consumption and pollution, which clearly demonstrates the global dimension of water. The United States tops the list as the largest gross importer of virtual waters, followed by Japan, Germany, and China. These countries, with large external water footprints, depend on fresh water resources in other countries. This highlights the need for scarce water countries to consider secure import of water intense commodities, in the governance of water. For instance, given its water stress in Northern China, China is securing its food supply by controlling land in Africa.

4.9.3 Global Threats to Human Water Security and Watercourse Convention, 1997

Water is widely recognized as the most essential natural resource for Earth's ecosystems and human society, yet human activities directly threaten the freshwater systems globally; it is likely to be further impacted by climate change. Unfortunately, the beneficial use of water for economic productivity also impairs the ecosystems and aquatic biodiversity. Rivers serve as the main source of renewable water supply for humans and freshwater or river ecosystems. These call for a diagnosis of primary threats to water security and mitigation at a global scale, including global treaties, such as the UN Convention on the Law of the Non-Navigational Uses of International Watercourses,[635] the Convention on the Protection and Use of Transboundary Watercourses and International Lakes,[636] and the Convention on Biological Diversity.[637]

The two Conventions, UN Convention on the Law of the Non-Navigational Uses of International Watercourses and Convention on the Protection and Use of Transboundary Watercourses and International Lakes, are fully compatible with no contradiction between them and are in many ways mutually complementary. The 1997 Watercourses Convention complements the 1992 Water Convention by detailing the factors relevant to equitable and reasonable utilization, providing the procedures for consultations on planned measures and describing the consequences of the occurrence of transboundary impact. The 1992 Water Convention complements the 1997 Watercourses Convention by prescribing the content of specific agreements and tasks of joint bodies, detailing the information subject to joint assessment and exchange, and providing detailed guidance on the water quality objectives and best available technology.

The primary purpose of the UN Watercourses Convention, a global framework instrument that sets out rules and principles for governing international watercourses, is to supplement existing regional (multi-basin) basin and sub-basin agreements. The Convention has potential in addressing the existing legal architecture for international watercourses, which is often described as fragmented. However, this 1997 Convention came into force only in 2014; its ratification process has been slow.

The Transboundary Watercourse Convention was driven by the concern over the existence and threats of adverse effects of the changes in the conditions of the transboundary watercourses and international lakes on the environment, economies and well-being of the member countries of the Economic Commission for Europe (ECE). Transboundary waters mean any surface or ground waters that mark, cross or are located on boundaries between two or more ECE states. Transboundary impact means any significant adverse effect on the environment resulting from a change in the conditions of transboundary waters caused by a human activity. The key requirements of this convention are

1. To prevent, control, and reduce pollution of waters causing or likely to cause transboundary impact.
2. To ensure that transboundary waters are used with the aim of ecologically sound and rational water management, conservation of water resources, and environmental protection.

3. To ensure that transboundary waters are used in a reasonable and equitable way, taking into particular account their transboundary character, in the case of activities that cause or are likely to cause transboundary impact.
4. To ensure conservation and, where necessary, restoration of ecosystems.

The Convention on ECE Water further states that "measures for the prevention, control, and reduction of water pollution shall be taken, where possible, at source and these measures shall not directly or indirectly result in a transfer of pollution to other parts of the environment."[638]

Convention on Biological Diversity is briefly described under the biodiversity section. A recent study[639] considering all anthropogenic stresses and their aggregate impact, often under competing demands for human water security and biodiversity, finds that 80% of the world's population lives in areas where either the incident threat to human water security or biodiversity exceeds the 75th percentile.

Fortunately, high levels of threat to human water security and aquatic biodiversity occur only when threat scores are high for multiple stressors. Because of common impact sources, catchment and pollution stresses occur concomitantly, such as cropland for catchment and pesticides, organic load and nutrients contributing to pollution load. Multiple factors determine the threat; for instance, China's arid areas with sparse population are not greatly affected despite arid weather and low dilution potential, but densely populated eastern China, despite greater rainfall, faces substantial threat. Barring some remote areas like Alaska or Northern Australia, very few rivers are unaffected by human activities.

Built capital or water infrastructure improves human water security. A study taking into account the investment benefit factor to adjust the human water security shows significant positive impact in reducing the threat. For instance, in the United States and Western Europe, massive investments in water infrastructure show much lower adjusted human water security threat, valued at trillions of dollars. Investments in water infrastructure for the Organization for Economic Cooperation and Development (OECD) and Brazil, Russia, India, China and South Africa (BRICS) is estimated to be US$800 billion per year for human water security. However, the very same factors, stabilization of human water security, impact biodiversity negatively through flow distribution and habitat loss. Remote areas may be best served by preserving biodiversity and ecosystems; interestingly, the question of who will pay for such preservation, especially at the cost of development of dams resonates of the very issues faced in climate change. Why should it not? Water, and ecosystems, like carbon, are omnipresent in human lives.

4.9.4 CLIMATE CHANGE IMPACT ON WATER SECURITY

As climate change warms the atmosphere, altering the hydrologic cycle, amount, timing, form, and intensity of precipitation continue to change. Also, the flow of water in watersheds, as well as the quality of aquatic and marine environments change. These impacts are likely to affect the ways to protect water quality, public health, and safety and to build resiliency[640] to respond to climate change.

Likely Climate change impacts[641]

1. *Increase water pollution*: Warm air could cause wild fire related soil erosion and sedimentation and water pollution. Also, as air and consequently water warm, less dissolved oxygen will be retained in water, which is detrimental to the aquatic ecosystem. Warm water could also lead to detrimental algal bloom and altered toxicity of some pollutants. These will impair water quality. Warmer temperatures will lead to increase in water loss through evapotranspiration and increase in demand for cooling water for power and other plants—all leading to greater stress on the water infrastructure.

2. *More extreme weather events*: Heavier precipitation will increase flood risk, expand flood areas, affect the variability and velocity of streams, increasing erosion. There will be an increase in nutrients and toxins discharged to waterbodies affecting human water and ecosystem. In effect, stress on water infrastructure and wetlands management will rise rapidly.

3. *Climate*: Climate change also adds to the risk of water energy conflict; water and energy industries are competing for the same resources. Its impact on alternative energy production processes and carbon sequestration could increase groundwater and surface water withdrawal requirements. In addition, higher water temperatures from cooling processes and reduced assimilative capacity, increased pollution from rise in agricultural production, effects of carbon sequestration on ground water or ocean environment, are some of the related impacts on water from climate change response.

4. *Water availability*: Distribution and level of rain and snow may affect availability and quality of drinking water and water for agriculture, power plants, and industry. These may stress the human water infrastructure demands greatly.

5. *Sea level rise*: Storm surge and waterbody boundary movement and displacement may move shorelines, displace wetlands, inundate areas, and alter tides. These are likely to affect relocation of water intakes, wastewater and water treatment facility upgrades, and stress coastal management efforts.

6. *Coastal area*: Combination of impacts from sea level rise, floods and storm surges, erosion, ocean acidification, and warming will make coastal management difficult.

4.9.5   THE WATER AND ENERGY NEXUS

Water and energy are closely linked and are highly interdependent. Water is required to produce energy, and energy is needed to manage water from underground aquifer extraction, transport through canals and pipes for treatment and distribution, desalination of brackish and seawater for producing fresh water, and pollution control. Per Hoffman, "The energy security of the United States is closely linked to the state of its water resources… at the same time, US water security cannot be guaranteed without careful attention to related energy issues."[642] For instance, huge amounts of cooling water are required for thermal power plants; hydropower generation depends on water, and fossil fuel production for energy use requires a lot of water and generates a lot of wastewater. On the other hand, treatment and transport of water and wastewater in urban areas require significant amounts of energy; pumping of the influent wastewater and aeration are the two biggest energy consumers in the wastewater treatment plants.

Water requirements for primary energy has been estimated to be as follows:

Primary Energy/Electricity	Water Requirement	Source Reference
Crude oil	1.058 cubic meter per gigajoule (GJ)	Gerbens-Leenes[643]
Coal	0.164 cubic meter per GJ	Gerbens-Leenes[644]
Natural gas	0.109 cubic meter per GJ	Gerbens-Leenes[645]
Uranium	0.086 cubic meter per GJ	Gerbens-Leenes[646]
Biomass and biofuel	Variable depending on feedstock	Not applicable
Hydropower	2.6–5.4 cubic meter per MW·h	Gleick[647]
Thermal electricity	Variable based on technology and fuel	Not applicable
Wind, solar, and photoelectric	Not available	Not applicable

Water is an essential and non-substitutable resource, and there is a high price inelasticity of demand. When supplies are abundant, there is a misperception of infinite supply and the value of water and energy are low, but small move toward depletion could destabilize markets with very significant price variations.[648] Demand side management of water, similar to that now being practiced by utility companies for electricity, may be invaluable. Energy savings in home use will impact energy demand, and, consequently, water demand very significantly.

4.9.6 METHODS OF VALUATION OF WATER

Water has a value, cost, and price.[649] Price is a financial or fiscal transaction between the user and the provider. Because of regulatory control it does not represent either the cost or its economic value. Different uses of water have different economic values. Household consumption is commonly valued using a stated value willingness to pay method from direct surveys. The revealed preference can be inferred from user preferences in consumption change following a tariff change. Irrigation water use can be valued by marginal productivity, the extra value obtained from additional water use. One way to value industrial-use water may come from cost of recycling. Hydropower water could be valued based on the cost advantage of hydro over thermal power. Water markets are characterized by externalities, uncertainty, imperfect competition, asymmetric information, and distributional impacts, making valuation a challenge.

4.10   OIL SPILLS

4.10.1   Oil Spills—Issue and Impact

Oil tankers transport nearly two billion tons of crude oil and refined products around the world by sea. Waste oil is generated from several systems in a ship, such as the sludge, slop, bilge, and ballast water system. Generally, ship-generated oily waste can either be delivered to shore, incinerated onboard, and (legally or illegally) discharged to sea. Today, the majority of oil tankers are built and operated to minimize the amount of oil spilled in the event of an accident, as well as during routine tank cleaning operations. Yet some accidental discharges or spills do occur. The size of the spill and the sensitivity of the environment where the spill occurrs determine the impact. For instance, the Exxon Valdez oil spill on the Prince Williams Sound on the shores of Alaska, an extremely sensitive ecosystem, an enormous the impact on the environment.

Oil impacts marine life by either its physical nature (physical contamination and smothering) or by its chemical components (toxic effects and biomagnification in the food chain). Cleanup operations may cause physical damage to the habitats in which plants and animals live. Physical smothering by the persistent residues of spilled oils and water-in-oil emulsions (*mousse*) is the main threat to living resources. The animals and plants, which could come into contact with a contaminated sea surface, are the most at risk, such as marine life on shorelines, marine mammals and reptiles and birds that feed by diving or form flocks on the sea. The most toxic components in oil tend to be those lost rapidly through evaporation when oil is spilt. Most toxic compounds in oil evaporate, hence large-scale mortalities of marine life because of lethal concentrations of toxic components are relatively rare, local, and short-lived. Other effects such as the ability to reproduce, feed, or perform other functions can be caused by prolonged exposure to a much lower concentration. Bioaccumulation occurs in sedentary, shallow-water animals, such as oysters, mussels, and clams all of which extract food by filtering large quantities of sea water.

4.10.2   The MARPOL Convention, 1973/1978

International Convention for the Prevention of Pollution from Ships, 1973, as modified by the Protocol of 1978 relating thereto, also known as the MARPOL Convention,[650] includes regulations to prevent and minimize the pollution of the marine environment by ships from operational or accidental causes. It combines two treaties: the International Convention for the Prevention of Pollution from Ships (MARPOL) adopted in 1973, which covers pollution by oil, chemicals, and harmful substances in packaged form, sewage, and garbage; and the Protocol of 1978 relating to the 1973 International Convention for the Prevention of Pollution from Ships (1978 MARPOL Protocol). It was adopted in response to a spate of tanker accidents. Measures relating to tanker design and operation were also incorporated into the Protocol of 1978.

Finally, any violation of the MARPOL 73/78 within the jurisdiction of any party to the Convention is punishable either under the law of that party or under the law of the flag state. Ships engaged on international voyages must carry onboard valid international certificates as prima facie evidence that the ship complies with the requirements of the Convention.

MARPOL has six technical Annexures:

1. *Annex I: Regulations for the prevention of pollution by oil*: Covers prevention of pollution by oil from operational measures and accidental discharges; the 1992 amendments to Annex I made it mandatory for new oil tankers to have double hulls and a phase-in schedule for existing tankers to fit double hulls.
2. *Annex II: Regulations for the control of pollution by noxious liquid substances in bulk*: Details the discharge criteria and pollution control measures for noxious liquid substances carried in bulk and lists about 250 substances. Discharge of residues containing noxious substances is not permitted within 12 miles of the nearest land.
3. *Annex III: Prevention of pollution by harmful substances carried by sea in packaged form*: Contains general requirements for standards on packing, marking, labeling, documentation, stowage, quantity limitations, exceptions, and notifications.
4. *Annex IV: Prevention of pollution by sewage from ships*: Provides requirements to control pollution of the sea by sewage; prohibits the discharge of sewage into the sea.
5. *Annex V: Prevention of pollution by garbage from ships*: Bans the disposal of all forms of plastics. The latest revision also prohibits the discharge of all garbage into the sea, except under certain circumstances.
6. *Annex VI: Prevention of air pollution from ships*: Sets limits on sulfur and nitrogen oxide emissions from ship exhausts, prohibits deliberate emissions of ozone depleting substances, and mandates technical and operational energy efficiency measures to reduce the amount of greenhouse gas emissions.

4.10.3 MAJOR OIL SPILLS—ASSESSMENT AND BP DEEPWATER HORIZON OIL SPILL

The BP Deepwater Gulf of Mexico oil spill in 2010 was a clear example of the all-around loss of immense proportions: regional ecosystems as well as the oil, fishing, and tourism industries were all big losers, but the biggest losers were the shareholders and bond holders of BP. BP has estimated the total cost of the catastrophe at ~$62 billion, including the $20 billion liability settlement. For comparison, at more than 600,000 tons[651] (more than 200 million gallons of oil) it killed 11 people, injured 17, and was six times larger than the Exxon Valdez spill[652] in 1989, and almost 75% as much as the 1991 Iraq Persian Gulf War oil spill.[653] Stout cites the events leading to the spill as the *dumbest idea in the world*.[654] Investigating government authorities identified the root cause to be chronic safety lapses, which could be traced to a systemic lack of environmental ethics culture. BP and its contractors cut short-term costs and thereby ignored standard safety procedures. However, when BP CEO Tony Hayward, under whose watch the spill occurred, tried to palm off the blame on contractors, he lost his job due to public outrage.

In 2012, BP was indicted by the US federal grand jury for corporate manslaughter and was sentenced to a criminal penalty of four billion dollars, and the two highest-ranking supervisors were also indicted on charges of involuntary manslaughter. Furthermore, BP will pay about half a billion dollars for having calculatingly lied to investors about the amount of oil spilling into the gulf. The US Government temporarily banned BP contracts to supply fuel to the US Department of Defense. The EPA said it imposed the ban because the company's conduct during the 2010 Deepwater Horizon disaster showed a *lack of business integrity*.[655] This event highlights several factors pertinent to environmental professionals and the cost of unethical culture. The leaders and the managers indulged in promoting short-term cost cutting related to safety and environmental maintenance expenses. Even the CEO indulged in lying to the regulatory authorities and in passing the blame on to contractors. And yes, BP paid a heavy price in fines and settlements, cost of cleanup, loss of reputation, loss of market capital, and indictment of personnel. In addition, intangibles such as employee morale and investor and lender distrust were some of the other factors.

4.10.4   Major Oil Spills and Response

Some of the large accidental oil spills[656] include Amoco Cadiz (1978, 22,700 tons) caused by a ship running aground and splitting; PEMEX IXTOC I (1979, 480,000 tons) triggered by fire and explosion of a well; Exxon Valdez (1989, 38,400 tons) due to a shipwreck in a sensitive ecosystem; and BP Deepwater Horizon (2010, 62,000) because of a wellhead blowout that also killed 11 people. In addition, as a military tactic to block enemy ships from docking, Iraqi troops fleeing Kuwait opened the valves on the Sea Island oil rig in the Persian Gulf causing a major oil spill (1991, 1.5 million tons).[657] No one could access the rig to plug the leak because of the ongoing war.

4.10.4.1   Double-Hull Tankers

The 1992 amendments to Annex I of Regulations for the Prevention of Pollution by Oil made it mandatory for new oil tankers to have double hulls and a phase-in schedule for existing tankers to fit double hulls. A double hull strengthens the hull of ships, reducing the likelihood of oil disasters in low-impact collisions and groundings over single-hull ships, especially, in port areas when the ship is under pilotage. Double-hulled tankers are unlikely to perforate both hulls in a collision, preventing oil from seeping out.[658]

4.10.4.2   Collaborative Cleanup

International Convention on Oil Spill Preparedness, Response, and Cooperation (OPRC Convention)[659] entered into force in 1995, requires parties to establish measures for dealing with pollution incidents, either nationally or in cooperation with other countries. These measures include a shipboard oil pollution emergency plan. A protocol to extend the Convention to cover hazardous and noxious substances was adopted in 2000: The Protocol[660] on Preparedness, Response, and Cooperation to Pollution Incidents by Hazardous and Noxious Substances entered into force in 2007. The Protocol provides a global framework for international cooperation in combating major incidents or threats of marine pollution.

4.10.4.3 Chemical Dispersants

Dispersants[661] are often a necessary part of an effective major oil spill response. Naturally occurring microbes, present in most environments, feed on and break down crude oil. Dispersants are chemicals sprayed onto spills by suitably equipped planes or boats. They break oil into tiny droplets spread out into water and increase their surface area, which makes them more available for microbial degradation. Turbulence from wind and current waves provide further assistance. Smaller oil particles are easier to biodegrade, prevent significant oiling, and provide a layer of protection to sensitive shoreline habitats from oil slick. The unexpected restriction on dispersant use in Deepwater Horizon oil spill response still remains unclear today.

ENDNOTES

510 Ramanan, R., Corporate carbon risk management—A strategic framework, *EM*, October 2010, 20.

511 The Canada Research Chair in International Political Economy, TIPEA (Trade and Investment Provisions in International Environmental Agreements) database, Accessed October 2017 and available at http://www.chaire-epi.ulaval.ca/en/publications-english.

512 NOAA, A Paleo perspective on global warming, Accessed December 2012 and available at http://www.ncdc.noaa.gov/paleo/globalwarming/paleolast.html.

513 Pope Francis encyclical on climate change, *On Care for Our Common Home*, Accessed June 24, 2015 and available at http://w2.vatican.va/content/francesco/en/encyclicals/documents/papa-francesco_20150524_enciclica-laudato-si.html.

514 NOAA, A Paleo perspective on global warming, Accessed December 2012 and available at http://www.ncdc.noaa.gov/paleo/globalwarming/paleolast.html.

515 Frankel, J., Global environmental policy and global trade policy—Harvard project on international climate agreements, Accessed December 2012 and available at http://belfercenter.ksg.harvard.edu/publication/18647.

516 European Capacity Building Initiative, A pocket guide to the Paris agreement, Accessed April 2017 and available at http://www.eurocapacity.org/downloads/PocketGuide-Digital.pdf.

517 European Commission, PESETA project, Accessed April 2017 and available at http://ftp.jrc.es/EURdoc/JRC55391.pdf.

518 Proceedings of the National Academy of Sciences (PNAS), Physical and economic consequences of climate change in Europe, Accessed April 2017 and available at http://www.pnas.org/content/108/7/2678.full.pdf.

519 US EPA, Climate change in the United States, benefits of global action, Accessed April 2017 and available at https://www.epa.gov/sites/production/files/2015-06/documents/cirareport.pdf.

520 Ibid.

521 Oelkers, E.H. and D.R. Cole, Carbon dioxide sequestration a solution to a global problem, Accessed May 2017 and available at https://ic.ucsc.edu/~mdmccar/ocea213/readings/15_GeoEngineer/C_sequestration/oelkers_2008_Elements_CO$_2$_sequestration_overview.pdf.

522 Gorte, R.W., Carbon sequestration in forests specialist in natural resources policy, August 6, 2009, Accessed May 2017 and available at https://fas.org/sgp/crs/misc/RL31432.pdf.

523 UNCTAD, International rules for GHG emissions trading—Defining the principles, modalities, rules and guidelines for verification, reporting and accountability, Accessed May 2017 and available at http://unctad.org/en/Docs/pogdsgfsbm6.en.pdf.

[524] UNFCC, International emissions trading UNFCC, Accessed May 2017 and available at http://unfccc.int/kyoto_protocol/mechanisms/emissions_trading/items/2731.php.

[525] EU, Adaptation to climate change, Accessed May 2017 and available at https://ec.europa.eu/clima/policies/adaptation_en.

[526] EU, Climate change adaptation mainstreaming, Accessed May 2017 and available at https://ec.europa.eu/clima/policies/adaptation/what/mainstreaming_en.

[527] UNFCC, Framework for action on adaptation, Accessed May 2017 and available at Communication Toward a comprehensive climate change agreement in Copenhagen, COM (2009) 39, 28.01.2009.

[528] US EPA, Climate change adaptation strategies, Accessed May 2017 and available at https://www.epa.gov/arc-x/strategies-climate-change-adaptation.

[529] Purdue University, How fuel ethanol is made from corn? Accessed May 2017 and available at https://www.extension.purdue.edu/extmedia/id/id-328.pdf.

[530] Yale University, Accessed May 2017 and available at https://www.yaleclimateconnections.org/2015/01/pros-and-cons-of-ethanol-in-motor-vehicle-gas-explored/.

[531] Texas Agriculture, ExxonMobil algae overview, Accessed May 2017 and available at https://www.texasagriculture.gov/Portals/0/Bioenergy/ExxonMobil_Algae_Overview%20_2010-02-16.pdf.

[532] Swarup, V., ExxonMobil, ".." Accessed May 2017 and available at https://energyfactor.exxonmobil.com/perspectives/exxonmobils-advanced-biofuels-research/.

[533] US Office of Fossil Energy, Shale gas, Accessed May 2017 and available at https://energy.gov/fe/shale-gas-101.

[534] US EPA, Accessed May 2017 and available at https://www.epa.gov/hydraulicfracturing/process-hydraulic-fracturing.

[535] EPRI, US energy efficiency potential through 2035, Accessed May 2017 and available at https://publicdownload.epri.com/PublicDownload.svc/product=00000000000 1025477/type=Product.

[536] Ibid.

[537] Alliance to Save Energy, Alliance commission on national energy efficiency policy: The history of energy efficiency, January 2013, available at https://www.ase.org/sites/ase.org/files/resources/Media%20browser/ee_commission_history_report_2-1-13.pdf.

[538] IECC, Building energy codes program, Accessed May 2017 and available at https://www.energycodes.gov/sites/default/files/becu/2015_IECC_residential_requirements.pdf.

[539] ASHRAE, Energy standards for buildings, Accessed May 2017 and available at https://www.ashrae.org/resources--publications/bookstore/standard-90-1.

[540] Consortium for Energy Efficiency, 2015 state of the efficiency program industry: Budgets, expenditures, and impacts, CEE Annual Industry Report, March 18, 2016, available at https://library.cee1.org/sites/default/files/library/12628/CEE_2015_Annual_Industry_Report.pdf.

[541] EPRI, The third wave of energy efficiency, Accessed May 2017 and available at https://publicdownload.epri.com/PublicDownload.svc/product=000000003002009354/type=Product.

[542] Kolios, A., Comparative study of multiple-criteria decision making methods under stochastic inputs, *Energies*, 2016, Accessed April 2017 and available at www.mdpi.com/1996-1073/9/7/566/pdf.

[543] Saelee, S., B. Paweewan, R. Tongpool, T. Witoon, J. Takada, and K. Manusboonpurmpool, Biomass type selection for boilers using TOPSIS multi-criteria model, *International Journal of Environmental Science and Development*, 2014, 5, 181–186.

544 Kantas, A.B., H.I. Cobuloglu, and I.E. Büyüktahtakın, Multi-source capacitated lot-sizing for economically viable and clean biofuel production, *Journal of Cleaner Production*, 2015, 94, 116–129.

545 Cobuloglu, H.I. and I.E. Büyüktahtakın, Food vs. biofuel: An optimization approach to the spatio-temporal analysis of land-use competition and environmental impacts, *Applied Energy*, 2015, 140, 418–434.

546 Mohsen, M.S. and B.A. Akash, Evaluation of domestic solar water heating system in jordan using analytic hierarchy process, *Energy Conversation and Management*, 1997, 38, 1815–1822.

547 Adinolfi, G., G. Graditi, P. Siano, and A. Piccolo, Multi-objective optimal design of photovoltaic synchronous boost converters assessing efficiency, reliability, and cost savings, *IEEE Transactions on Industrial Informatics*, 2015, 11, 1038–1048; Graditi, G., G. Adinolfi, and G.M. Tina, Photovoltaic optimizer boost converters: Temperature influence and electro-thermal design, *Applied Energy*, 2014, 115, 140–150.

548 Datta, A., D. Saha, A. Ray, and P. Das, Anti-islanding selection for grid-connected solar photovoltaic system applications: A MCDM based distance approach, *Solar Energy*, 2014, 110, 519–532.

549 Mardani, A., A. Jusoh, E.K. Zavadskas, F. Cavallaro, and Z. Khalifah, Sustainable and renewable energy: An overview of the application of multiple criteria decision-making techniques and approaches, *Sustainability*, 2015, 7, 13947–13984.

550 Pope Francis encyclical on climate change, On Care for Our Common Home, Accessed June 24, 2015 and available at http://w2.vatican.va/content/francesco/en/encyclicals/documents/papa-francesco_20150524_enciclica-laudato-si.html.

551 Ramanan, R., Carbon-share—A contentious distributive justice issue globally, *AWMA International Conference on Addressing Climate Change: Emerging Policies, Strategies, and Technological Solutions*, Chicago, IL, September 9–10, 2015.

552 Preston, I., N. Banks, K. Hargreaves, A. Kazmierczak, K. Lucas, R. Mayne, C. Downing, and R. Street, Climate change and social justice: An evidence review, Summary, Accessed June 18, 2015 and available at http://www.jrf.org.uk/sites/jrf/climate-change-social-justice-summary.pdf.

553 Global Reporting Initiative, "G-4 56-58 ethics and integrity" within Governance metrics of Global Reporting Initiative, available at https://g4.globalreporting.org/general-standard-disclosures/governance-and-ethics/ethics-and-integrity/Pages/default.aspx.

554 Waite, M., SURF framework for a sustainable economy, *Journal of Management and Sustainability*, 2013, 3(4), 25. doi:10.5539/jms.v3n4p25.

555 McKeown, C., Interpreting the quadruple bottom line, Accessed June 16, 2015 and available at http://futureconsiderations.com/2013/05/quadruple-bottom-line/; also see Pope Francis.

556 Lawler III, E.E., The quadruple bottom line: Its time has come, Accessed June 16, 2015 and available at http://www.forbes.com/sites/edwardlawler/2014/05/07/the-quadruple-bottom-line-its-time-has-come/.

557 Ramanan, R., Ethics—The fourth bottom-line of sustainability, *Volume 10 of Compendium—Spirituality for Corporate Social Responsibility, Good Governance and Sustainable Development*, ISOL (Integrating Spirituality & Organizational Leadership) Foundation, under publication by Bloomsbury to be released at the 5th International Conference on Integrating Spirituality & Organizational Leadership at Chicago Art Institute, September 10, 2015.

558 Taback, H. and R. Ramanan, *Environmental Ethics and Sustainability* (Boca Raton, FL, CRC Press, 2013), 4.

[559] UN, The role of the aarhus convention in promoting good governance and human rights, September 2012, …, Accessed April 2017 and available at http://www.ohchr. org/Documents/Issues/Development/GoodGovernance/Corruption/ECONOMIC_COMMISSION_FOR_EUROPE.pdf.

[560] UN, Convention on access to information, public participation in decision-making and access to justice in environmental matters, 2nd edition 2014, Accessed April 2017 and available at http://www.unece.org/fileadmin/DAM/env/pp/Publications/Aarhus_Implementation_Guide_interactive_eng.pdf.

[561] United Nations Economic Commission for Europe (UNECE), Accessed April 2017 and available at http://www.unece.org/fileadmin/DAM/env/pp/documents/cep43e.pdf.

[562] Protocol on Pollutant Release and Transfer Registers (PRTR), Protocol on pollutant release and transfer registers to the Aarhus convention, Accessed April 2017 and available at http://www.unece.org/fileadmin/DAM/env/pp/prtr/Protocol%20texts/PRTR_Protocol_e.pdf.

[563] US EPA, National Environmental Policy Act, Accessed April 2017 and available at https://www.epa.gov/environmentaljustice/environmental-justice-and-national-environmental-policy-act.

[564] US EPA, Environmental justice guidance under the National Environmental Policy Act, December 1997, Accessed April 2017 and available at https://www.epa.gov/environmentaljustice/ceq-environmental-justice-guidance-under-national-environmental-policy-act.

[565] US EPA, Equitable development, Accessed April 2017 and available at https://www.epa.gov/environmentaljustice/equitable-development-and-environmental-justice.

[566] US EPA, Policy on environmental justice for working with federally recognized tribes and indigenous peoples, Accessed April 2017 and available at https://www.epa.gov/environmentaljustice/environmental-justice-tribes-and-indigenous-peoples.

[567] US EPA, Environmental justice 2020, Accessed April 2017 and available at https://www.epa.gov/environmentaljustice/ej-2020-action-agenda-epas-environmental-justice-strategy.

[568] Miceli, T.J., The economic approach to law, 2nd ed., Ch. 6, *The Economics of Property Law: Fundamentals*, Stanford University Press, Palo Alto, CA, 2008, Accessed December 2013 and available at http://www.sup.org/economiclaw/?d=Key%20Points&f=Chapter%206.htm.

[569] Ekardt, F., Climate change and social distributive justice, Accessed June 18, 2015 and available at http://www.sustainability-justice-climate.eu/files/texts/KAS-ClimateJustice-engl.pdf.

[570] Hyams, K., A just response to climate change: Personal carbon allowances and the normal-functioning approach, *Journal of Social Philosophy*, 2009, 40(2), 237–256.

[571] Heikkinen, N., climatewire: MIT competition uses crowdsourcing to find climate change solutions, Accessed June 5, 2015 and available at http://climatecolab.org/community/-/blogs/mit-competition-uses-crowdsourcing-to-find-climate-change-solutions.

[572] Global Reporting Initiative (GRI) is an international independent organization that helps businesses, governments and other organizations understand and communicate the impact of business on critical sustainability issues. Accessed June 16, 2015 and available at https://www.globalreporting.org.

[573] Brown, D.A., Why contraction and convergence is still the most preferable equity framework for allocating national GHG targets, http://blogs.law.widener.edu/climate/tag/distributive-justice-and-climate-change/.

[574] Maximum Achievable Control Technology (MACT), US EPA regulatory approach to control hazardous air pollutants.

[575] Nuclear Regulatory Commission (NRC), Chernobyl nuclear power plant accident, Accessed April 2017 and available at https://www.nrc.gov/reading-rm/doc-collections/fact-sheets/chernobyl-bg.pdf.

576 IAEA, Safety conventions, Accessed April 2017 and available at https://www.iaea.org/topics/nuclear-safety-conventions.

577 UK Government's Health, Safety and Environment, Major hazard incidents, Accessed April 2017 and available at http://www.hse.gov.uk/news/buncefield/major-hazard-incidents.htm.

578 FP Lees, Loss prevention in the process industries—Hazard identification, assessment and control, Volume 3, Appendix 5, Butterworth Heinemann, 1996.

579 Business Insurance, Bhopal disaster changed handling of industrial risks, Accessed April 2017 and available at http://www.businessinsurance.com/article/20041205/ISSUE01/100015848/bhopal-disaster-changed-handling-of-industrial-risks.

580 US EPA, The emergency planning and community right-to-know act, Accessed April 2017 and available at https://www.epa.gov/epcra/what-epcra.

581 US EPA, Risk management program rule, Accessed April 2017 and available at https://www.epa.gov/rmp/risk-management-plan-rmp-rule-overview.

582 US EPA, General guidance on risk management program rule, Accessed April 2017 and available at https://www.epa.gov/rmp/guidance-facilities-risk-management-programs-rmp#general.

583 NRC, Chernobyl nuclear power plant accident, Accessed April 2017 and available at https://www.nrc.gov/reading-rm/doc-collections/fact-sheets/chernobyl-bg.pdf; IAEA, Safety conventions, Accessed April 2017 and available at https://www.iaea.org/topics/nuclear-safety-conventions; UK Government's Health, Safety and Environment, Major hazard incidents, Accessed April 2017 and available at http://www.hse.gov.uk/news/buncefield/major-hazard-incidents.htm; What is biodiversity? United Nations Environment Programme, World Conservation Monitoring Centre, Accessed October 2017 and available at http://www.unesco.pl/fileadmin/user_upload/pdf/BIODIVERSITY_FACTSHEET.pdf.

584 International Union for Conservation of Nature, Improving knowledge on biodiversity and ecosystems, Accessed December 2012 and available at http://www.iucn.org/about/work/programmes/business/?6230.

585 Convention on Biological Diversity, December 29, 1993, 1760 U.N.T.S. 79, Accessed December 2012 and available at http://www.cbd.int/doc/legal/cbd-en.pdf.

586 Convention on Wetlands of International Importance Especially as Waterfowl Habitat, *Ramsar*, December 21, 1975, 996 U.N.T.S. 245; The Convention on International Trade in Endangered Species of Wild Fauna and Flora, July 1, 1975, 993 U.N.T.S. 243, Accessed December 2012 and available at http://www.cites.org/; The Convention on the Conservation of Migratory Species of Wild Animals, June, 23, 1979, 1459 U.N.T.S. 362, Accessed December 2012 and available at http://www.cms.int/.

587 Ibid.

588 USEPA, Considering ecological processes in environmental impact assessments, Accessed April 2017 and available at https://www.epa.gov/sites/production/files/2014-08/documents/ecological-processes-eia-pg.pdf.

589 Ecosystem and human well-being, Millennium ecosystem assessment, 29, Accessed December 2012 and available at http://www.millenniumassessment.org/documents/document.354.aspx.pdf.

590 Colwell, R.K. Biodiversity: Concepts, patterns and measurement. In Levin, S.A. (Ed.), *The Princeton Guide to Ecology* (Princeton, NJ: Princeton University Press, 2009), 257–263.

591 European Communities, *The Economics of Ecosystems and Biodiversity* (Cambridge, UK: Banson, 2008), Accessed December 2012 and available at http://ec.europa.eu/environment/nature/biodiversity/economics/pdf/teeb_report.pdf.

592 Business and biodiversity programme, International Union for Conservation of Nature, Accessed February 17, 2009 and available at http://www.iucn.org/about/union/secretariat/offices/rowa/iucnwame_ourwork/business___biodiversity_programme_/.

593 International Union for Conservation of Nature, The red list, Accessed December 2012 and available at http://www.iucnredlist.org/.

594 Cavalieri, P. and Singer, P. (Eds.), *The Great Ape Project: Equality beyond Humanity* (New York: St. Martins Griffin, 1993), 152.

595 "Greed is good" is a catchphrase by Gordon Gekko, a fictional character, from the 1987 film *Wall Street*.

596 Visser, W., Ages and stages of CSR. In *The Age of Responsibility*, 1st ed. (Hoboken, NJ: John Wiley & Sons, 2011), 37.

597 Bruner, R.F., The economic climate's impact on corporate culture and ethics, *Presented at Business Roundtable*, Institute for Corporate Ethics, Charlottesville, VA, November 2009.

598 Forbes, Four banks plead guilty to foreign exchange collusion, Accessed April 2017 and available at https://www.forbes.com/sites/antoinegara/2015/05/20/four-banks-plead-guilty-to-foreign-exchange-collusion-ubs-pleads-guilty-to-wire-fraud/#3f9f2d0f5108.

599 Ibid.

600 ISO, ISO 370000 Anti Bribery, Accessed May 2017 and available at https://www.iso.org/files/live/sites/isoorg/files/archive/pdf/en/iso_37001_anti_bribery_mss.pdf.

601 Fombrun, C.J., *Reputation: Realizing Value from the Corporate Image* (Boston, MA: Harvard Business School Press, 1996), presented in Sims, Corporate Social Responsibility, 234.

602 Sims, R.R., *Ethics and Corporate Social Responsibility—Why Giants Fall* (Santa Barbara, CA: Greenwood Publishing Group, 2003), 221.

603 Ibid, 220.

604 Ibid, 223.

605 Ibid.

606 Investopedia, Sarbanes-oxley act, Accessed May 2017 and available at http://www.investopedia.com/terms/s/sarbanesoxleyact.asp.

607 Investopedia, Dodd-Frank Wall Street reform and consumer protection act 2010, Accessed May 2017 and available at http://www.investopedia.com/terms/d/dodd-frank-financial-regulatory-reform-bill.asp.

608 UN, Convention against corruption, Accessed May 2017 and available at https://www.unodc.org/unodc/en/corruption/uncac.html.

609 ISO, ISO 370000 Anti Bribery, Accessed May 2017 and available at https://www.iso.org/files/live/sites/isoorg/files/archive/pdf/en/iso_37001_anti_bribery_mss.pdf.

610 Summary of the toxic substance control act, United States Environmental Protection Agency, Accessed August 23, 2012 and available at http://www.epa.gov/lawsregs/laws/tsca.html.

611 United States Environmental Protection Agency, Toxic release inventory, Accessed November 29, 2012 and available at http://www.epa.gov/TRI/.

612 United States Environmental Protection Agency, Hazardous waste regulations, Accessed December 2012 and available at http://www.epa.gov/osw/laws-regs/regs-haz.htm; and FIFRA http://www.epa.gov/lawsregs/laws/fifra.html.

613 United States Environmental Protection Agency, Cleaning up the Nation's hazardous waste sites, Accessed November 16, 2012 and available at http://www.epa.gov/superfund/.

614 REACH, European Commission, Accessed September 14, 2012 and available at http://ec.europa.eu/environment/chemicals/reach/reach_intro.htm.

615 PRTR, Protocol on pollutant release and transfer registers to the Aarhus convention, Accessed April 2017 and available at http://www.unece.org/fileadmin/DAM/env/pp/prtr/Protocol%20texts/PRTR_Protocol_e.pdf.

616 Basel Convention, Basel convention on the control of transboundary movements of hazardous wastes and their disposal, Accessed December 2012 and available at http://www.basel.int/.

617 UNEP, The International Programme on chemical safety—IPCS (IPSC, 2010), Accessed June 2017 and available at http://chm.pops.int/TheConvention/POPsReviewCommittee/Meetings/POPRC8/MeetingDocuments/tabid/2801/ctl/Download/mid/9135/Default.aspx?id=12&ObjID=14593.

618 WHO, IPCS risk assessment terminology, Geneva, International Programme on Chemical Safety (Harmonization Project Document No. 1) 2004, Accessed April 2017 and available at http://www.inchem.org/documents/harmproj/harmproj/harmproj1.pdf.

619 van Leeuwen, C.J. and T.G. Vermeire (Eds.), *Risk Assessment of Chemicals: An Introduction*, 2nd ed. (Dordrecht, the Netherlands: Springer, 2007).

620 World Health Organization, Who human health risk assessment toolkit: Chemical hazards, Accessed April 2017 and available at http://www.inchem.org/documents/harmproj/harmproj/harmproj8.pdf.

621 USEPA, *Supplemental Guidance for Conducting Health Risk Assessment of Chemical Mixtures as a Supplement to the EPA's Guidelines for the Health Risk Assessment of Chemical Mixtures* (USEPA, 1986), 2002.

622 COT, UK committee on toxicity of chemicals in food, consumer products and the environment, *Risk Assessment of Mixtures of Pesticides and Similar Substances*, 2002.

623 Boobis, A.R., B.C. Ossendorp, U. Banasiak, P.Y. Hamey, I. Sebestyen, and A. Moretto, Cumulative risk assessment of pesticide residues in food, *Toxicol Letters*, 2008, 180, 137–150.

624 European Commission, Toxicity and assessment of chemical mixtures, Accessed April 2017 and available at http://ec.europa.eu/health/scientific_committees/environmental_risks/docs/scher_o_155.pdf.

625 Linder, M. and M. Williander, Circular business model innovation: Inherent uncertainties. *Business Strategy and the Environment*. doi:10.1002/bse.1906; McKinsey, Moving toward a circular economy, Accessed April 2015 and available at http://www.mckinsey.com/business-functions/sustainability-and-resource-productivity/our-insights/moving-toward-a-circular-economy.

626 Accenture, Circular advantage, Accessed April 2017 and available at https://www.accenture.com/t20150523T053139__w__/us-en/_acnmedia/Accenture/Conversion-Assets/DotCom/Documents/Global/PDF/Strategy_6/Accenture-Circular-Advantage-Innovative-Business-Models-Technologies-Value-Growth.pdf.

627 Greenbiz, Circular economy, Accessed April 2015 and available at https://www.greenbiz.com/article/5-business-models-put-circular-economy-work.

628 Einstein, A., *Speech at the California Institute of Technology*, Pasadena, CA, February 16, 1931, as reported in *The New York Times*, February 17, 1931, 6.

629 US EPA, Green chemistry's roots in the pollution prevention act of 1990, Accessed April 2017 and available at https://www.epa.gov/greenchemistry/basics-green-chemistry#ppa.

630 US EPA, Green chemistry, Accessed April 2017 and available at https://www.epa.gov/greenchemistry/basics-green-chemistry#bookmarks.

631 Ibid.

632 ACS, Green engineering, Accessed April 2017 and available at https://www.acs.org/content/acs/en/greenchemistry/what-is-green-chemistry/principles/12-principles-of-green-engineering.html.

633 Hoekstra, A.Y. and M.M. Mekonnen, The water footprint of humanity, *Proceedings of the National Academy of Sciences*, 2012, 109(9), 3232–3237.

[634] Hoekstra, A.Y., A.K. Chapagain, M.M. Aldaya, and M.M. Mekonnen, The *Water Footprint Assessment Manual: Setting the Global Standard* (London, UK: Earthscan, 2011).

[635] UN, Convention on the law of the non-navigational uses of international watercourses, Accessed May 2017 and available at https://www.unece.org/env/nyc.html.

[636] UN, Convention on the protection and use of transboundary watercourses and international lakes (ECE Water Convention), Helsinki, 1992, Accessed May 2017 and available at https://www.unece.org/fileadmin/DAM/env/water/pdf/watercon.pdf.

[637] UN, Convention on biological diversity, Accessed May 2017 and available at Convention on Biological Diversity http://www.biodiv.org/convention/articles.aspæ (2004).

[638] UN, Convention on the protection and use of transboundary watercourses and international lakes (ECE Water Convention), Helsinki, 1992, Accessed May 2017 and available at https://www.unece.org/fileadmin/DAM/env/water/pdf/watercon.pdf.

[639] Vörösmarty, C.J., P. McIntyre, M.O. Gessner, D. Dudgeon, A. Prusevich, P. Green, S. Glidden, S.E. Bunn, C.A. Sullivan, and C.R. Liermann, Global threats to human water security and river biodiversity, *Nature*, 2011, 467(7315), 555–561.

[640] Water Governance Facility, Water adaptation in national adaptation programmes for action—Freshwater in climate adaptation planning and climate adaptation in freshwater planning, 2009, available at http://www.watergovernance.org/documents/WGF/Reports/Water_Adaptation_in_NAPAs.pdf.

[641] US EPA, Water program to climate change, Accessed April 2017 and available at https://www.epa.gov/climate-change-water-sector; Office of Water Climate Change Adaptation Plan, Accessed April 2017 and available at https://www3.epa.gov/climatechange/Downloads/OW-climate-change-adaptation-plan.pdf.

[642] Hoffman, A., The connection: Water and energy security, Institute for the analysis of global security, 2004, Accessed August 2004 and available at http://www.iags.org.

[643] Gerbens-Leenes, P.W., A.Y. Hoekstra, and T.H. Van Der Meer, Water footprint of bioenergy and other primary energy carriers, Accessed May 2017 and available at http://waterfootprint.org/media/downloads/Report29-WaterFootprintBioenergy.pdf.

[644] Ibid.

[645] Ibid.

[646] Ibid.

[647] Gleick, P.H., Water and energy, *Annual Review of Energy and Environment*, 1994, 19, 267–299.

[648] Olsson, G., Water and energy nexus. In *Encyclopedia of Sustainability Science and Technology* (Springer, 2011), available at http://ac4ca.eie.ucr.ac.cr/data/uploads/Gustaf%20Olsson%20Water%20and%20Energy%20nexus.pdf.

[649] United Nations, Managing water under uncertainty and risk, The United Nations World Water Development Report 4 (2012), Vol. 1, Ch. 1, 22–42, Ch. 2, 44–52, and Ch. 10, 276–88.

[650] Marpol Convention, International convention for the prevention of pollution from ships (MARPOL 73/78) PRACTICAL GUIDE, Accessed April 2017 and available at https://maddenmaritime.files.wordpress.com/2015/08/marpol-practical-guide.pdf.

[651] US Coast Guard, US Geological Survey, Deepwater Horizon MC252 Gulf incident oil budget. National Oceanic and Atmospheric Administration, 2010.

[652] Major oil spills, International Tanker Owners Pollution Federation, Accessed November 2, 2008.

[653] Khordagui, H. and D. Al-Ajmi, Environmental impact of the Gulf War: An integrated preliminary assessment, *Environmental Management*, 1993, 17(4), 557–562.

654 Stout, L., *The Shareholder Value Myth—How Putting Shareholders First Harms Investors, Corporations and the Public* (San Francisco, CA: Berrett-Koehler Publications, 2012), 1.

655 Ivory, D., BP temporarily banned from contracts with US Government, *Bloomberg News*, November 28, 2012, available at http://www.bloomberg.com/news/print/2012-11-28/bp-temporarily-suspended-from-new-contracts-with-u-s-government.html.

656 Fortune, Six big oil spills and what they cost, Accessed April 2017 and available at http://archive.fortune.com/galleries/2010/fortune/1005/gallery.expensive_oil_spills.fortune/index.html

657 Ibid.

658 Chircop, A. E. and O. Lindén, *Places of Refuge for Ships: Emerging Environmental Concerns of a Maritime Custom* (Martinus Nijhoff Publishers, 2006), 194. After the Exxon Valdez oil spill disaster in Alaska in 1989, the US Government required all new oil tankers built for use between US ports to be equipped with a full double hull.

659 IMO, International convention on oil pollution preparedness, Accessed April 2017 and available at http://www.imo.org/en/About/Conventions/ListOfConventions/Pages/International-Convention-on-Oil-Pollution-Preparedness%2c-Response-and-Co-operation-%28OPRC%29.aspx.

660 IMO, Protocol for oil pollution preparedness, Accessed April 2017 and available at http://www.imo.org/en/About/Conventions/ListOfConventions/Pages/Protocol-on-Preparedness%2c-Response-and-Co-operation-to-pollution-Incidents-by-Hazardous-and-Noxious-Substances-%28OPRC-HNS-Pr.aspx.

661 Oil spill response, Dispersants, Accessed April 2017 and available at http://www.oil-spillprevention.org/oil-spill-cleanup/oil-spill-cleanup-toolkit/dispersants.

5 Sustainability Analytics Applications

5.1 VALUATION OF STATISTICAL LIFE

Estimating a monetary value of human life is considered by many to be an abomination, an unethical and immoral act. However, valuation of projected reduction in the risk of premature mortality is an evolving area in the community of economic and public policy analysis, and a considerable body of academic literature has been emerging on the subject. Although sensitive, the value of a statistical life is an unavoidable input in policy option evaluation.

5.1.1 VALUE OF STATISTICAL LIFE OR VALUE OF PREVENTING A FATALITY

1. The value of statistical life (VOSL) is a measure of society's willingness to pay for statistical reductions in small risks of premature death. It has no application to an identifiable individual or to anything other than small variations in individual risks. It is not intended to suggest that any individual life can be reduced to a mere monetary value. The change in risk is a statistical case, not for any identifiable individual. Its sole purpose is to help describe the likely benefit of an intervention. Health and Safety Executive (HSE) UK uses the term value of a prevented fatality (VPF) as equivalent to VOSL. VOSL or VPF is not the value the society or the legal system might put on the life of a real person or on the compensation appropriate to the loss of that person.

2. In establishing the value of VOSL (or VPF), US policies rely on the revealed preference method based on wage risk studies, while Europe, Canada, and Australia use the stated preference methods, eliciting people's willingness to pay (WTP) for small changes in mortality risk. Society's willingness to pay is different from a human capital approach, which is based on lost earnings. WTP is the preferred approach to estimate VOSL, over the human capital approach that uses avoided lost earnings, because it most closely conforms to economic theory. Contingent value (CV) studies solicit WTP directly from respondents. Hedonic wage (HW) or wage risk studies base WTP on estimate of additional compensation demanded in the labor markets for riskier jobs. Hedonic price (HP), on the other hand, bases WTP on price differentials for housing environment or product safety features.

3. The value placed on mortality risk reduction also depends on the nature of the risk. For instance, premature mortality risks from air pollution are experienced on an involuntary basis and are generally uncompensated,

while job-related risks are assumed by individuals who presumably have some choice as to occupation and are compensated for taking a riskier job. A stark manifestation of this can be seen in the safe exposure standards for toxins in the environment for the general public under US Environmental Protection Agency (EPA) standards versus those for workers within plant premises under US Occupational Safety and Health Administration (OSHA) standards. Similarly, the US Food and Drug Administration (FDA) used a VOSL of US$2.5 million for its tobacco rule and US$5 million for the mammography rule, a 100% premium, once again, likely driven by the voluntary versus non-voluntary nature of the risk.

4. Likewise, HSE UK VPF also varies depending on the particular hazardous situation. HSE takes the view that when death is caused by cancer, people are willing to pay a premium for the benefit of preventing a cancer-caused fatality. European commission states that, although evidence is minimal, cancer premium, for long period of illness before death, needs to be captured. While HSE UK uses a 100% cancer premium, European Commission recommends a 50% cancer premium.

5. VOSL does not take into account the demographics of the population, in terms of age, income, or health status. The value of statistical life-years extended (VOSLE) method does address age differences. For instance, if regulatory action protects individuals whose average remaining life expectancy is 30 years, a reduction of one fatality is expressed as *30 years life extended*. However, this approach is politically sensitive and does not account for health status.

6. The Organisation for Economic Co-operation and Development (OECD)[662] recommends that the VOSL base value be adjusted as follows:

    a. *Population characteristics*: Adjust between countries based on gross domestic product (GDP) per capita. Also, adjust by a factor of 1.5–2.0 if the intervention is targeted at reducing risk to children. No adjustments within a country based on income or health status and strata.

    b. *Risk characteristics*: No adjustment for perception, morbidity prior to death or magnitude of risk. Recalculation of VOSL base value in case of risk greater than one in ten thousand annually.

    c. *Economic characteristics*: Adjust for inflation based on CPI and for increase in real income over time based on percentage increase in GDP per capita.

    d. *Other considerations*: No adjustment between private versus public, potential bias in stated preference studies, as well as for altruism.

5.1.2 Normalizing Benchmarked Value of Statistical Life or Value of Preventing a Fatality Globally—A Human Rights Issue

Figure 5.1, Benchmarking value of statistical life for equivalency, presents the value of VOSL adopted and recommended by various agencies in the United States, United Kingdom, Europe, Canada, and the New Zealand. References for the source of data in Figure 5.1 are A,[663] B,[664] C,[665] D,[666] E,[667] F,[668] G,[669] H,[670] and I.[671] These values

Source	Country/region	Social discount policy	Uprate policy	Currency basis	Benchmark benefit	Exchange rate 2003	Safety road 2003 US $ per year	Environment cancer 2003 US $ per year	Safety road 2003 US $ over 30 years	Environment cancer 2003 US $ over 30 years	Average metro population density
A	Australia/New Zealand	4%		2001 NZ $	2,490,000	1.47	$1,758,916		$30,415,240		4,023
B	Canada	4%		1999 CND	5,000,000	1.27		$4,345,833		$75,148,286	2,959
C	Europe	4%		2000 Euro	1,400,000	0.79	$1,892,559		$32,726,201		12,226
D	Europe	4%		2000 Euro	2,100,000	0.79		$2,838,839		$63,579,935	12,226
E	Europe Max	4%		2000 Euro	3,500,000	0.79		$4,731,399		$81,815,502	12,226
F	UK	6%	4%	2001 UK Pounds	1,000,000	0.55	$1,887,993		$42,284,361		4,348
G	UK	6%	4%	2001 UK Pounds	2,000,000	0.55		$3,775,987		$84,568,722	4,348
H	USA	7%		1996 US $	2,700,000	1.00	$3,164,627		$39,269,989		2,918
I	USA	7%		1997 US $	5,900,000	1.00		$6,760,187		$83,887,438	2,918
						VOSL	$3,400,000	$6,800,000	$42,190,740	$84,381,480	

Year	1990	1991	1992	1993	1994	1995	1996	1997
CPI US	130.7	136.2	140.3	144.5	148.2	152.4	156.9	160.5
Year	1998	1999	2000	2001	2002	2003	2004	2005
CPI US	163.0	166.6	172.2	177.1	179.9	183.9	188.9	195.3

FIGURE 5.1 Benchmarking value of statistical life for equivalency. (From A,[672] B,[673] C,[674] D,[675] E,[676] F,[677] G,[678] H,[679] and I.[680])

fall into two broad categories: (1) voluntary risks, such as *road safety*; and (2) involuntary risks, such as *cancer*. Present value of risk reduction benefit over a thirty-year assessment period and relevant agency US Office of Management and Budget (OMB) and UK HSE recommended factors for discounting/updating is computed to be US 2003 $ ~ 80 million for involuntary risks and $ ~ 40 million for voluntary risks. The benchmarked VOSL of US 2003 $ ~ 80 million for involuntary risks is equivalent to an annualized stream of US 2003 $6.8 million per year for averting one premature statistical fatality per year. Likewise, VOSL of US 2003 $ ~ 40 million for voluntary risks is equivalent to an annualized stream of US 2003 $3.4 million per year for averting one premature statistical fatality per year.

5.1.3 APPLICATION OF VALUE OF STATISTICAL LIFE OR VALUE OF PREVENTING A FATALITY

Government stewards, for policy decision analytics, depend on small changes in statistical risk to health at the individual level. Statistical life or averted fatality is arrived at by aggregating these small changes in risk across large groups of individuals. For instance, in the United States, the annual likelihood of death increases each year with age; from 1 in 1,000 at age 20 to 10 in 1,000 at age 65, if one survives to that age.[681]

If the WTP reveals an average value of say $700 for a reduction of 1 in 10,000 in the individual's risk of premature death in the current year, then the VOSL equals $700 average individual WTP multiplied by 10,000 = $7,000,000.

If an intervention option decreases this annual risk in a population of 4,000 individuals by 0.5 in 1,000, the risk reduction is equal to 2.0 statistical cases (0.5/1,000)*4,000 = 2.0 and the monetized value of the benefit equals 2*$7,000,000— that is, $14,000,000. This benefit value could be one factor in cost benefit analysis or cost effectiveness evaluation, or in comparing intervention options.

5.2 VALUATION OF AIR POLLUTANTS

Valuation of pollution reduction benefits, globally (from recent benefit analyses by various Government agencies in the United States, United Kingdom, Europe, Canada, and World Bank), helps relate change in pollutant emissions directly to monetized environmental benefits. The use of government data provides a defensible scale, because it conforms to current economic principles that society will raise company cost to equal perceived benefit. For opportunity prioritization purposes, it is more prudent to use benchmarked pollution reduction benefit unit values from government benefit analyses. A representative set[682] of these are presented in Figure 5.2. All monetary values have been converted to US dollars uses exchange rates as of December 2003 and US consumer price index of relevant years. Data availability may be limited to a set of *select pollutants*, namely CO – Carbon Monoxide, CO_2 – Carbon Dioxide, NO_x – Nitrogen Oxides, SO_2 – Sulfur Dioxide, PM – Particulate Matter, BOD – Biological Oxygen Demand, TSS – Total Suspended Solids, noise, and some toxins. For all other pollutants, relative potency or toxic factors must be used to convert them into select pollutant equivalents.

Basis: Annualized benefit = Cost of externality Averted, 90% from averted mortality	2003 US $s US short tons (2,000 #s)	Particulate PM 2.5 μ	Sulfur dioxide	Nitrogen oxide	Volatile organic compounds	Voluntary transport risk VOSL	Involuntary cancer risk VOSL
Annualized Benefit per EU	2000 €/year/tonnes per year	40,000	8,700	6,600	1,400		
Convert to 2003 US $/Short ton/Year = (2240/2000)*(VOSL!I17/VOSL!F17)*(VOSL!G11/VOSL!G7)	1.514	$60,562	$13,172	$9,993	$2,120		
Selected VOSL and annualized benefit	2003 US $s/year/TPY	$60,000	$13,000	$10,000	$2,100	$3,400,000	$6,800,000
Pollution reduction or averted equivalent = Averting 1 fatality/year; VOSL = $6,800,000	Pollution reduction of TPY per Year	113	523	680	3,238		
Methodology for the cost-benefit analysis	http://ec.europa.eu/environment/archives/cafe/pdf/cba_methodology_vol1.pdf						
Assessment of externalities of air pollution	http://ec.europa.eu/environment/archives/cafe/activities/pdf/cafe_cba_externalities.pdf						
Update of air pollution externalities	http://ec.europa.eu/environment/archives/cafe/pdf/cba_update_nov2006.pdf						

FIGURE 5.2 Air pollutant reduction benefits—benchmarked values.

5.2.1 Air Pollutant Reduction Benefits—Benchmarked Values

The most common approach for benefit analyses involves modeling changes in exposure to different emission levels, quantifying impacts using exposure-response functions, and valuing human health and welfare (e.g., averted impact on visibility, productivity, and property damage) benefits based on willingness to pay. Emission trade and/or cost effectiveness of a regulation or negotiated permit are alternate sources that yield market value of a pollutant. These, however, are often location and time specific and may vary considerably.

Figure 5.2 shows the benchmarked values of Air Pollutant Reduction Benefits. The benchmarking is done with published government agency or academic information.[683–685]

Quantification of benefits from pollutant reduction and are conducted by the European Commission using the *impact pathway* or *damage cost* approach.[686] It involves (1) quantification of emissions (by the Regional Air Pollution INformation and Simulation [RAINS][687] model); (2) description of pollutant dispersion across Europe (by the RAINS and European Monitoring and Evaluation Program [EMEP] models); (3) quantification of exposure of people, environment, and buildings that are affected; (4) quantification of the impacts of air pollution; (5) valuation of the impacts; and (6) description of uncertainties.

Health impacts and damage to crops and properties because of air pollution can be quantified. Damage to ecosystems and cultural heritage can be quantified on a relative basis. Other impacts such as reduced visibility from air pollution and the social dimensions of health impacts may not currently be quantifiable in terms of impact or monetary value; they permit only a qualitative analysis.

Multiple cost-benefit analyses have shown that human health impacts will generate the largest quantified monetary benefits when air pollution is reduced. The pollutants of most concern here are fine particles and ground-level ozone both of which occur naturally in the atmosphere. Fine particle concentration is increased close to ground level by emissions from human activity. This may be through direct emissions of so-called *primary* particles, or indirectly through the release of gaseous pollutants (especially SO_2, NO_x, and NH_3 – Ammonia) that react in the atmosphere to form so-called *secondary* particles. Ozone concentrations close to ground level are increased by anthropogenic emissions, particularly of VOCs and NO_x. The quantification of health impacts addresses the impacts related to both long-term (chronic) and short-term (acute) exposures.

The quantification deals with both mortality and morbidity impact on adults and infants. Both methods, value of statistical life (VOSL, applied to the change in number of deaths) and value of life year extended (VOSLE, applied to changes in life expectancy), are used to show transparently the inherent uncertainty.

5.2.2 Potency Factors to Convert to Equivalent Select Pollutants

Pollution reduction benefit (PRB) of the opportunity is the sum of PRBs of all pollutants. The PRB for any pollutant equals the change in emission level multiplied by its benchmarked value from recent government benefit analyses. Estimation of PRB requires benchmarked values of individual pollutants. These values are available only for a limited set of pollutants. These pollutants are termed *select pollutants*.

It is prudent to limit the number of these select pollutants to a small set, because of resource constraints in establishing the benchmarked benefit values. However, this requires a way to convert all other pollutants into this set of *select pollutants*. All other pollutants have to be converted to equivalent select pollutants. The use of Relative Potency, or Toxic Weight factors, is one feasible and defensible way to accomplish this. To convert other pollutants into equivalent *select pollutants*, the change in pollution level is multiplied by its relative potency or toxic weight factor.[688]

Relative potency, or Toxic Weight factors, is required for converting all pollutants to a limited set of equivalent select pollutants. For most common pollutants, relative potency factors for air, such as photochemical oxidation potential (PCOP), global warming potential (GWP), ozone depletion potential (ODP), and toxic weight factors for water (TWF), are available in *The Sustainability Metrics* from the Institute of Chemical Engineers, UK.[689] Atmospheric pollutants are categorized into five groups and the environmental burden (EB) of each group is calculated in terms of a select pollutant. The five categories are (1) Atmospheric acidification EB is ton/y sulfur dioxide equivalent, (2) global warming EB is te/y carbon dioxide equivalent, (3) human health (carcinogenic) effects EB is ton/y benzene equivalent, (4) stratospheric ozone depletion EB is ton/y CFC-11 equivalent, and (5) photochemical ozone (smog) formation EB is ton/y ethylene equivalent. A representative set of relative potency factors are presented in Figure 5.3. Documents such as the Sustainability Metrics of the Institute of Chemical Engineers, UK[690] and those listed in the following provide detailed listing of potencies. The toxicity weight for each chemical, as well as a sample calculation, can be found in the toxicity weighting spreadsheet available on the Risk Screening Environmental Indicators (RSEI)[691] website. Several other scientific and regulatory bodies (US EPA,[692–694] UK Environment Agency (EA),[695–697] and University of California, Berkeley,[698] among others) list comparable data.

A hazardous air pollutant's (HAP) potential impact on public health depends on two significant factors: mass of emissions and toxicity. In the toxicity-weighted emissions approach, toxicity-weighted emissions of a given HAP, in toxicity-weighted tons per year, equals the weight of the given HAP, which is emitted to the air in tons of HAP per year (TPY) multiplied by the toxicity factor for the given HAP.[699]

For *select* pollutants (i.e., for the set of pollutants for which PRB value from government benefits analyses is readily available and is included in Figure 5.2), conversion is not required. For pollutants other than those in the *select set*, conversion to an equivalent select pollutant is necessary. To convert other pollutant discharge level changes into equivalent *select pollutants*, multiply it by its relative potency or toxic weight factor. For example, to convert 10 TPY of butylene emission into the equivalent select pollutant *ethylene*, first find relative potency factor for butylene from the table, which is 0.7. Therefore, 10 TPY butylene equals 10 * 0.7 Butylene/ethylene equal 7 TPY VOC ethylene equivalent.

Atmospheric acidification potency			Ecotoxicity to aquatic life (vales for sea water conditions)	
Sulfur dioxide	1.00		Arsenic	0.20
Ammonia	1.88		Cadmium	2.00
Hydrogen chloride	0.88		Chromium	0.33
Nitrogen dioxide	1.60		Copper	1.00
Hydrogen flouride	0.70		Iron	0.01
Sulfuric acid mist	0.65		Lead	0.20
			Manganese	0.10
			Mercury	16.67
			Nickel	0.17
			Vanadium	0.13
The unit of Environmental Burden is te/y sulphur dioxide equivalent.			The unit of Environmental Burden is te/y copper equivalent.	
			The potency factor is equal to the reciprocal of the Environmental Quality Standard (EQS) divided by the reciprocal of the EQS of copper.	

FIGURE 5.3 Potency factors for atmospheric acidification and aquatic ecotoxicity. (From Institute of Chemical Engineers, UK.[700] Source: http://www.icheme.org/communities/subject_groups/sustainability/resources//~/media/Documents/Subject%20Groups/Sustainability/Newsletters/Sustainability%20Metrics.pdf.)

5.3 VALUATION OF ENVIRONMENTAL RISKS

5.3.1 ENVIRONMENTAL OPPORTUNITY PRIORITIZATION FRAMEWORK

Manufacturing facilities need a screening framework to focus on the highest environmental performance improvement opportunities and to monetize benefits to prioritize environmental investments. It is often a challenge to find diverse environmental benefits at the same time when allocating resources among opportunities with comparable financial returns or compliance accomplishments. For instance, an opportunity to reduce sulfur dioxide may increase carbon dioxide emissions and effluent discharges. The screening framework presented here is designed to help prioritize environmental investment opportunities with conflicting benefits on a relative basis, in a consistent manner. It describes how environmental investments, credits, and benefits are evaluated in the prioritization framework, shows how relative potency or toxic weight factors serve to convert relevant pollutants into a limited set of *select pollutants*, and defines a prioritization metric for environmental opportunities.

5.3.2 ENVIRONMENTAL OPPORTUNITY PRIORITIZATION METRIC

Relative ranking of competing opportunities requires the development of an appropriate criteria. The environmental opportunity prioritization metric, *environmental investment efficiency*, is defined to be a ratio of the environmental benefits to some defined measure of cost or investment or both. Environmental investment opportunities reduce either the risk of an environmental incident or the level of environmental pollution. Environmental opportunity value comes from two very distinct components—one from lowering environmental impacts from a mix of probabilistic unplanned incidents and the second from deterministic releases and discharges including planned maintenance. Because of the inherent difference between them, the proposed prioritization framework recommends separate approaches for the valuation of benefits and ranking of deterministic pollution reduction and probabilistic pollution averted opportunity components.

5.3.2.1 Environmental Investment Opportunity Valuation

Environmental opportunity investments are associated with averting pollution impacts from event-related releases or discharges by reduction of risk and pollution reduction from regular operations including planned maintenance.

1. Pollution reduction investments may involve commitment of assets to
   a. Reduce emissions, waste, or both to levels below legal compliance requirements.
   b. Create offsets to allow for expansion potential.
   c. Make it a de minimus source to avoid applicability of unnecessary regulations.
   d. Comply early to improve public perception.
   e. Reduce opposition to operations/regulatory permitting.
2. Pollution averting or risk reduction investments involves resource allocation to
   a. Increase risk mitigation capacity to buffer against exceedance.
   b. Reduce episodic releases and impacts.
   c. Improve reliability and maintenance.
   d. Reduce community concerns/complaints.
3. Economic credits from environmental investments

Conventional opportunity evaluation often overlooks economic credits directly attributable to environmental investments. They should be included just like other costs and benefits in conventional opportunity economics. Further, for multiple year cash flows, US EPA guidelines recommend use of discount rate of 7% to obtain present value (PV). These components include emission fees, waste disposal, energy usage, environmental investment tax credit, and emission credits with a binding sales agreement or to avoid purchase of required credits. Cost increases, if any, are treated as negative economic credits. Some examples of cost changes directly attributable to the environmental investment opportunity include

1. Energy pollution control may require additional energy and increase cost.
2. Emission fees emission reduction could lead to lower emission fees. (Note: there could be a simultaneous increase in emission of some other pollutant: e.g., NO_x or carbon dioxide and the fee increase if any should be accounted for.)

3. Waste disposal waste reduction may reduce disposal cost, while additional discharges from an air pollution control equipment, for instance a scrubber, may increase treatment or discharge cost.
4. Reporting labor elimination of a pollution source may reduce reporting.
5. Environmental investment tax credit some national regulations may allow tax credit for investment in pollution control measures.

5.3.2.2 Pollution-Reduction Opportunity Valuation

The deterministic pollution reduction opportunities, which are definitive, can be prioritized using a benefit cost ratio approach. The prioritization metric, environmental investment efficiency is defined to be a ratio of the PV of the pollution reduced environmental benefits to the PV of the investments less environmental economic credits. US EPA guidelines recommend use of discount rate of 7% to obtain present value (PV).

Environmental Investment Efficiency Metric for Pollution Reduction is a benefit to cost ratio

$$= \frac{\text{(Pollution Reduction Benefit)}}{\text{(Environmental Investments} - \text{Environmental Economic Credits)}}$$

The PRB of the opportunity is the sum of PRBs of all pollutants. The PRB for any pollutant equals the change in emission level multiplied by its benchmarked value from recent government benefit analyses. These values are available only for a limited set of pollutants. These are termed *select pollutants*, and the values are included in Figure 5.2. All other pollutants are converted to equivalent select pollutants using relative potency/toxic weight factors.[701] Section 5.2 provides a detailed discussion of the PRB valuation methodology.

5.3.2.3 Pollution-Averted Opportunity Valuation

The stochastic pollution averted opportunities are more effectively prioritized by using the risk matrix for screening. Typically, process safety risk-management methods address benefits from averting consequences of fire, explosion, and acute pollutant and flammables released through risk reduction. Most incidents involve a diverse set of consequences with varying degrees of severity. Consequences include potential impact on human health, other species health, property value, aesthetic, and recreational value of natural resources, and so on (e.g., sulfur dioxide may contribute to human health effect, and damage to plants and to property). Aggregation of effects to a single attribute is difficult and depends on multiple factors such as atmospheric conditions, receiving body characteristics, and interaction among a broad set of compounds released to the environment. Typically, benefits of averting human fatality dominate.

Current risk matrix screening approaches are event focused and do not address non-acute pollution. Release of non-acute pollutants is usually overlooked. Also, prediction of non-acute pollutant release potential is incident specific and is difficult to correlate to changes in levels of risk. So, in addition to assessing the risk reduction benefits, the quantity of event specific release of non-acute pollutants averted has to be assessed. Once the release quantity of non-acute pollutants averted is estimated, it can be multiplied by the benchmarked (monetized) value of the non-acute emissions and discharges averted, because of that specific risk reduction, and added to the risk reduction benefits.

5.3.3 RISK-REDUCTION OPPORTUNITY EVALUATION—ISO-RISK SCREENING MATRIX

Risk screening matrix is a qualitative risk assessment tool that enables risk-based prioritization into a few categories. Section 3.1.5.4 describes the risk ranking and the probability/consequence screening matrix for risk prioritization. However, additional resolution is often needed for comparing risks and informing the decision-making process for resource allocation. This is particularly so when risks being compared or risks prior to and post mitigation are positioned within the same category; and hence not discernable. Risk reduction opportunity prioritization calls for application of benefit cost or cost effectiveness analysis, and both require more granular, preferably quantitative, differentiation between small changes in risk.

Iso-Risk matrix, a simple modification of the standard risk screening matrix, is a semiquantitative approach that narrows some of these gaps.[702] Iso-Risk contours are lines of equal risk. The Iso-Risk matrix is built by overlaying the standard matrix with these lines. One such Iso-Risk line is chosen and assigned a numerical value to represent risk significance. All other contours, called Iso-Risk lines, automatically take on a relative risk significance. It neither measures absolute risk, nor does it indicate the tolerability of risk. Consistent numerical characterization of risk and risk reduction can be gleaned from this method and used in the relative ranking of risk reduction opportunities and conduct cost-effectiveness analysis. A simple but useful way is to depict the center of the cell value in discrete risk units. This discrete risk unit matrix allows comparing risks or set of prior to and post mitigation risks, plotted on the matrix, even within the same cell, based on its position relative to Iso-Risk line (Figure 5.4).

5.3.4 SERIATIM OF OPPORTUNITIES

In the case of a large number of opportunities, investment seriatim could be constructed with cumulative returns and cumulative costs, and a cut off chosen for selecting projects to go forward within a limited investment available. Figure 5.5 shows an example seriatim of ranked investment opportunities.

				Probability in %				
Immediate action weekly review				Almost certain	Likely	Possible	Unlikely	Rare
Urgent action monthly review				30	3	0.3	0.03	0.003
Timely action quarterly review				**A**	**B**	**C**	**D**	**E**
General action yearly review								
No action eliminated document								
Extreme significant long-term	45	I		1350	135	14	1	0.1
Major significant medium term	4.5	II		135	14	1	0.1	0.01
Moderate minor local for days	0.45	III		14	1	0.1	0.01	0.001
Insignificant very minor within facility	0.045	IV		1	0.1	0.01	0.001	0.0001
Almost none very small area	0.0045	V		0.1	0.01	0.001	0.0001	0.00001

(Left axis label: Consequence in severity of impact on organization)

FIGURE 5.4   Pollution averted benefits ranking Iso-Risk screening matrix.

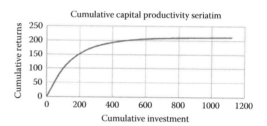

NPV	INV	Cum INV	Cum NPV
		0	0
100	100	100	100
60	120	220	160
30	150	370	190
15	200	570	205
5	250	820	210
2	300	1120	212

FIGURE 5.5   Example seriatim of ranked investment opportunities. Net Present Value (NPV), Investments (INV).

5.4   PRIORITIZATION OF ENVIRONMENTAL OPPORTUNITIES

5.4.1   ISSUE

A major corporation with several operating facilities plans to optimize its investments to enhance its global environmental performance. Environmental opportunity investments are associated with averting pollution from event-related releases or discharges by the reduction of risk or pollution reduction from regular operations including planned maintenance. Environmental investments include both capital and operating costs, and investment tax credits. Additional cost changes, directly attributable to the environmental investment opportunity, include energy for pollution control, emission fees, waste disposal, and compliance reporting labor. The corporation needs a capital productivity-based ranking for prioritization of these investment opportunities.

5.4.2 Analytics

5.4.2.1 Definition of Terms

EC = Economic credits, say 10% of investment (e.g., investment tax subsidies)

EE = Environmental investment efficiency metric = (PAB + PRB)/(INV − EC)

ERDU = Event risk discrete units (e.g., moving from one cell to another)

ERRU = Policy defined equivalent risk reduction units to avert one fatality (e.g., 250 to represent the equivalent of averting one fatality per year)

INV = Opportunity investment, annualized capital and operating cost

Opp PAB = Pollution-averted benefit = Opp ERDU * VOSL/ERDU

Opp PRB = Opportunity pollution reduction benefit = $\sum$PRUi * ($/year/PRBUi)

Opportunity ERDU = Opp ERDU = (Pre-project ERDU—post project ERDU)

Opportunity PV = Opp PV = Present value of multiyear cash-flow of (Opp PAB + Opp PRB + EC − INV), all in current dollars, discounted @ 7.0% per US EPA guidelines for relative ranking purposes

PRBUi = $/year per TPY PRUi (e.g., $ benefit per year per TPY reduction of pollutant i)

PRUi = Pollution reduction in TPY of a select pollutant i (e.g., VOC as ethylene)

PRVj = Pollution reduction in TPY of a non-select pollutant j (e.g., butylenes)

Relative Potency = Factor to convert non-select pollutant to equivalent select pollutants

VOSL = Policy defined value of statistical life, per government benchmarks

5.4.2.2 Computation of Capital Productivity

1. *Pollution reduction benefit (PRB) estimation*

   The PRB of the opportunity is estimated as follows:

   a. Estimate the change in pollution (emission/discharge) level for all pollutants including each *select pollutant*. Estimate *pre* and *post* opportunity emission levels. Also, note if region exceeds ozone or other ambient air quality standards. This may impact the benchmarked benefits. For instance, the opportunity may decrease ethylene emissions by 100 TPY, but increase butylene emissions by 40 TPY in a region that exceeds ozone ambient air quality standard.

   b. Convert pollution-level changes for all the pollutants to the equivalent select pollutants using relative potency/toxic weight factor. Multiply the change in the pollution level by its relative potency, or toxic weight factor, from Figure 5.3 or from the references. (e.g., relative potency factor for butylene, is 0.7 in ethylene equivalent, therefore, 40 TPY Butylene = 40 TPY butylene * 0.7 butylene/ethylene = 28 TPY ethylene).

c. For each *select pollutant* estimate aggregate change in pollution level, add a and b to obtain the aggregate change in pollution (emission/discharge/release) level for that select pollutant (e.g., for select pollutant VOC (as ethylene), PRUi = (100 − 28) = 72 TPY decrease).

d. Obtain benchmarked value of PRBUi for the relevant select pollutant(s) from Figure 5.2 (e.g., value in 2003 US $/year of reducing one TPY of VOC [as ethylene]) $/year/PRBUi = $2,100. This has to be adjusted for being in a region that exceeds the ozone ambient air quality standard, and current (2017) dollars. EU suggests an increase of one-third for the non-attainment area, yields $2,100 * 1.33 = $2,800. The US CPI of June 2017 is 243.79. Adjusting for = $2,800 * (US CPI 2017/ US CPI 2003) = $2,800 * (243.79/183.9) = $3,712.

e. Calculate PRB for each *select pollutant*. The pollution reduction benefit for each *select pollutant* equals this aggregate change in pollution level from c above multiplied by its benchmarked value from recent government benefit analyses from d above.

$$= PRUi*(\$/year/PRBUi),\ for\ example$$

$$= 72\ TPY\ VOC * \$3,712/year\ per\ TPY\ of\ VOC$$

$$= \$267,254/year$$

f. PRB is the sum of PRB of all pollutants; that is, all *select pollutants* (including the converted equivalents). Calculate opportunity PRB = $\sum$PRB for all select pollutants (e.g., VOC, PM, BOD, and so on).

2. *Pollution averted benefit assessment*

The PAB opportunity prioritization, in this proposed prioritization framework, for probabilistic incident risk reduction, is accomplished through a risk screening matrix and adding computed benefits of related non-acute pollution averted.

a. Develop, compile, and include discrete risk units derived from Iso-Risk contours in the matrix. In many common 4 × 5 risk matrices with four categories of risk, a knee may occur at some of the transition points because the consequence levels span differing orders of magnitude. One simple way is to adjust the consequence level cells for consistency in their relative sizes, thus removing the knee in the Iso-Risk contours.[703] Another way is to use a 5 × 5 risk matrix as shown in the example.

b. Establish, as a policy, the benchmarked value of statistical life (e.g., the benchmarked value of VOSL in 2003 US $ is 6.8 million per fatality per year for involuntary risks). Convert that to the current year, using the ratio of consumer price indices. The US CPI of June 2017 is 243.79. Adjusting for US CPI = $6,800,000 * (US CPI 2017/ US CPI 2003) = $6,800,000 * (243.79/183.9) = $9 million.

c. Establish, as a policy, risk reduction units, for example, 250 as equivalent of averting one fatality per year. This number represents a basis for relative valuation of opportunities.

d. Assess and assign discrete risk units to the pre- and post-situation of the opportunity. Review pre- and post-*opportunity* potential environmental event scenarios, identify what potential events it averts, and find the number of Risk units reduced using the risk screening matrix.

e. Locate and place the pre- and post-*opportunity* potential environmental event scenarios in the appropriate *cells* in the matrix, for example, B ll (pre) to C lll (post). For instance, an opportunity reduces incident risk from cell B II 14 to cell C III 0.1 in the risk screening matrix; the opportunity units of annual risk reduction is $= 14 - 0.1 = 13.9$. Then the value of the opportunity benefit from risk reduction equals $\$500,400 (= 13.9*(\$9,000,000/250) = 13.9*\$36,000)$.

f. The benefits of event-related non-acute pollution averted can be computed using the same method proposed for PRB evaluation. Estimate non-acute pollution averted in tons of emissions and releases by the initiative preventing that event. Benefits of related non-acute pollution averted can be computed following the same method as proposed for PRB evaluation.

5.4.3 Insight

5.4.3.1 Opportunity-Capital Productivity Computation and Ranking

1. Rank opportunities by environmental investment efficiency metric $= EE = (PAB + PRB)/(INV - EC)$. From Figure 5.4, option A and option C are nearly equal and both dominate option B. However, this metric does not reckon multiyear cash-flow situations.

2. Alternatively, opportunities could be ranked by opportunity PV. Com-pute, for each opportunity, the present value of multiyear cash flow of $(PAB + PRB + EC - INV)$, all in current dollars, discounted @ 7.0% per US EPA guidelines for relative ranking purposes.

3. Then rank them in the order of highest to lowest opportunity PV (Figure 5.6).

5.5 BASF ECO-EFFICIENCY ANALYSIS INFORMS CORPORATE STRATEGIC DECISION[704]

5.5.1 Issue

Polyol, a raw material for furniture foam, is conventionally derived from petroleum. Foams are derived from a reaction of the polyol and toluene diisocyanate. The primary reaction during the production of polyurethanes[705] is of the form where polyols (compounds containing multiple –OH groups) and diisocyanates (compounds containing –NCO groups) form polyurethanes $-NCO + HO- \rightarrow -NH-CO-O-$. Alternatively, polyol could be produced from bio-based natural oils, such as soy

Step	Description	Term	Input	Option A	Option B	Option C
1	Opportunity Investment = INV, annualized capital and operating cost	INV	Input	$1,200,000	$8,000,000	$7,000,000
2	Economic Credits = EC, say 10% of Investment	EC	10%	$120,000	$800,000	$700,000
3	Policy defined Equivalent Risk Reduction Units to avert one fatality	ERRU	250			
4	Policy defined Value of Statistical Life, per Government benchmarks	VOSL	$9,000,000			
5	Event Risk Discrete Units e.g., moving from one cell to another	ERDU	Input	CI to CIII	AII to BIII	BI to CI
6	Opportunity ERDU = (Pre-project ERDU – Post Project ERDU).	Opp ERDU	Calculated	13.9	134	121
7	Pollution Averted Benefit = PAB = Opp ERDU * VOSL/ERDU	Opp PAB	Calculated	$500,400	$4,824,000	$4,356,000
8	Pollution Reduction in TPY of a non-select pollutant j e.g., Butylene	PRVj	Input	−40	10	15
9	Conversion of non-select pollutant to equivalent select pollutants	Potency	0.7	−28	7	10.5
10	Pollution Reduction in TPY of a select pollutant I e.g., VOC as Ethylene	PRUi	Input	100	20	30
11	$/year per TPY PRUi e.g., $ benefit per year per TPY reduction of pollutant i	PRBUi	$3,712			
12	Opp Pollution Reduction Benefit = PRB = ΣPRUi * ($/year/ PRBUi)	Opp PRB	Calculated	$267,254	$100,220	$150,331
13	Environmental Investment Efficiency Metric = (PAB + PRB)/(INV − EC)	EE	Calculated	71.1%	68.4%	71.5%
14	Alternatively Opportunity PV = (PAB + PRB + EC − INV) all in current dollars, discounted @ 7.0% per US EPA Guidelines for relative ranking	Opp PV				

FIGURE 5.6 Computations for capital productivity-based prioritization of opportunities.

or castor. Are the bio-based polyol manufacturing routes more sustainable than the traditional petroleum route? The objective is to compare sustainability of polyol production processes from petroleum and natural bio-based oils.

5.5.1.1 Process Basis

1. Polyols are formulated such that there are no differences in the foam manufacturing process, manufacturing equipment, amounts of catalyst, additives, and isocyanate or the scrap generation rates.
2. Polyols are delivered by railcar, and energy supplied to maintain proper polyol viscosity is included.
3. Performance and durability of the all finished foams (the use phase of the life cycle) are identical and ability to recycle, reuse, or recover are the same.

5.5.1.2 Alternative Processes

1. Polyol, a raw material for furniture foam, is conventionally derived from petroleum. Alternatively, polyol could be produced from bio-based natural oils such as soy or castor.
2. Products under evaluation must have same functional unit or customer benefit and must cover 90% of the relative market.

5.5.2 ANALYTICS

BASF eco-efficiency analysis (EEA) contributes to strategic and informed decision making. It is a comparative analysis—compares sustainability of alternative products or processes, more like cost-effectiveness and not cost-benefit analysis. EEA measures entire lifecycle environmental impacts and lifecycle costs for product or process alternatives for a defined level of output for its entire supply chain. SEEBALANCE®, a SocioEcoEfficiency Analysis developed by BASF social metrics into eco-efficiency analysis. BASF has recently introduced the Revised SEEBALANCE in 2017 integrates social metrics into eco-efficiency analysis. BASF EEA,[706] established in 1996, has been validated by NSF International in 2016. The assessment of life-cycle costs and aggregation to an overall eco-efficiency is based on ISO 14045:2012 and environmental life-cycle assessments follows ISO 14040:2006 and 14044:2006. The BASF anniversary book[707] reports on the development history over two decades in addition to presenting numerous example studies.

5.5.2.1 Define Customer Benefit or Functional Unit

1. Functional unit provides a reference point for comparison: production, use, and disposal of one-million board-feet of 1.8 lb/cubic feet density high-quality furniture foam
2. Develop clear performance criteria and boundaries
3. Consider as many alternatives to perform same function. Furniture foam produced from polyol derived from fossil or renewable sources (i.e., castor oil or soy oil)
4. Life cycle analysis (LCA) of production, use, and disposal considering system boundaries
5. Identical life-cycle stages could be eliminated in the relative comparison

5.5.2.2 Determine Economic Impacts

1. LCA to determine overall total cost of ownership from customer perspective
2. Basis-defined customer benefit and system boundaries
3. Include all initial and all future cost impacts or benefits
4. Use either constant (real) or nominal monetary values, but no mixing
5. Calculate at a point of time or use Net Present Value (NPV) to account for time value of money

5.5.2.3 Determine Environmental (Risk) Impacts

1. *Inventory analysis*: This step involves the compilation of all economic data on manufacturing and use costs, as well as the collection of relevant environmental data on a range of categories.
2. *Impact assessment*: This step comprises of assessing range of impacts of the product or process on the environment.
3. Characterize
   a. Energy consumption
   b. Resource consumption
   c. Emissions—air, water, land

    d. Land use

    e. *Toxicity potential*: risk phrases per Annexure III of the EU Directive 67/548/EEC or estimate from Safety Data Sheet (SDS)

    f. Risk potential

4. Normalize

    a. *Data normalizing* is required to make data understandable and comparable; applies to all environmental impact data categories except emissions.

    b. The summed life-cycle impact of each category is normalized relative to the alternative with the highest impact (or least favorable) in that area, which is assigned a value of 1 and others are valued proportionately.

5. Aggregate

    a. Aggregate score is required to obtain an overall environmental impact score by combining subcategories; which in turn requires a weighting process that incorporates both environmental relevance factors and societal significance factors.

    b. Environmental relevance is required to allow high environmental impact burdens to be more heavily weighted. These are unique to each EEA and region specific and reflect the contribution of that impact. It is calculated by dividing the alternative's impact by the total burden that impact category imposes on that region. This data is obtained from published regional environmental data.

    c. Societal significances are required to account for society's opinion on the importance of each environmental impact. These are based on third-party market surveys; while constant for each analysis, have to be periodically updated. Figure 5.7 shows an example of social weighting factors.

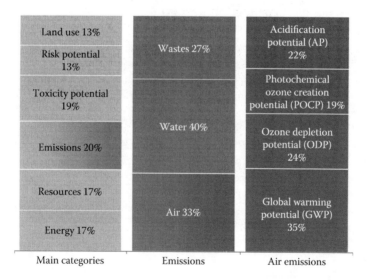

FIGURE 5.7 Social weighting factors.[708]

d. Societal significance factors are used in conjunction with the environmental relevance factors. The geometric mean of the environmental relevance factor and the social weighting factor is calculated as an overall weighting factor for each impact category.

e. Normalized environmental impacts for each of the six categories is multiplied by these geometric means and summed to get final environmental impact score for each alternative product or process.

5.5.2.4   Generate Eco-Efficiency Portfolio

1. Weighting of the environmental and cost impacts are used to account for relative importance of influence level.[709]

2. Each alternative's environmental score, in conjunction with its normalized economic impact, is captured on a biaxial plot, as illustrated in Figure 5.8, which is called the eco-efficiency portfolio. The plot reveals the relative positions of the eco-efficiency of the alternative products and processes evaluated. The most eco-efficient alternative is the alternative with the largest perpendicular distance from the diagonal line in the direction of the upper-right quadrangle, given equal importance to both the impacts.

5.5.3   INSIGHT

The petroleum- and castor-based polyol have similar eco-efficiency, equal perpendicular distance from the diagonal in the biaxial plot. Figure 5.9 provides additional comments on the three options.

For further information, please see Dr. Saling's book and BASF.[710]

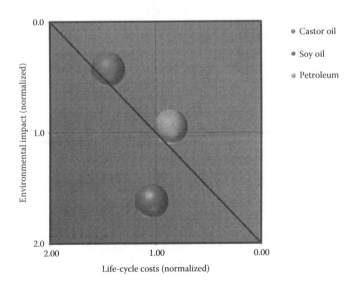

FIGURE 5.8   BASF eco-efficiency portfolio.

Polyol type	Cost	Environmental	Comments
Petroleum	Lowest	Intermediate	Lowest costs. Lowest risk potential and land use
Castor oil	Highest	Lowest	Lowest energy usage, fewest emissions, lowest toxicity potential, but high land use. Highest costs
Soy oil	Intermediate	Highest	High energy consumption and land use. Moderate cost

FIGURE 5.9 Eco-efficiency comparison.

5.6 CMU GUIDES GOVERNMENT METHANE-REDUCTION POLICY WITH DATA ANALYTICS

5.6.1 Issue

Methane is a high GWP greenhouse gas (GHG). Methane emissions from the natural gas sector account for about a quarter of US anthropogenic methane emissions. Several recent studies of methane emissions from natural gas transmission and storage compressor stations show a skew; a majority of emissions come from only a few emission sources, termed super-emitters. In the Carnegie Mellon University (CMU) study,[711] abating emissions from the highest emitting components or facilities, based on defined thresholds, was hypothesized to be a potentially efficient strategy for reducing emissions. However, locating the highest emitting components or facilities is a challenge. Detection across all facilities is necessary under all policy options, especially if emissions across facilities are skewed.

5.6.1.1 Policy Trade-Off Dimensions

1. Private costs and benefits (detection, abatement, or tax on unabated abatable emissions versus market value of recovered natural gas) and social costs and benefits (averting climate impacts of release of methane emissions)
2. Detection and emission reduction potential abatement technologies (sensitivity and efficacy)
3. Government policy instruments: emission standards or Pigouvian tax on emissions

5.6.1.2 Alternative Intervention Options

1. Unconstrained system-wide policy with detection and abatement of methane emissions at all facilities to minimize net social costs or maximize net benefits, including private costs of abatement and detection and methane social costs.
2. System-wide policy with a facility-level emissions reduction target to minimize private costs, including abatement and detection costs subject to an emission reduction target per facility at 10%, 50%, and 75%.
3. Super-emitter policy with absolute emission threshold calls for detection at all facilities and abatement only at a targeted subset with the highest absolute methane emissions, where the objective is the same as unconstrained

system-wide policy, but abatement only at a subset of facilities with absolute annual emissions above a specified threshold.
4. Super-emitter policy with proportional loss threshold, same as the one above under (3), but abatement only at a subset with proportional loss above a specified threshold.

5.6.2 Analytics

1. Establish baseline emissions for each component and facility by implementing a Monte Carlo simulation framework to model annual methane emissions. This was done based on Subramaniam et al.[712] and Zimmerle et al. data sets comprising 22,000 components, under 27 categories at 45 Transmission and Storage (T&S) systems and operating hours based on Zimmerle et al.[713] study of 24,000 components at 514 T&S systems. By assigning emissions for each component to a facility, each facility is described by a profile. Facility profile, emission estimates, and operating hours are used to develop estimates of throughput and proportional loss rate for each facility. Proportional loss rate equals quotient of annual emissions and throughput.
2. Develop abatement costs for each emission category and component as a function of simulated baseline emissions, abatement efficiency, abatement costs, and natural gas prices. Baseline estimate, the abatement efficiency is about 90% for most components based on published ICF and EPA estimates. However, for sensitivity analysis, these were conservatively reduced by 50% because they do not incorporate covariation of efficiency and baseline and variation after the first abatement. Abatement costs are computed as PV of incremental capital and labor over abatement interval in equal annual payment. Costs and sensitivity of existing and emerging detection method are based on optical gas imaging or EPA method 21 per New Source Performance Standards (NSPS). Henry Hub Spot Prices are used to address uncertainties in natural gas prices. Marginal benefits are computed based on the social costs of methane. If saved natural gas is marketed, greater private benefits are generated.
3. Formulate integer linear programs with different objective functions, simulated inputs and constraints as the optimization model to maximize net social benefits or minimize private costs to evaluate abatement strategies for the T&S systems. One year was chosen as the analysis period. Optimal level of abatement occurs where marginal benefits equals marginal costs.

5.6.3 Insight

1. Total methane emissions from 1,758 T&S compressor stations is about 440,000 tons per year. About 57% of the methane emissions emanate from a few categories, such as reciprocating compressors isolation valves and rod packing, and 30% of facilities emit 60% of the methane emissions. Nearly 83% of methane emissions can be abated using existing abatement technology.

2. Optimal abatement occurs at 80% of the emission reduction. Marginal abatement costs are under \$50 per ton of CO_2—assuming no savings from recovered gas. Optimal cost is \$62 million for the system-wide options and \$74 million for the super-emitter options. To provide a context, for the T&S systems, the total operating costs and income are \$6.2 billion and \$4.1 billion, respectively. This suggests that the operators cost of abatement are relatively trivial. Detection costs across all facilities are not trivial.

3. Optimal benefits, based on a social cost of \$1,330 per ton of methane (= \$46 per ton of CO_2), across the T&S sector compressors is about \$420 million without accounting for savings of natural gas and about \$600 million with savings of natural gas at \$5 per Million British Thermal Units (MMBTU).

4. Annual net benefits across the system range from \$160 million at \$600 per ton of methane emissions averted (=\$13 per ton of CO_2) to \$1.2 billion at \$3,600 per ton of methane emissions (=\$136 per ton of CO_2). Also, net benefits from abating low emitting sources are non-trivial.

5. System-wide policies with emissions reduction targets are not necessarily irrelevant, even if facilities choose the least cost abatement strategy. Requiring all facilities to reduce emissions by an equivalent percentage may be equitable in a sense; however, some facilities may be required to conduct abatement at low-emitting components with high marginal abatement costs that exceed the social cost of methane. Also, for the unconstrained system-wide option, both the intervention policies, standards, and emission tax, lead to the same optimal results. Applying intervention policy, the unconstrained system-wide option dominates both the super-emitter policies. The super-emitter policy with absolute annual threshold dominates the super-emitter policy with proportional loss threshold. In effect, these options are ranked as follows: (1) unconstrained system-wide, (2) super-emitter policy with absolute annual threshold, and (3) super-emitter policy with proportional loss threshold.

6. While based on efficiency for the entire T&S, this ranking is true, when equity and ease of political acceptability are considered, the aforementioned ranking may not hold. Also, we modeled just a single year snapshot of the T&S system. The baseline emissions model may not capture low probability or catastrophic leaks from facilities; thus, if emissions across the system are more skewed than what has been modeled, there may be greater efficiency gains in targeting super-emitters.

5.7 ACUMEN PIONEERS VENTURE PHILANTHROPY—SOCIAL INVESTMENT METRICS

5.7.1 Issue

Similar to venture capital that bets on disruptive technologies or business models to seek disproportionate return opportunities, social venture investors focus on building thriving enterprises that serve a vast number of underserved by leveraging finance, innovation, and scale-up through replication.

How to assess if the mission to create financially sustainable enterprise that delivers broad social impact is being accomplished? How to quantify intangible social benefits that allows comparison of social impact investment options? How to evaluate social impact and cost-effectiveness of investments to allocate resources? For instance, is it better to make a risk-free grant to UNICEF versus providing a loan to a private-sector organization for an innovative technology for bed nets to protect from malaria? Acumen executives were challenged to find a consistent method to answer these questions. They were on a quest for metrics.

5.7.2 ANALYTICS

Metrics are needed to measure the success of a mission-driven social investment and comparison of investments options for decision making.

1. At the individual investment level, the metric measures—is it relatively cost effective, is it sustainable, is it scalable, is it broadly accessible, and is it risky?
2. At the portfolio of investment level, the metric measures financial performance as the net asset value of the portfolio year over year but weighs this factor directly with the social impact performance. The social impact is defined by *breadth* (the number of lives impacted), *depth* (the meaningfulness of that impact on the lives touched), and the *poverty focus* (the percent of the portfolio companies' consumers who live below the poverty line).
3. At the organizational level, the metric tracks efficiency, growth, and risk management.

The metric informs where and how to achieve the greatest impact. Success of each investment is measured by its ability to deliver against social impact, financial viability, and breakthrough insights. For each aspect, definitions are developed to assess success and failure.

Acumen fund does not seek to beat an absolute standard for social return, especially across a very vastly diverse portfolio of social venture investment opportunities. Instead, it seeks to compare opportunities to current charitable options to accomplish the very same social mission.

"For each dollar invested, how much social impact will this generate over the life of the investment relative to the best available charitable option (BACO)?"[714] Figure 5.10

Rating	Social Impact	Financial Viability	Breakthrough Insights
Poor	Less cost effective than other charities	Lower than expected return on capital	No additional insights
Good	Greatest impact per $ invested	Just adequate return on capital with subsidies	Gained insights influenced fund
Better	Proven to be at least ten times scalable	Operations yield positive cash flow without subsidies	Unique insights gathered and broadly shared
Best	Materially improved one million plus people	Enterprise able to tap into commercial capital markets	Insights stimulated external replication

FIGURE 5.10 Social investment success screening matrix.

provides one way to screen on a relative basis. BACO is based on whether the proposed investment outperforms a plausible alternative—existing charities providing similar functions or services that lead to comparable social impacts. In other words, success is measured against other similar investments; impacts are benchmarked across subsectors. For instance, the benchmark for breadth of people who can be reached by energy product companies is observed to be much higher than that of vertically integrated agriculture businesses, while the depth of impact of the latter might be higher.

The breakthrough insight Acumen seeks to gain through their investments is stated in their *Theory of Change*, which is included in each investment memo prior to investment. This theory explicitly lays out the hypothesis(es) Acumen believes to be true, which are validated or invalidated by the financial and social metric performance. These metrics can include operational efficiency, impacts such as reliability or time savings, but also broader ecosystem impacts such as attracting commercial capital to the sector or facilitating or catalyzing the expansion of an entire sector (i.e., mini-grids in East Africa).

5.7.3 INSIGHT

Acumen[715] plans to allocate $325,000 toward the mission of malaria protection of people in the BoP. One option is to provide a loan at an annual interest of 6% to a net maker along with adopting an innovative technology for long-lasting insecticide-treated bed nets with a five-year product life. Instead of seeking an absolute standard for social return across a very diverse portfolio, Acumen opts to compare the investment relative to the BACO. Acumen identifies BACO to be an outright grant to UNICEF to purchase standard bed nets that protect two people with a 2.5-year product life at $3.25 each and distribute them only to people in the BoP. The Social Impact Cost Effectiveness of Acumen lending the money to the net maker is about fifty times greater than UNICEF BACO. Please see Figure 5.11 below.

	BACO	Acumen	Comments
Social Impact Computation			
Investment	$325,000	$325,000	
Price per bed net	$3.25		
Purchased # of Bed Nets	100,000		
Scale-up additional # of Bed Nets		400,000	Additional production driven by private enterprise productivity that is attributable to investment
People Protected per bed net	2	2	
Product life in Years per bed net	2.5	5.0	Longer product life driven by superior technology
Penetration into Base of the Pyramid	100%	50%	Reduced BOP penetration because of distribution
Total Social Impact	$500,000	$2,000,000	= Number of Bed Nets* Number of People Protected per Bed Net*Bed Net Life in Years * BOP Penetration
Net Cost Impact Computation			
Investment	($325,000)	($325,000)	
Management Costs	($65,000)	($130,000)	Management Cost Grant @ 20% and Loan @ 40%
	0	$422,500	Principal + Interest Return @ 6% for 5 years
Net Cost Impact	($390,000)	($32,500)	
Social Impact Cost Effectiveness	1.28	61.54	Social Impact per $ Invested = Total Social Impact/Net Cost Impact

FIGURE 5.11   Social impact cost effectiveness computation.

5.8   IBM BOOSTS INSIGHTS AND ENERGY CONSERVATION WITH REAL-TIME ANALYTICS

5.8.1   Issue

IBM has a longstanding and comprehensive global energy management program, overseen by more than 50 energy engineers, which delivers consistent, substantial year on year energy conservation savings. The energy management team utilizes the full repertoire of energy conservation and efficiency tools: time of day management of lighting and heating ventilating and air conditioning (HVAC) systems; retrofit to more efficient heating and cooling equipment; high efficiency LED lighting systems; variable speed systems that adjust output to match demand; building recommissioning programs; and other standard energy efficiency and conservation strategies and tools. Since 2010, the IBM energy management team has developed or procured real-time, analytics-based monitoring and management systems to track, report, and optimize energy consumption in IBM's HVAC, cooling, and chiller plant infrastructure supporting buildings and data centers. These systems operate on an IBM cloud platform and check performance against defined rules and enable corrective action to be taken either automatically or through operator intervention.

5.8.2   Analytics

5.8.2.1   Building Monitoring

IBM has developed and deployed an analytics-based building monitoring and management system (MMS), which continually collects data from roughly 16,500 field data sources, storing more than 500 million data points each year. The effort began in 2009 as a collaboration across the company's real estate, research, and software development organizations at a single IBM site, collecting a few hundred points of available field data sources at the site and writing analytics software, which took established operating rules for system operations and identified when system performance/operations fell outside those operating parameters. An *out of specification* condition triggers an alert, which is evaluated by a member of the engineering and maintenance team. If the alert is found to be valid, a maintenance work order is generated and the condition is corrected. The analytics system can then be used to validate the correction of the out-of-spec condition.

The analytics system has evolved over time and is now based on Skyspark Analytics,[716] a widely used fault detection and diagnostic tool. Today the MMS collects the field data sources at 27 major IBM campus, which encompass 155 buildings with more than 24 million square feet of space and represent 40% of IBM's energy consumption. Where necessary, additional sensors have been installed to capture key unmonitored data points. Data from the field sources is collected into the database and analyzed against the operating rules contained in the analytics engine. The analytics engine runs a broad range of IBM developed algorithms that enables a comprehensive, robust analysis of the dataset.

Over time, the IBM energy management team has established 116 unique rules applied to 15 different types of equipment involving air conditioning systems, chilled water systems, air compressors, boilers, heat exchangers, and data center equipment. Example rules include

1. Heating valve and cooling valve are open at the same time. The temperature in the space is fine, but there is a high energy input to the system indicating that energy is being wasted to maintain that temperature.
2. Finding leaking heating valves when the unit is off at night. When the temperature of the Variable Air Volume (VAV) box warms while the system is off, a heating valve is leaking.
3. If it is wintertime and the temperature of the unit goes down unexpectedly, it indicates that outside air dampers could be leaking.
4. The room temperature exceeds a specific maximum temperature for more than 3 hours in a 24-hour period.
5. The supply valve to a specified system was more than 97% open for more than 4 hours in a 24-hour period.

New rules, based on operating experience, continue to be proposed and adopted periodically, driving advances in the system's capabilities. One important feature of the system is that as new rules are verified and added, they can be applied to the full data set, immediately identify operating anomalies, and enable corrective actions to be taken at all the buildings being monitored (Figure 5.12).

The MMS has enabled the energy management team to reduce energy consumption by more than 175,000 MW·h and save $8 million in energy expense over a 6-year period. In addition, the system enables real-time capture of out-of-spec conditions, preventing them from remaining undiscovered for months or years depending on the nature of the anomaly and the frequency and quality of preventative maintenance checks on a given system.

Savings	Asset	Description
1,800 MW·h $200,000	Air Handler	Building was cooling more air than needed and was drawing excess air. Air flow meters were added to improve control.
2,600 MW·h $300,000	Chiller	Upon investigation of a chiller alert it was found that the condenser pipes were fouled by silica (SiO_2). Cleaning of the pipes resulted in a significant improvement in the tons of cooling delivered per kwh consumed, generating the noted savings.
2,100 MW·h $190,000	Air Handler	A *hot call* alert investigation found heating was occurring in vacant space. The heating setting was turned down to minimum generating significant savings.

FIGURE 5.12 Specific savings project examples.

5.8.2.2   Data Center Cooling Optimization

IBM has installed real-time monitoring and management systems using temperature, humidity and air flow sensors at more than 40 data center facilities to optimize and minimize cooling delivery into these spaces. The system utilizes the sensor data and fluid flow algorithms to create three-dimensional heat maps of the data center floor. The heat map identifies the impact of information technology (IT) equipment changes and air flow rebalancing efforts, enabling the data center staff to maintain the optimum temperature balance. By stabilizing the temperature across the data center and eliminating hot spots, it has been possible to raise the average data center temperature at data centers equipped with these systems by 2.5°C, optimize the chilled water system efficiency, and reduce the quantity of energy required to cool the data center. These projects have saved an annualized 200,000 MW·h and $20 million over the 6-year period.

The monitoring systems are now being upgraded with an analytics-based artificial intelligence (AI) engine directly adjusting the air flow output on the variable speed computer room air conditioning (CRAC) units to optimize and balance cooling delivery. The AI-enhanced system has been installed at three data centers, with excellent results. At one of the three data centers, energy consumption of the CRAC units was reduced by 69% over a 12-month period, saving 990 MW·h per year in energy consumption. Additional savings were realized through reduced chiller loads and more efficient chiller operation. The system continually monitors data center conditions and adjusts the CRAC system airflows as IT equipment is idled, removed, and added, enabling the data center to maintain operation at the optimum temperature.

5.8.2.3   Chilled Water Optimization

IBM has deployed chilled water optimization (CWO) software that integrates the management of the chiller units and free cooling systems, and distribution system to optimize the overall efficiency of the chiller operation by analyzing sensor and control inputs across the system. The system considers the efficiency characteristics of the individual units and the availability of free cooling. By balancing the operation of all the system components in accordance with a defined set of rules and the performance characteristics of the active and passive cooling systems, the cooling delivery is maximized while the energy use is minimized. IBM has realized annualized savings of 29,000 MW·h and more than $2.2 million at the seven sites where CWO has been implemented.

5.8.3   Insight

The results are substantial. These systems collectively produced an annualized total savings of 256,000 MW·h of electricity and fuel and $15.5 million from 2010 through 2016.

Cloud-based analytics systems enable the centralization of all of the available sensor data for continuous assessment of the operating performance of energy consuming systems. Analysis of the data quickly captures the obvious problems, which would have been identified, albeit belatedly, through routine preventative maintenance (PM) activities. The analytics tools go a step deeper, revealing problems and inefficiencies that would remain undiscovered without a detailed analysis of the data. On all three examples described earlier, the use of analytics surfaces anomalies and inefficiencies quickly so they can be investigated and resolved. Energy use reductions of 10%–20% are common across the system types and in some cases substantially more savings can be found and sustained. These systems represent the next generation of technology to drive efficiency into the operation of building infrastructure.

5.9 CHAKR INNOVATES PROCESS TO CAPTURE BLACK CARBON AND CONVERT TO BLACK INK

5.9.1 ISSUE

Nano carbonaceous particulate matter, black carbon as particulate matter <2.5 μm (PM2.5), has adverse impacts on human health, ecosystems, and visibility. Black carbon particulates in the atmosphere is also a major contributor to global warming and climate change.

5.9.1.1 Human Health and Ecosystem Impact

Black carbon is a primary participant in both indoor and outdoor particulate air pollution; which together, per a World Health Organization (WHO) 2011 study,[717] was estimated to cause more than three million deaths annually, and 88% of these deaths would occur in low- and middle-income countries. A more recent study by Health Metrics Institute reports that "In 2015, long-term exposure to PM2.5 contributed to 4.2 million deaths and to a loss of 103 million years of healthy life. China and India together accounted for 52% of the total global deaths attributable to PM2.5."[718] According to WHO, a reduction in the particulate matter levels from 70 μg/m^3 of PM10 and 35 μg/m^3 of PM2.5 to 20 μg/m^3 of PM10 and 10 μg/m^3 of PM2.5 respectively would result in ~15% reduction in deaths caused due to air quality.

Short- and long-term exposure can lead to penetration into the human body—the lungs with inhalation, the gastrointestinal tract via ingestion, and skin and mucosa through dermal pathways. Black carbon is known to cause lung cancer, chronic obstructive pulmonary disease (COPD), ischemic heart disease, stroke, bronchitis, and aggravated asthma, as well as premature death, and has been classified as a class I carcinogen by the International Agency for Research on Cancer (IARC), WHO.

Black carbon particles impact visibility significantly. Visibility or visual range is inversely proportional to the sum of light scattered and absorbed by particles, and light scattered and absorbed by gases. Submicron size black carbon is an efficient light-absorbing species. Its absorption efficiency has been estimated[719] to be 7–14 m^2/g and its scattering efficiency has been estimated[720] to be about 3.2 m^2/g. Fraction of the particulate light extinction due to particle absorption is reported to range from 0.35 to 0.50 in urban areas, 0.13 to 0.27 in rural and residential areas, and 0.05 to 0.11 in remote areas.

Black carbon particles are also linked to reduced crop yields and damage to materials and buildings.

5.9.1.2   Global Warming and Climate Change Impact

Black carbon, or soot, does not behave like GHGs and does not become well-mixed in the atmosphere. Black carbon particulates remain suspended in the air until they settle back on the surface, become washed out by rain, or participate in cloud formation. The average lifetime of a single soot particle in the atmosphere is only two or three weeks. Because black carbon has a short atmospheric lifetime, emission reductions are more effective near-term in mitigating global warming, and because of the albedo effect, especially impactful in climate-sensitive regions, including the Arctic and the Himalayas.

Black carbon particles absorb solar radiation and settle on icecaps, reducing its reflectivity and altering cloud formation patterns.

1. *Radiation absorption effect*: Black carbon particles absorb significant amounts of energy, trapping heat and warming the planet. Despite lasting in the atmosphere for less than one month, one ton of black carbon has a warming effect equal to 1,000–2,000 tons of CO_2 over a 100-year period, and an impact greater than 4,000 tons of CO_2 over a 20-year span.[721] Black carbon's radiative forcing is an area of active research. A 2013[722] study estimated that black carbon has a net climate radiative forcing of +1.1 W/m^2. While black carbon radiative forcing estimates have large uncertainty ranges, there is growing evidence that black carbon has one of the largest impacts of all global warming pollutants, second only to CO_2.

2. *Snow albedo effect*: Some black carbon emissions settle on snow, glaciers, and sea ice, and darken their surfaces. This significantly reduces the reflectivity, or albedo, of the surface, increasing absorption of more solar energy and accelerating ice melt. While globally the average effect of this process is about 0.1 W/m^2, in reality, this snow albedo impact is much higher in very climate-sensitive regions; the Arctic and the Himalayas are particularly vulnerable to black carbon emissions.

3. *Cloud interaction effect*: Black carbon particles participate in climate change by affecting cloud stability, precipitation, and reflectivity.

5.9.2   ANALYTICS

5.9.2.1   Selecting Black Carbon Source for Emission Control Innovation

Black carbon, a major component of soot, is a product of incomplete combustion of coal, diesel, biofuels, and biomass. Closed combustion makes up 59% of emissions; open burning is responsible for the rest. Major sources of black carbon include inefficient biomass cooking stoves, diesel and two-stroke engines, and open-air-vented coal furnaces. Potential mitigation options include better power plant efficiency to reduce demand for power from diesel generators, improved efficiency of diesel engines, more stringent new engine emission standards or enhanced fuel standards to reduce emissions from mobile sources, and replacing or retrofitting diesel generators, diesel fired boilers, and diesel automobiles.

Black carbon accounts for about 75% of particle emissions from diesel combustion, while particles emitted in biomass burning are primarily organic carbon, which is generally more reflective than black carbon.[723] Also, it is important to consider the effects of co-emitted particles and gases when evaluating mitigation opportunities. Some particulate matter emissions co-emitted with black carbon have significant cooling impacts that offset a portion of black carbon's full warming impact. The emissions ratio of black carbon to cooling particulates depends on the source, making some mitigation strategies (i.e., cleaner diesel engines) more attractive because they yield a greater potential global warming mitigation impact.

Urban regions in developing nations are densely populated, have a large concentration of other polluting industries, and have significant diesel truck traffic for transportation of goods. Furthermore, the urban areas also utilize diesel generating sets for power back up because of unreliable power supply. In effect, controlling black carbon from diesel engines and generating sets also yields one of the highest health protection impact per unit of investment.

The Chakr innovation effort naturally gravitated toward controlling black carbon emissions from diesel generating sets and other diesel combustion units, as the first step toward their mission.

5.9.2.2 Applying Circular Economy Principles in the Capture and Recycling of Black Carbon

Economies have been living on borrowed time for more than 250 years, enjoying abundant and inexpensive natural resources and ignoring the impact on the environment. Circular economy is conceptually[724] a transformation of this two-century old mind-set, from a linear take-make-throw economy to a circular Cradle-to-Cradle model that

1. Applies closed-loop recycling to eradicate waste systematically throughout the lifecycles and uses of products and their components.
2. Innovates business models that build on the interaction between products and services, as well as product design that utilizes the economic value retained in products after use in the production of new offerings.

In circular economy models, companies become highly involved in the use and disposal of products, find ways to generate revenue from selling the functionality, and dematerializes and optimizes performance along the entire value chain. Per Accenture,[725] one of the circular supplies model recovers and reuses resource outputs.

Traditional diesel combustion particulate matter emission control technologies either burn off the captured carbon (like in diesel particulate filters) or dump the soot particles (like in wet scrubbers), which in turn converts one form of pollution into another. These technologies, however, miss the opportunity to utilize the waste and create a circular economy to enhance sustainability. This is a classic example of circular economy principles being overlooked in technology innovation.

In the spirit of circular economy, and the name Chakr, which means circle in Hindi, not burning off the black carbon, saving it for reuse, and converting it into a useful product became the mantra.

5.9.2.3   Review of Traditional Black Carbon Emission Control Technologies

1. *Flow-through and wall-flow diesel particulate filters*: Diesel particulate filters (DPF) is the most dominant proven technology to remove particulates from diesel exhaust by filtering exhaust from the mobile or stationary diesel engines.[726,727] Flow-through DPFs employ catalyzed metal wire mesh structures or tortuous flow, metal foil-based substrates with sintered metal sheets to reduce diesel PM. They are coated with catalyst materials to assist in oxidizing the soot or used in conjunction with an upstream diesel oxidation catalyst to oxidize diesel soot. Flow-through filters, with no active regeneration or ash removal necessary, can achieve PM reduction of about 30%–75% depending on the engine operating characteristics. The most common high efficiency filter, the wall-flow DPF, is based on a porous wall, square cell, honeycomb design where every alternate channel is plugged on each end. These wall-flow filters can be made from a variety of ceramic materials (cordierite and silicon carbide) or sintered metal fibers. As the gas passes through the porous walls of the filter cells (thus the wall-flow filter designation), the particulate matter is deposited on the upstream side of the cell wall. Over time, the soot deposited on the cell wall leads to saturation of filter and a build-up of back-pressure. It requires a means of burning off or removing accumulated particulate matter and reducing the backpressure of the exhaust stream.

2. *Passive and active filter regeneration*: The most common method of removing accumulated particulate matter from the filter is to burn or oxidize it within the filter. By burning off trapped material, the filter is cleaned or *regenerated*. Filters may be passively regenerated using available exhaust heat or actively regenerated using some kind of energy input, like injection of diesel fuel into an upstream unit. In general, new applications of DPFs employ a combination of passive and active regeneration strategies to ensure that filter regeneration occurs under all operating conditions. In passive regeneration, the ceramic or metal filter substrate is coated with a high surface area oxide and precious or base metal catalyst. The catalyst acts to reduce the ignition temperature of the accumulated particulate matter, and it is burnt or oxidized on the filter when exhaust temperatures are adequate. The most commonly applied method of active regeneration is to introduce a temporary change in engine operation or an oxidation catalyst to facilitate an increase in exhaust temperature. Catalytic oxidation of unburnt HC and CO in exhaust provides the exothermic heat of combustion to raise the temperature of the exhaust gas. Alternatively, oxidizing NO_x in the exhaust to nitrogen dioxide (NO_2), which oxidizes carbon at a lower temperature than oxygen, facilitates filter regeneration at lower exhaust temperatures (Figure 5.13).

3. *Diesel generator black carbon retrofit emission control cost-benefit analysis*: US EPA, "Diesel Retrofit Technology—An Analysis of Cost-Effectiveness of Reducing Particulate Matter and Nitrogen Oxide Emissions from Heavy Duty Non-Road Diesel Engines Through Retrofits," EPA420 R-07-005 May 2007 provides a detailed cost-effectiveness analysis of retrofit particulate matter emission control of diesel generators (Figure 5.14).[729]

Product Name	Retrofit Technologies—Diesel Particulate Filter with 85% PM Reduction Applicability
Catalytic Exhaust Products Ltd. Dieselytic SXS-SC DPF	Stationary prime and emergency standby generators and pumps with Tier 1, Tier 2, or Tier 3 certified off-road engines meeting 0.2 g/bhp-hr or less diesel PM
Clariant Corporation EnviCat®- DPF	Stationary prime and emergency standby generators and pumps; CARB diesel; biodiesel.*
Cummins Pacific eMission DPF	Stationary emergency standby generators with a PM emission rate of 0.15 g/bhp-hr or less and between 23 to 78 liter displacements.
DCL International Inc.	Stationary prime and emergency standby generators, pumps, and compressors; Tier 1, Tier 2, or Tier 3 off-road engines certified to <0.15 g/bhp-hr PM; CARB diesel; biodiesel.*
Global Emissions Systems, Inc. (GESi) 6000DPF	Stationary prime and emergency standby generators and pumps with Tier 1, Tier 2, or Tier 3 certified off-road engines meeting 0.2 g/bhp-hr or less diesel PM
Johnson Matthey CRT	Stationary emergency/standby generators; conditionally verified for stationary prime generators. CARB diesel; biodiesel.*
MIRATECH® LTR™ DOC/DPF System	Stationary emergency standby generators with a PM emission rate of 0.22 g/bhp-hr or less.
Nett Technologies. NETT GreenTRAPTM DPF	Stationary prime and emergency standby generators and pumps with Tier 1, Tier 2, or Tier 3 certified off-road engines meeting 0.2 g/bhp-hr or less diesel PM
Rypos, Inc. HDPF/C™	1996–2007 stationary emergency standby generators and pumps with a PM emission rate of 0.2 g/bhp-hr or less and certified to Tier 1, Tier 2, or Tier 3 off-road diesel engine standards; CARB diesel; biodiesel.*
Universal Emissions Technologies GreenShield®	Stationary prime and emergency standby power generators and pumps with Tier 1, Tier 2, or Tier 3 certified off-road engines.

FIGURE 5.13   Select verified diesel engine particulate emission control equipment.[728] (* These systems have been verified for use with biodiesel blends subject to certain requirements. https://www.arb.ca.gov/diesel/verdev/reg/biodieselcompliance.pdf.)

Cost Effectiveness Diesel Generator PM Retrofit Controls		
Retrofit Technology	Range of $/ton PM Emission Reduced	
Diesel Oxidation Catalysts	$18,700	$46,100
Catalyzed Diesel Particulate Filters with Passive Regeneration	$20,800	$51,300
Capacity, in hp	100	
Annual Hours of Operation	338	
Load Factors	0.43	

FIGURE 5.14   Cost effectiveness of diesel generator PM retrofit controls.

The benefits of eliminating black carbon emissions may be evaluated by benchmarking European Commission (EC) data.[730]

= 40,000 Euros (base year 2000) per year per tonnes of PM2.5 averted per year

= 40,000 Euros*(CPI 2016/CPI 2000)*(US Dollars/Euro)*(2,000 pounds/ton)/

(2240 pounds/tonnes) = 40,000*(240.007/172.2)*(1.18/1.0)*(2,000/2,240)

= US$ 58,737 per year per ton of PM2.5 emission averted per year

Health impacts, damage to crops and properties because of air pollution, can be quantified, and are included in the benchmarked EC data above. Damage to eco-systems and cultural heritage can be quantified, but only on a relative basis. Other impacts such as reduced visibility from air pollution and the social dimensions of health impacts may not currently be quantifiable in terms of impact or monetary value; they permit only a qualitative analysis. This evaluation is conservative and does not include climate change mitigation benefits and visibility and social dimensions of health impact.

The previous assessment confirms a benefit to cost ratio in excess of 1.0 under both scenarios, justifying the retrofit control of black carbon emissions from diesel generators.

Now, in the spirit of circular economy, if the captured black carbon is converted into a useful product, then this results in a reduction in costs because of the gain in the market value of the useful product. Further benefits come from elimination of additional air pollution from the regeneration process, as well as from the elimination of environmental foot print of black carbon sourcing in the current processes to make black ink.

5.9.3 Innovation

5.9.3.1 Objectives

1. The emission control technology must achieve greater than 85% capture efficiency.
2. The impact on the diesel generator operations should not exceed the current methods.
3. The captured black carbon should not be burned off, to enable later conversion into ink.
4. The carbon footprint of the ink pigment manufacturing process should not exceed the current methods.

5.9.3.2 Chakr Shield

Chakr Shield (patent pending) is an award-winning innovative technology to reduce emissions from diesel generators. Chakr Shield technology captures particulate matter being emitted from diesel generators, without consuming too much energy or causing an adverse impact on the engine. Its performance has been verified by several accredited government agencies in India. It is a retrofit device, which offers several advantages—meets India's regulatory emission control mandates with minimal back pressure and efficiency loss, improves ambient air quality, and improves environment and health of the local population. It helps users achieve corporate sustainability goals and green building ratings. Rather significantly, by reducing black carbon, the short-term impact global warming pollutant, Chakr Shield also contributes to climate change mitigation efforts.

5.9.3.3 Chakr POINK

Chakr POINK (Ink from Pollution), in the spirit of circular economy, is made from captured particulate matter that produces a high-quality ink with rich black color. The Chakr POINK process converts the captured *soot* into a black pigment.

This pigment can be used as a coloring agent for paints and inks by adding binders and stabilizing agents. The ink pigment has been tested and certified to be safe to use.

5.10   IBM HELPS REDUCE AIR POLLUTION IN CHINA WITH IoT AND BIG DATA ANALYTICS

5.10.1   Issue

IBM's research division (IBM Research) is one of the largest and most influential private research institutions worldwide, with more than 3,000 researchers in 12 labs located across six continents.[731] IBM Research has developed an analytics/Internet of Things (IoT) tool that gathers weather, air pollutant, and chemical concentration, emissions, and traffic data across a network of sensors installed over a broad geographical area. The analysis of this data provides air pollution forecasts that can be used to implement mitigation measures in advance to reduce the impact of periodic, weather-induced air pollution events.

Beijing has been challenged by dangerous levels of air pollution for years. In January 2013, the fine particulate matter concentration in Beijing reached 40 times the exposure limit recommended by the WHO.[732] In 2015, the city issued two unprecedented red alerts indicating critical pollutant levels.[733]

Air pollution fluctuates greatly based on weather conditions such as humidity, temperature, and wind. Pollutants bond with water molecules to create smog, a denser, visible pollution that typically sits stagnant. While winds can help disperse pollutants, they can also bring new ones with them. When temperatures rise, so does the hot air, causing a greater mixture and movement of the pollution resulting in chemical reactions that can be particularly harmful. These factors make it extremely difficult to predict the rapid fluctuation of air pollution levels; this means that residents often do not receive ample time to protect themselves from the smog.

5.10.2   Analytics

5.10.2.1   IBM and Chinese Government Collaborate to Reduce Air Pollution Events

In September 2013, the Chinese government announced a plan to reduce air pollution in the whole country, including measures such as limiting the burning of coal and taking high-polluting vehicles off the roads with the objective to reduce particulate matter concentration in major cities.[734] Under this plan, Beijing would have to reduce its fine particulate matter (known as PM2.5) concentration by 25% (to less than 60 micrograms per cubic meter, or $\mu g/m^3$) by 2017, compared to the 2012 levels. PM2.5 is a measure of particulate emissions including carbon, nitrogen, sulfur, and heavy metals.[735]

In July 2014, IBM announced that it would deploy a 10-year initiative, in collaboration with Chinese government entities, to transform China's national energy systems and protect the health of citizens with its solution called Green Horizon. One of the first partners to this initiative was the Beijing Municipal Government. The collaboration, focused on tackling the city's air pollution challenges, leverages some of IBM's most advanced technologies, such as cognitive computing, optical sensors and IoT integrated into a big data and analytics platform and drawing upon IBM's deep experience in weather modeling and prediction. The project, led by the IBM Research lab in China, taps into IBM's network of 12 global research labs to create an ecosystem of partners from across government, academia, industry, and private enterprise.

5.10.2.2 IBM Green Horizon—Air Pollution Event Intervention

Green Horizon works by collecting vast amounts of data from environmental and weather monitoring stations and traffic cameras, as well as meteorological satellites. In the case of Beijing, 35 monitoring stations and cameras with hundreds of installed sensors collect the data on the monitored parameters, with a geographical resolution of a kilometer square. IBM IoT technology connects the devices to IBM's systems for subsequent analysis and interpretation. The solution is able to forecast air quality up to 10 days in advance.

The collected data is ingested by IBM's machine learning technologies to determine the highest probability forecast. Errors in model forecasting often result from uncertainties in initial conditions. The tool leverages data assimilation to combine the range of different data sources, including surface monitoring data, weather data, emissions data, satellite data, and geographical data, to estimate the initial state of a model (i.e., initial condition) and to set the stage to provide a high-accuracy air quality outlook. Once the optimal initial conditions are set, the data enters the *model blending* process, where different physics and chemistry predictive models are combined. As each model performs best in different conditions (i.e., temperature, wind speed/ strength, geographic and topographic factors, season of the year, etc.), IBM's machine learning techniques play a key role in finding just the right blending of the different models to deliver the best forecast given the actual parameters being provided by the recorded data set at any given point of time. The system uses adaptive machine learning mechanisms to train those models and adaptively adjust the parameters for each model, and selects the optimized one with best performance for each specific situation, based on the combination of the historical and current data recorded.[736]

IBM Research helped develop a system capable of generating predictive models showing pollution sources, flow direction, and its potential effect on pollutant concentrations. Known sources of emissions are tracked in real-time, compared to the modeled data and analyzed for anomaly detection. Green Horizon can also identify suspected unknown sources of air pollution based on type of pollutants and other factors. Using scenario modeling, the tool allows for simulation of hypothetical *what-if* scenarios— enabling city officials to try out the effectiveness of different action plans to achieve a balance between environmental and economic concerns. Provided with accurate forecasts, authorities can make better decisions on enforcement and inspection actions on critical emissions source areas or identify the location of suspected sources.

5.10.3 Insights

The purpose of IoT is to strengthen the understanding of our surroundings with measured connected data, but without cognitive computing and big data analytics, the potential of IoT is limited by the human ability to understand and process the complexity and scale of the data it gathers. Green Horizon is a great example of how different kinds of digital technologies are applied to drive business results that also deliver sustainability benefits. Systems such as Green Horizon augment human decision-making capacity, unearthing hidden insights in instrumental ways to address some of our generation's biggest challenges.

IBM Green Horizon reports an accuracy of more than 80% for its 3-day forecasts and around 75% for its 7- to 10-day forecasts. Since the city of Beijing launched its five-year campaign against air pollution, the city has seen a significant reduction in PM2.5 concentration. During the first seven months of 2017, the PM2.5 concentration was recorded to be 34.7% lower than during the same period in 2013, and the average PM2.5 concentration during 2016 was 73 µg/m^3, against the set target of 60 µg/m^3.[737]

IBM's modeling system is also being used in two other Chinese cities with pollution challenges: Baoding and Zhangjiakou. In addition, Green Horizon's geographical coverage is expanding: IBM has entered a research collaboration with the Delhi Dialogue Commission, under the Delhi government, to leverage Green Horizon's technology to help calculate the most effective and sustainable strategies for tackling air pollution in India's capital. In South Africa, IBM has forged an innovation partnership with the city of Johannesburg to model air pollution trends and quantify the effectiveness of the city's intervention programs in support of Johannesburg's air quality targets and sustainable development.[738]

5.10.3.1   IBM Green Horizon—Renewable Energy Integration and Industrial Energy Conservation

Beyond air quality management, Green Horizon also includes two additional capabilities: renewable energy integration and industrial energy conservation. Air pollution is a multifaceted problem, and the key to tackling it goes beyond monitoring emissions. It also includes adopting a *system-in-a-system* approach to air quality management to address the issues at their roots, such as mitigating measures that enable the shift from fossil fuels to renewable energy. Such a shift not only will improve air quality but also reduce CO_2 emissions, the main cause of climate change.

IBM Research has developed a renewable energy forecasting system, which combines weather modeling, IoT collection of sensor data, and cognitive computing to help utility companies predict the energy output of their generating capacity. Solar farms use sky-facing cameras to monitor cloud movement and calculate their potential blocking impact on solar radiation. Wind turbines are fitted with sensors 80 m above the ground to monitor wind speed, moisture, and air pressure. Assimilating those readings with weather forecasting data, the system is able to predict the performance of wind and solar energy farms with 90% accuracy several days ahead—providing the ability to effectively dispatch thousands of megawatts of power that could otherwise be lost. Such knowledge is key to integrating more renewable electricity into the grid and creating an optimum balance between supply and demand in energy markets. The system has already been rolled out to 30 wind, solar, and hydro power sources.

ENDNOTES

662 OECD, Mortality risk valuation in environment, health and transport policies, 2012, Accessed May 2017 and available at http://www.oecd.org/environment/tools-evaluation/49446853.pdf.

663 Land transport safety authority and transfund New Zealand, Environment Canterbury technical report # U01/89, November 2001, p. ii.

664 Health Canada, Paul de Civita, International perspective on valuing mortality risk for policy, a Canadian perspective, November 2001.

665 Environment Directorate General, Matti Vainio, Value of statistical life in Europe, November 6, 2001.

666 Ibid.

667 Environment Directorate General, Matti Vainio, Value of statistical life in Europe, November 6, 2001.

668 Health and Safety Executive (HSE), Decision making process, 2001, p. 65.

669 Ibid.

670 Office of Management and Budget (OMB), Report to congress on the costs and benefits of federal regulations 2000, 72, DOT Federal Railway Administration Worker Protection Rule, 61FR65973, September 1996.

671 Office of Management and Budget (OMB), Report to congress on the costs and benefits of federal regulations 2000, 25, EPA Clean Air Act Section 812, Cost Benefit Analysis, 1997.

672 Arias, E. 2010. United States life tables, *National Vital Statistics Reports*, 58(21). Center for Disease Control and Prevention, US Department of Health and Human Services.

673 Land transport safety authority and transfund New Zealand, Environment Canterbury technical report # U01/89, November 2001, p. ii.

674 Health Canada, Paul de Civita, International perspective on valuing mortality risk for policy, a Canadian perspective, November 2001.

675 Environment Directorate General, Matti Vainio, Value of statistical life in Europe, November 6, 2001.

676 Ibid.

677 Environment Directorate General, Matti Vainio, Value of statistical life in Europe, November 6, 2001.

678 Health and Safety Executive (HSE), Decision making process, 2001, p. 65.

679 Ibid.

680 Office of Management and Budget (OMB), Report to congress on the costs and benefits of federal regulations 2000, 72, DOT Federal Railway Administration Worker Protection Rule, 61FR65973, September 1996.

681 Office of Management and Budget (OMB), Report to congress on the costs and benefits of federal regulations 2000, 25, EPA Clean Air Act Section 812, Cost Benefit Analysis, 1997.

682 EC, Damages per tonne of emissions for EU-25, Accessed May 2017 and available at http://ec.europa.eu/environment/archives/cafe/activities/pdf/cafe_cba_externalities.pdf; http://ec.europa.eu/environment/archives/cafe/pdf/cba_update_nov2006.pdf.

683 EC, Methodology for the cost-benefit analysis, Accessed May 2017 and available at http://ec.europa.eu/environment/archives/cafe/pdf/cba_methodology_vol1.pdf.

684 EC, Assessment of externalities of air pollution, Accessed May 2017 and available at http://ec.europa.eu/environment/archives/cafe/activities/pdf/cafe_cba_externalities.pdf.

685 EC, Update of air pollution externalities, Accessed May 2017 and available at http://ec.europa.eu/environment/archives/cafe/pdf/cba_update_nov2006.pdf.

686 EC, Methodology for the Cost-benefit analysis, Accessed May 2017, available at http://ec.europa.eu/environment/archives/cafe/pdf/cba_methodology_vol1.pdf.

687 EU "RAINS Model" Accessed June 2017 available at http://ec.europa.eu/environment/archives/cafe/pdf/working_groups/021106rainsreview.pdf.

688 UK Institution of Chemical Engineers (IChemE), The sustainability metrics environmental burdens for emissions to air 2002, Available at https://www.scribd.com/document/51848632/IChemE-Metrics-sustainability.

689 IChemE UK, Sustainability metrics, Accessed April 2017 and available at http://nbis.org/nbisresources/metrics/triple_bottom_line_indicators_process_industries.pdf.

690 Ibid.

691 US EPA Risk screening environmental indicators model, Accessed May 2017 and available at https://www.epa.gov/rsei.

692 US EPA, Understanding global warming potential, Accessed May 2017 and available at https://www.epa.gov/ghgemissions/understanding-global-warming-potentials.

693 US EPA, Class I ozone depletion substances, Accessed May 2917 and available at https://www.epa.gov/ozone-layer-protection.

694 US EPA, Toxicity weights for TRI chemicals, Accessed May 2017 and available at https://www.epa.gov/sites/production/files/2015-12/documents/technical_appendix_a-toxicity_v2.3.4.pdf.

695 UK Environmental Agency, Possible environmental burden measures for water, R&D technical report P6-015/TR3, 2003.

696 UK Environmental Agency, Environmental burden measures for air, R&D technical report P6-015/TR2, 2003.

697 UK Environmental Agency, Possible environmental burden measures for hazardous substances, R&D technical report P6-015/TR4, 2003.

698 UC Berkeley, Carcinogenic potency project, Accessed May 2017 and available at https://toxnet.nlm.nih.gov/cpdb/.

699 Wright, D.W., Toxicity-weighting: A prioritization tool for quality assurance of air toxics inventories, Accessed May 2017 and available at https://www3.epa.gov/ttnchie1/conference/ei16/session6/wright.pdf.

700 IChemE UK, Sustainability metrics, Accessed April 2017 and available at http://nbis.org/nbisresources/metrics/triple_bottom_line_indicators_process_industries.pdf.

701 UK Institution of Chemical Engineers (IChemE), The sustainability metrics environmental burdens for emissions to air 2002, Available at https://www.scribd.com/document/51848632/IChemE-Metrics-sustainability.

702 Kimbril, D.R. and I. Clarke, A semi-quantitative method for evaluating risk reduction alternatives, Accessed May 2017 and available at https://www.icheme.org/~/media/Documents/Subject%20Groups/Safety_Loss_Prevention/Hazards%20Archive/LP2007/LP2007-001.pdf.

703 Ibid.

704 Uhlman, B.W. and Saling, P., Measuring and communicating sustainability through eco-efficiency analysis, *Chemical Engineering Progress*, December 2010, Accessed June 2017 and available at http://www2.basf.us/AcrylicsDispersions/asphalt/docs/MeasuringAndCommunicatingSustainabilityThroughEco-EfficiencyAnalysis.pdf.

705 http://polyurethanes.org/uploads/documents/eco_polyol.pdf.

706 BASF, SEEBALANCE® Measuring sustainable development on a product level, Accessed and available at https://www.basf.com/us/en/company/sustainability/management-and-instruments/quantifying-sustainability/seebalance.html.

707 Saling, P., The BASF eco-efficiency analysis–A 20-year success story, Available at https://www.basf.com/en/company/sustainability/management-and-instruments/quantifying-sustainability/eco-efficiency-analysis.html.

708 Uhlman, B.W. and P. Saling, Measuring and communicating sustainability through eco-efficiency analysis, *Chemical Engineering Progress*, December 2010, Accessed June 2017 and available at http://www2.basf.us/AcrylicsDispersions/asphalt/docs/Measurin gAndCommunicatingSustainabilityThroughEco-EfficiencyAnalysis.pdf.

709 Kicherer, A. et al., Eco-efficiency combining lifecycle assessment and lifecycle costs via normalization, *International Journal of Lifecycle Assessment*, 2007, 12(7), 537–543.

710 Saling, P., The BASF eco-efficiency analysis, Accessed April 2018 and available at https://www.basf.com/en/company/sustainability/management-and-instruments/ quantifying-sustainability/eco-efficiency-analysis.html and BASF, Accessed October 2017 and available at https://www.basf.com/en/company/sustainability/management-and-instruments/quantifying-sustainability/eco-efficiency-analysis.html.

711 Mayfield, E.N., A.L. Robinson, and J.L. Cohon, System-wide and superemitter policy options for the abatement of methane emissions from the U.S. natural gas system, *Environmental Science & Technology*, 2017, 51, 4772–4780. doi:10.1021/acs.est.6b05052.

712 Subramanian, R., L.L. Williams, T.L. Vaughn, D. Zimmerle, J.R. Roscioli, S.C. Herndon, T.I. Yacovitch et al., Methane emissions from natural gas compressor stations in the transmission and storage sector: Measurements and comparisons with the EPA greenhouse gas reporting program protocol, *Environmental Science & Technology*, 2015, 49(5), 3252–3261; Zimmerle, D.J., L.L. Williams, T.L. Vaughn, C. Quinn, R. Subramanian, G.P. Duggan, B. Willson et al., Methane emissions from the natural gas transmission and storage system in the United States, *Environmental Science & Technology* 2015, 49(15), 9374–9383.

713 Zimmerle, D.J., L.L. Williams, T.L. Vaughn, C. Quinn, R. Subramanian, G.P. Duggan, B. Willson et al., Methane emissions from the natural gas transmission and storage system in the United States, *Environmental Science & Technology* 2015, 49(15), 9374–9383.

714 Acumen, Fund Metrics, Accessed July 2017 and available at http://acumen.org/content/ uploads/2007/01/Metrics%20methodology1.pdf.

715 Acumen, Acumen fund concept paper, Accessed July 2017 and available at http://acumen. org/wp-content/uploads/2007/01/BACO%20Concept%20Paper_01.24.071.pdf.

716 IBM, Global solutions directory SkySparks analytics 3.0, Accessed October 2017 and available at http://www-304.ibm.com/partnerworld/gsd/solutiondetails. do?solution=48 500&expand=true&lc=en.

717 Air quality and health, World Health Organization, 2011.

718 Institute for Health Metrics and Evaluation, State of global air 2017, Accessed October 2017 and available at https://www.stateofglobalair.org/sites/default/files/SOGA2017_ report.pdf.

719 Wolff, G.T., Particulate elemental carbon in the atmosphere, *Journal of the Air Pollution Control Association*, 1981, 31(9), 935–938. doi:10.1080/00022470.1981.10465298.

720 Ibid.

721 Environmental and Energy Study Institute, Short lived climate pollutants: Why are they important, Accessed October 2017 and available at http://www.eesi.org/files/ FactSheet_SLCP_020113.pdf.

722 Bond T.C., S.J. Doherty, D.W. Fahey, P.M. Forster, T. Berntsen, B.J. DeAngelo, M.G. Flanner et al., Bounding the role of black carbon in the climate system: A scientific assessment, *Journal of Geophysical Research: Atmospheres*, 2013, 118, 5380–5552.

[723] GWU, Black carbon emissions: Impacts and mitigation, Accessed October 2017 and available at https://www2.gwu.edu/~ieresgwu/assets/docs/Evans_paper.pdf and EPA. Report to congress on black carbon, Washington DC, U.S. Environmental Protection Agency. Accessed 2012 and available at: http://www.epa.gov/blackcarbon/.

[724] Linder, M. and M. Williander, Circular business model innovation: Inherent uncertainties, *Business Strategy and the Environment*, 2015. doi:10.1002/bse.1906; McKinsey, Moving toward a circular economy, Accessed April 2015 and available at http://www.mckinsey.com/business-functions/sustainability-and-resource-productivity/our-insights/moving-toward-a-circular-economy.

[725] Greenbiz, Circular economy, Accessed April 2015 and available at https://www.greenbiz.com/article/5-business-models-put-circular-economy-work.

[726] MECA, Emission control technologies for diesel powered vehicles, Accessed October 2017 and available at http://www.meca.org/galleries/files/MECA_Diesel_White_Paper_12-07-07_final.pdf.

[727] California Air Resources Board, Summary of verified diesel emission control strategies Accessed October 2017 and available at https://www.arb.ca.gov/diesel/verdev/vt/cvt.htm.

[728] Ibid.

[729] US EPA, Diesel retrofit technology—An analysis of cost-effectiveness of reducing particulate matter and nitrogen oxide emissions from heavy duty non-road diesel engines thru retrofits, EPA420 R-07-005 May 2007. Accessed October 2017 and available at https://nepis.epa.gov/Exe/ZyPDF.cgi/P10023OA.PDF?Dockey=P10023OA.PDF.

[730] EC, Assessment of externalities of air pollution, Accessed May 2017 and available at http://ec.europa.eu/environment/archives/cafe/activities/pdf/cafe_cba_externalities.pdf.

[731] IBM, Accessed October 2017 and available at http://research.ibm.com/.

[732] New York Times, Accessed October 2017 and available at http://www.nytimes.com/2013/09/13/world/asia/china-releases-plan-to-reduce-air-pollution.html.

[733] British Broadcasting Corporation, Accessed October 2017 and available at http://www.bbc.com/news/world-asia-china-35026363.

[734] New York Times, Accessed October 2017 and available at http://www.nytimes.com/2013/09/13/world/asia/china-releases-plan-to-reduce-air-pollution.html?mcubz=0.

[735] IBM, Accessed October 2017 and available at http://www.research.ibm.com/green-horizons/#fbid=dmqN9ooWQdQ.

[736] YouTube, Accessed October 2017 and available at https://www.youtube.com/watch?v=qcJz6nkYbkQ.

[737] China Daily, Accessed October 2017 and available at http://www.chinadaily.com.cn/china/2017-08/22/content_30944199.htm.

[738] Technology Review, Accessed October 2017 and available at https://www.technologyreview.com/s/600993/can-machine-learning-help-lift-chinas-smog/.

Appendix A: List of SASB Material Sustainability Components[739]

1. Environment
   a. GHG emissions
   b. Air quality
   c. Energy management
   d. Fuel management
   e. Water and wastewater management
   f. Waste and hazardous materials management
   g. Biodiversity impacts
2. Social capital
   a. Human rights and community relations
   b. Access and affordability
   c. Customer welfare
   d. Data security and customer privacy
   e. Fair disclosure and labeling
   f. Fair marketing and advertising
3. Human capital
   a. Labor relations
   b. Fair labor practices
   c. Diversity and inclusion
   d. Employee health, safety, and well-being
   e. Compensation and benefits
   f. Recruitment, development, and retention
4. Business model and innovation
   a. Life-cycle impacts of products and services
   b. Environmental and social impacts on assets and operations
   c. Product packaging
   d. Product quality and safety
5. Leadership and governance
   a. Systemic risk management
   b. Accident and safety management
   c. Business ethics and transparency of payments
   d. Competitive behavior

    e.   Regulatory capture and political influence
    f.   Materials sourcing
    g.   Supply-chain management

ENDNOTE

[739] SASB, List of material sustainability components, Accessed June 2017 and available at https://library.sasb.org/sasb-conceptual-framework-2/.

Appendix B: Definition of Terms—US OMB Circular No. A-94 Revised

Source: US Office of Management and Budget Circular No. A-94 Revised[740]

MEMORANDUM FOR HEADS OF EXECUTIVE DEPARTMENTS AND ESTABLISHMENTS

Subject: Guidelines and Discount Rates for Benefit-Cost Analysis of Federal Programs

1. The Circular applies specifically to
   a. Benefit-cost or cost-effectiveness analysis of federal programs or policies.
   b. Regulatory impact analysis.
2. Specifically exempted from the scope of this Circular are decisions concerning:
   a. Water resource projects (guidance for which is the approved *Economic and Environmental Principles and Guidelines for Water and Related Land Resources Implementation Studies*).
   b. For small projects that share similar characteristics, agencies are encouraged to conduct generic studies and to avoid duplication of effort in carrying out economic analysis.

DEFINITION OF TERMS—APPENDIX A OF CIRCULAR NO. A-94 REVISED

Benefit-cost analysis: A systematic quantitative method of assessing the desirability of government projects or policies when it is important to take a long view of future effects and a broad view of possible side effects.

Capital asset: Tangible property, including durable goods, equipment, buildings, installations, and land.

Certainty-equivalent: A certain (i.e., nonrandom) outcome that an individual values equally to an uncertain outcome. For a risk-averse individual, the certainty-equivalent for an uncertain set of benefits may be less than the mathematical expectation of the outcome; for example, an individual may value a 50–50 chance of winning $100 or $0 as only $45. Analogously, a risk-averse individual may have a certainty equivalent for an uncertain set of costs that is larger in magnitude than the mathematical expectation of costs.

Cost-effectiveness: A systematic quantitative method for comparing the costs of alternative means of achieving the same stream of benefits or a given objective.

Consumer surplus: The maximum sum of money a consumer would be willing to pay to consume a given amount of a good, less the amount actually paid. It is represented graphically by the area between the demand curve and the price line in a diagram representing the consumer's demand for the good as a function of its price.

Discount rate: The interest rate used in calculating the present value of expected yearly benefits and costs.

Discount factor: The factor that translates expected benefits or costs in any given future year into present value terms. The discount factor is equal to $1/(1 + i)t$ where i is the interest rate and t is the number of years from the date of initiation for the program or policy until the given future year.

Excess burden: Unless a tax is imposed in the form of a lump sum unrelated to economic activity, such as a head tax, it will affect economic decisions on the margin. Departures from economic efficiency resulting from the distorting effect of taxes are called excess burdens because they disadvantage society without adding to Treasury receipts. This concept is also sometimes referred to as deadweight loss.

External economy or diseconomy: A direct effect, either positive or negative, on someone's profit or welfare arising as a byproduct of some other person's or firm's activity. Also referred to as neighborhood or spillover effects, or externalities for short.

Incidence: The ultimate distributional effect of a tax, expenditure, or regulatory program.

Inflation: The proportionate rate of change in the general price level, as opposed to the proportionate increase in a specific price. Inflation is usually measured by a broad-based price index, such as the implicit deflator for the gross domestic product or the consumer price index.

Internal rate of return: The discount rate that sets the net present value of the stream of net benefits equal to zero. The internal rate of return may have multiple values when the stream of net benefits alternates from negative to positive more than once.

Life-cycle cost: The overall estimated cost for a particular program alternative over the time period corresponding to the life of the program, including direct and indirect initial costs plus any periodic or continuing costs of operation and maintenance.

Multiplier: The ratio between the direct effect on output or employment and the full effect, including the effects of second order rounds or spending. Multiplier effects greater than 1.0 require the existence of involuntary unemployment.

Net present value: The difference between the discounted present value of benefits and the discounted present value of costs.

Nominal values: Economic units measured in terms of purchasing power of the date in question. A nominal value reflects the effects of general price inflation.

Nominal interest rate: An interest rate that is not adjusted to remove the effects of actual or expected inflation. Market interest rates are generally nominal interest rates.

Opportunity cost: The maximum worth of a good or input among possible alternative uses.

Real or constant dollar values: Economic units measured in terms of constant purchasing power. A real value is not affected by general price inflation. Real values can be estimated by deflating nominal values with a general price index, such as the implicit deflator for the gross domestic product or the consumer price index.

Real interest rate: An interest rate that has been adjusted to remove the effect of expected or actual inflation. Real interest rates can be approximated by subtracting the expected or actual inflation rate from a nominal interest rate. (A precise estimate can be obtained by dividing one plus the nominal interest rate by one plus the expected or actual inflation rate, and subtracting one from the resulting quotient.)

Relative price: A price ratio between two goods as, for example, the ratio of the price of energy to the price of equipment.

Shadow price: An estimate of what the price of a good or input would be in the absence of market distortions, such as externalities or taxes. For example, the shadow price of capital is the present value of the social returns to capital (before corporate income taxes) measured in units of consumption.

Sunk cost: A cost incurred in the past that will not be affected by any present or future decision. Sunk costs should be ignored in determining whether a new investment is worthwhile.

Transfer payment: A payment of money or goods. A pure transfer is unrelated to the provision of any goods or services in exchange. Such payments alter the distribution of income, but do not directly affect the allocation of resources on the margin.

Treasury rates: Rates of interest on marketable Treasury debt. Such debt is issued in maturities ranging from 91 days to 30 years.

Willingness to pay: The maximum amount an individual would be willing to give up in order to secure a change in the provision of a good or service.

ENDNOTE

[740] US Office of Management and Budget Circular No. A-94 Revised (Transmittal Memo No. 64), October 29, 1992, Accessed April 2017 and available at https://obamawhitehouse.archives.gov/omb/circulars_a094#8.

Appendix C: Pentagon Presentation on Measuring Performance, 1998

Measuring Performance

Essential Activities:
- information architecture
- technology in place to support architecture
- activity-based management
- incentives
- design processes to ensure the above activities occur

Information architecture: Umbrella term for the categories of information needed to *manage* an organization's businesses, the methods the organization uses to generate this information, and the rules regulating its flow. The information architecture should be implicitly defined by the accounting system. The design for this begins with the data that *management* needs to pursue the organization's strategy.

Hardware
Management systems $\Big\}$ Mobil's approach to
Culture $\quad$ improve performance

Activity-based management: The integration of activity-based costing into day-to-day decision-making. ABM then becomes a powerful tool for continuous rethinking and improving business processes. The implementation of ABM is a major organizational-change effort that involves a significant amount of work.

Environmental cost accounting
Value tracking $\Big\}$ Mobil's
Achieve acceptable EHS risk levels at least cost $\quad$ Cost Tracking

References:
Doll, B.E.: Mobil EHS Management System. Brief presented 22 Sep 98 to SAF/MIQ.
Eccles, R.G.: The Performance Measurement Manifesto. Harvard Business Review, Jan-Feb 1991.
Ness, J.A. & Cucuzza, T.G.: Tapping the Full Potential of ABC. Harvard Business Review, Jul-Aug 1995.
Ramanan, R.: EHS Costs & Value Tracking. Brief presented 1 Apr 98 to Conference Board.
Source: Pentagon, Pentagon Staff Slide Presentation on Measuring Performance, 1998.

Pentagon
1998

Appendix D: Green Chemistry Principles

From US EPA, https://www.epa.gov/sites/production/files/documents/green-chemistry-bookmark_1.pdf.

Appendix E: Assessment of Climate Change Impact

Source: US EPA

	In the year 2050, global GHG mitigation is projected to result in...	In the year 2100, global GHG mitigation is projected to result in...
HEALTH		
AIR QUALITY	An estimated 13,000 fewer deaths from poor air quality, valued at $160 billion.*	An estimated 57,000 fewer deaths from poor air quality, valued at $930 billion.*
EXTREME TEMPERATURE	An estimated 1,700 fewer deaths from extreme heat and cold in 49 major U.S. cities, valued at $21 billion.	An estimated 12,000 fewer deaths from extreme heat and cold in 49 major U.S. cities, valued at $200 billion.
LABOR	An estimated avoided loss of 360 million labor hours, valued at $18 billion.	An estimated avoided loss of 1.2 billion labor hours, valued at $110 billion.
WATER QUALITY	An estimated $507-$700 million in avoided damages from poor water quality.[†]	An estimated $2.6-$3.0 billion in avoided damages from poor water quality.[†]
INFRASTRUCTURE		
BRIDGES	An estimated 160-960 fewer bridges made structurally vulnerable, valued at $0.12-$1.5 billion.[‡]	An estimated 720-2,200 fewer bridges made structurally vulnerable, valued at $1.1-$1.6 billion.[*]
ROADS	An estimated $0.56-$2.3 billion in avoided adaptation costs.[†]	An estimated $4.2-$7.4 billion in avoided adaptation costs.[†]
URBAN DRAINAGE	An estimated $56 million to $2.9 billion in avoided adaptation costs from the 50-year, 24-hour storm in 50 U.S. cities.[†]	An estimated $50 million to $6.4 billion in avoided adaptation costs from the 50-year, 24-hour storm in 50 U.S. cities.[†]
COASTAL PROPERTY	An estimated $0.14 billion in avoided damages and adaptation costs from sea level rise and storm surge.	An estimated $3.1 billion in avoided damages and adaptation costs from sea level rise and storm surge.
WATER RESOURCES		
INLAND FLOODING	An estimated change in flooding damages ranging from $260 million in damages to $230 million in avoided damages.[†]	An estimated change in flooding damages ranging from $32 million in damages to $2.5 billion in avoided damages.[†]
DROUGHT	An estimated 29%-45% fewer severe and extreme droughts, with corresponding avoided damages to the agriculture sector of approximately $1.2-$1.4 billion.[‡]	An estimated 40%-59% fewer severe and extreme droughts, with corresponding avoided damages to the agriculture sector of $2.6-$3.1 billion.[†]
WATER SUPPLY AND DEMAND	An estimated $3.9-$54 billion in avoided damages due to water shortages.[†]	An estimated $11-$180 billion in avoided damages due to water shortages.[†]

289

	In the year 2050, global GHG mitigation is projected to result in...	In the year 2100, global GHG mitigation is projected to result in...
AGRICULTURE & FORESTRY		
AGRICULTURE	An estimated $1.5-$3.8 billion in avoided damages.	An estimated $6.6-$11 billion in avoided damages.
FORESTRY	Estimated damages of $9.5-$9.6 billion.	An estimated $520 million to $1.5 billion in avoided damages.
ECOSYSTEMS		
CORAL REEFS	An estimated avoided loss of 53% of coral in Hawaii, 3.7% in Florida, and 2.8% in Puerto Rico. These avoided losses are valued at $1.4 billion.	An estimated avoided loss of 35% of coral in Hawaii, 1.2% in Florida, and 1.7% in Puerto Rico. These avoided losses are valued at $1.2 billion.
SHELLFISH	An estimated avoided loss of 11% of the U.S. oyster supply, 12% of the U.S. scallop supply, and 4.6% of the U.S. clam supply, with corresponding consumer benefits of $85 million.	An estimated avoided loss of 34% of the U.S. oyster supply, 37% of the U.S. scallop supply, and 29% of the U.S. clam supply, with corresponding consumer benefits of $380 million.
FRESHWATER FISH	An estimated change in recreational fishing ranging from $13 million in avoided damages to $3.8 million in damages.[†]	An estimated $95-$280 million in avoided damages associated with recreational fishing.[†]
WILDFIRE	An estimated 2.1-2.2 million fewer acres burned and corresponding avoided wildfire response costs of $160-$390 million.[†]	An estimated 6.0-7.9 million fewer acres burned and corresponding avoided wildfire response costs of $940 million to $1.4 billion.[†]
CARBON STORAGE	An estimated 26-78 million fewer metric tons of carbon stored, and corresponding costs of $7.5-$23 billion.[†]	An estimated 1-26 million fewer metric tons of carbon stored, and corresponding costs of $880 million to $12 billion.[†]

Appendix F: Criteria—Pollutant Health and Welfare Benefits in 2010

Source: US EPA, "The Benefits and Costs of Clean Air Act 1990–2010"[741]

TABLE 8.1

Criteria Pollutant Health and Welfare Benefits in 2010

	Monetary Benefits (in millions 1990$)[a]		
	Primary		
Benefits Category	**Low**	**Central**	**High**
Mortality			
Ages 30+	14,000	100,000	250,000
Chronic Illness			
Chronic Bronchitis	360	5,600	18,000
Chronic Asthma	40	180	300
Hospitalization			
All Respiratory	76	130	200
Total cardiovascular	93	390	960
Asthma-related ER visits	0.1	1.0	2.8
Minor illness			
Acute bronchitis	0.0	2.1	5.2
URS	4.2	19	39
LRS	2.2	6.2	12
Respiratory illness	0.9	6.3	15
Mod/worse asthma[b]	1.9	13	29
Asthma attacks[b]	20	55	100
Chest tightness, shortness of breath, or wheezing	0.0	0.6	3.1
Shortness of breath	0.0	0.5	1.2
Work loss days	300	340	380
MRAD/Any-of-19	680	1,200	1,800

(Continued)

TABLE 8.1 (*Continued*)
Criteria Pollutant Health and Welfare Benefits in 2010

	Monetary Benefits (in millions 1990$)[a]		
	Primary		
Benefits Category	Low	Central	High
Welfare			
Decreased worker productivity	710	710	710
Visibility—recreational	2,500	2,900	3,300
Agriculture (net surplus)	7.1	550	1,100
Acidification	12	50	76
Commercial timber	180	600	1,000
Aggregate range of benefits[c]	26,000	110,000	270,000

Note:

[a] The estimates reflect air quality results for the entire population in the United States.

[b] Moderate-to-worse asthma, asthma attacks, and shortness of breath are endpoints included in the definition of MRAD/Any of 19 respiratory effects. Although valuation estimates are presented for these categories, the values are not included in total benefits to avoid the potential for double-counting.

[c] The Aggregate Range reflects the 5th, mean, and 95th percentile of the estimated credible range of monetary benefits based on quantified uncertainty, as discussed in the text.

ENDNOTE

[741] US EPA, The benefits and costs of clean air act 1990–2010, Accessed May 2010 and available at https://www.fda.gov/ohrms/dockets/dockets/07n0262/07n-0262-bkg0001-36-(Ref35)-EPA-Report-to-Congress-1999-vol6.pdf.

Index

Note: Page numbers followed by f refer to figures respectively.